T0283152

Before They Vanish

BEFORE THEY VANISH

Saving Nature's Populations—and Ourselves

Paul R. Ehrlich, Gerardo Ceballos, Rodolfo Dirzo

Illustrations by Darryl Wheye and and Mattias Lanas

Foreword by Jared Diamond

JOHNS HOPKINS UNIVERSITY PRESS

Baltimore

© 2024 Johns Hopkins University Press
All rights reserved. Published 2024
Printed in the United States of America on acid-free paper
9 8 7 6 5 4 3 2 1

Johns Hopkins University Press
2715 North Charles Street
Baltimore, Maryland 21218
www.press.jhu.edu

The photos that appear in plates 1–16 are courtesy of Gerardo Ceballos.

Library of Congress Cataloging-in-Publication Data for this book is available.

Names: Ehrlich, Paul R., author. | Ceballos, Gerardo, author. | Dirzo,
    Rodolfo, author.
Title: Before they vanish : saving nature's populations—and ourselves /
    Paul R. Ehrlich, Gerardo Ceballos, and Rodolfo Dirzo.
Description: Baltimore : Johns Hopkins University Press, [2024] |
    Includes bibliographical references and index. |
Identifiers: LCCN 2023058852 | ISBN 9781421449692 (hardcover ; alk.
    paper) | ISBN 9781421449708 (ebook)
Subjects: LCSH: Endangered species. | Extinction (Biology) |
    Plants—Extinction.
Classification: LCC QL88 .E39 2024 | DDC 591.68—dc23/eng/20240312
LC record available at https://lccn.loc.gov/2023058852

A catalog record for this book is available from the British Library.

*Special discounts are available for bulk purchases of this book. For more information,
please contact Special Sales at specialsales@jh.edu.*

To Guille, who we wish were still with us.

Contents

Foreword

New Vanishings of Old Creatures

Jared Diamond

As of the year 1981, New Guinea's largest and highest mountain range still remaining unexplored was the Foja Mountains, rising steeply out of northwest New Guinea's lowland swamps. In February 1981, a helicopter dropped me, three New Guinean field associates, a month's worth of supplies, and an emergency radio in a natural clearing near the summit of those mountains, which were completely uninhabited by humans. I hoped to rediscover there New Guinea's mystery bird, the Golden-fronted Bowerbird, known only from four specimens that had turned up in 1895 in a Paris feather shop, with no hint of where in New Guinea they had been collected. The Foja Mountains had been a reasonable guess as to where that mystery bird might be lurking. In fact, I did find the long-lost bowerbird there: it was the first bird species that I encountered when I stepped into the jungle from our camp clearing.[1]

But I also nourished wilder hopes when I was planning that expedition into the Foja Mountains. Until humans reached New Guinea and Australia around 50,000 years ago, across the water gaps separating that double continent from Indonesia's islands and Asia, New Guinea and Australia had been inhabited by many species of big marsupials distantly related to kangaroos and to the other small marsupials still surviving in New Guinea and Australia today. Those former giants, known today only from their fossilized bones, included carnivorous kangaroos three meters (almost ten feet) tall, rhinoceros-like marsupial herbivores called diprotodonts, and marsupial equivalents of leopards and wolves. Their disappearance following human arrival 50,000 years ago was presumably due to effects of those first humans, such as overhunting and fires. I dreamed: perhaps the Foja Mountains, uninhabited by humans today because they are so isolated, steep, and surrounded by swamps, were also not penetrated by the first-arriving ancestral New Guineans of 50,000 years ago. Perhaps I'd rediscover a lost world not only of bowerbirds, but also of big flesh-eating kangaroos, diprotodonts, and marsupial leopards.[2]

Alas, those wilder dreams of mine proved to be unfounded. The Foja Mountains did have little bowerbirds, but no big flesh-eating kangaroos. Ancient New Guineans evidently did explore everywhere in their quest for easily hunted big animals that had never experienced humans.

But my fantasies of astonishing the world by rediscovering creatures believed extinct had precedents. The world's largest rail, a chicken-like family of ground birds, is New Zealand's takahe, known in modern times only from four specimens shot between 1850 and 1898. Although the takahe was not as impressive as a carnivorous kangaroo three meters (almost ten feet) tall, it was still striking: blue plumage, long red legs, an enormous red bill, flightless, and weighing about three kilograms (around six and a half pounds). The takahe was also known from fossil bones. Hence it was presumed to be one of the many New Zealand bird species reduced in numbers by Polynesian settlers and then exterminated by Europeans and their introduced mammals.[3] But in 1948, New Zealanders were astonished and delighted when a search rediscovered a takahe population in a remote mountain valley. Today, thanks to a captive breeding program, there are hundreds of takahe in that valley and more introduced to mammal-free offshore islands.

Far more astonishing than the rediscovery of the takahe was that of coelacanths, a group of primitive fishes distantly related to lung-fishes. They were formerly widely distributed in oceans hundreds of millions of years ago, until they apparently vanished from the fossil record around 66 million years ago, at the same time as the extinction of the dinosaurs and three-quarters of the world's other animal and plant species. Coelacanths were not "mere" three-kilogram (around six and a half pounds) flightless birds: they included fish two meters long (nearly six and a half feet) and weighing seventy kilograms (154 pounds).[4]

It was therefore the biggest shock in modern paleontology when, 66 million years after the last coelacanth had supposedly vanished, one turned up in 1938 in the catch of a fishing trawler off the coast of South Africa. Since then, coelacanths have been caught regularly and observed swimming off Africa's east coast. A further shock came with the discovery of an Indonesian coelacanth population 8,000 kilometers (almost 5,000 miles) to the east of the African population, and initially recognized by one specimen in an Indonesian fish market.

The coelacanth exemplifies what is called a "living fossil": a creature known only from ancient fossils, until a surviving living population is discovered unexpectedly. The coelacanth's rediscovery has encouraged the burgeoning of a field termed "cryptozoology": the claimed discovery or rediscovery of other big animals, such as dragons, the claimed Abominable Snowman (alias Yeti) of the Himalayas, the claimed Bigfoot (alias Sasquatch) of northwestern North America, and Scotland's claimed Loch Ness Monster. The existence of the latter, supposedly inhabiting Scotland's largest lake, is suggested by sightings and photos dismissed by skeptics as hoaxes, but interpreted by cryptozoologists as a living fossil plesiosaur—a group of large marine reptiles otherwise believed to have gone extinct at the same time as the dinosaurs and the ancient coelacanths. It seems unlikely, though, that plesiosaurs could have survived for 66 million years in Loch Ness, which was completely frozen during the recent ice ages. Nevertheless, the Loch Ness Monster is regarded as at least worthy of consideration by numerous British people—including by one Cambridge scientist colleague of mine who made serious but unsuccessful searches for the monster, and who shall remain unnamed.

That ends my good news about real or hoped-for reappearances of old creatures previously thought extinct. There are just a few of them, but they make heartwarming stories. Now comes the bad news about the thousands or millions of old creatures that really have turned out to be extinct. They make miserable, sickening stories.[5]

Many creatures are known with certainty to be extinct because someone saw the last individual and described how he or she killed it or watched it eke out its last days. For example, the Great Auk was the largest species of the auk family, seabirds of Arctic and sub-Arctic oceans convergent on but unrelated to Antarctic penguins. The Great Auk, a powerful swimmer that could not fly, nested in colonies on offshore rocky islands of the North Atlantic Ocean. Its large size and down feathers made it a preferred target for aboriginal hunters, then for modern European hunters, and finally (as it became rare and as specimens became increasingly valuable) for European specimen collectors. The last individual seen in Britain was captured and killed in July 1840 by three men who considered it a witch and beat it to death with a stick. The last three individuals of the species, a nesting pair of adults incubating one egg on an island off Iceland, were killed on June 3, 1844, by three men acting for a merchant who wanted specimens: Jon Brandsson strangled one of the adults, Sigurdur Isleifsson strangled the other adult, and

Ketill Ketilsson smashed the egg. That ended millions of years of Great Auk evolution.[6]

Another species whose decline and end can be followed bird by bird is Bachman's warbler, a small yellow and green songbird. It formerly bred in swampy woods of the southeast United States. Then, in the late summer, it migrated south over the Gulf of Mexico and Strait of Florida to spend the winter in forests on the island of Cuba. Already uncommon when first recorded in 1832, its numbers declined as its breeding and wintering habitats were destroyed, until the last specimen was collected in 1941. But breeding warblers were reported from South Carolina until 1953, then only about a dozen scattered individuals elsewhere, until the last confirmed sighting in 1988.

My college classmate John Terborgh was lucky to see one of those last individuals. One evening in 1954, in his final year of high school, John received a phone call from a friend reporting that one lonely male Bachman's warbler had been discovered on a creek off the Potomac River south of our national capital of Washington, DC. John raced off in his car before dawn the next morning, reached the creek—and there, sure enough, was a male Bachman's warbler singing incessantly for the next two hours! In the following two years, a male warbler, probably that same individual, returned to that same creek in the spring. Then, nothing. What had happened?[7]

Males and females of migratory songbirds find each other in the spring on their breeding grounds, pair off, and nest. Then, they separate and fly to their wintering grounds. In the following spring, the male returns first to the breeding grounds, sets up a nesting territory, and sings his heart out to attract a female—in the hopes that there are any females around to find him. But as the already uncommon Bachman's warbler declined towards extinction, the likelihood that any of the few remaining females would find any of the few remaining singing males scattered over their breeding range declined toward zero. Probably the last female never found the last singing male.

What happened to the Great Auk and the Bachman's warbler has happened to thousands, perhaps millions, of other animal and plant and microbe species within the last century or two. Their extinctions, after millions of years of evolution, have been due to the effects of modern humans exploding in numbers, destroying habitats, hunting large or prized species, inadvertently spreading animal and plant diseases, and intentionally or unintentionally spreading pest species that exterminated other species with no evolutionary experience of the pest. Among the first of these modern extinctions to attract notice was that of the dodo, a huge flightless pigeon confined to the remote Indian Ocean island of Mauritius, discovered there in 1598, and last sighted in 1662.[8]

We now know, from discoveries of recent fossil bones by archaeologists, that arrivals of humans in every part of the world previously uninhabited by humans—North and South America, Australia and New Guinea, Pacific and Atlantic and Indian Ocean islands, and Mediterranean islands—were followed by waves of extinctions.[9] In some cases, our discovery of a species was followed by the species' extinction within a few years—such as in the case of Australia's gastric-brooding frogs, which had evolved a unique method of protecting and rearing their offspring. (The fertilized eggs or embryos were swallowed by the mother and developed safely in her stomach![10])

For those of you who don't care about extinctions of warblers and frogs, I'll now catch your attention by telling you about extinctions of humans. Many non-European human populations have been exterminated within the last few centuries by Europeans expanding around the world. The 5,000 native Tasmanians (Aboriginal Australians who used to live on the Australian island of Tasmania) were hunted down and killed following permanent European settlement of Tasmania in 1803, until the 200 surviving

Tasmanians were evacuated to nearby Flinders Island, where the last full-blooded Tasmanian died in 1876. Most of the 400 Yahi, a Native American tribe of Northern California, were killed by European settlers in four massacres between 1861 and 1865; the forty survivors went into hiding; and the last survivor, known as Ishi, emerged starving from hiding in 1911 as the last "wild Indian" in the United States, spent his remaining years in San Francisco as an employee of the University of California, and died in 1916.[11]

Europeans were not the only people to eliminate other human groups. Most San people of sub-equatorial Africa disappeared in the course of the Bantu expansion. Hoabinhian people of mainland and island Southeast Asia disappeared in the course of the Chinese expansion. The original Jomon people of Japan disappeared in the course of the Yayoi modern Japanese expansion. Much earlier, somewhat after 40,000 years ago, Neanderthals, a recent human species distinct from us modern humans (*Homo sapiens*), were somehow replaced/eliminated by our own expansion out of Africa. Before vanishing, all of these groups, even the Neanderthals, interbred with the human populations that replaced them.

What are the major risks that this book's potential readers, and that all contemporary humans, whether literate or not, face today? We face two risks that could kill many or all of us quickly, plus at least three sets of risks that are already slowly undermining the quality of life for all of us.

Of the two risks that would kill us quickly if they materialized, one is a large-scale nuclear war, which would transform the atmosphere and lead to a so-called nuclear winter. That may or may not happen. But nine countries already have enough nuclear weapons to make it happen, if even just one of their leaders so decided. Some of the current and recent leaders of those nine countries are famous for their dangerous ideas and poor judgment.

The other risk that could kill us quickly is an asteroid collision. The big collision of 66 million years ago, involving an asteroid with an estimated diameter of nearly sixteen kilometers (ten miles), is the one that exterminated the dinosaurs, plesiosaurs, and most other living plant and animal species. Smaller meteors and asteroids of various sizes have often been observed to hit our atmosphere or the Earth's surface in modern times.

The biggest recent such collision happened on the morning of June 30, 1908, in the remote Tunguska region of eastern Siberia.[12] That Tunguska object appears to have been a small stony asteroid about sixty meters (almost 200 feet) in diameter, which exploded in our atmosphere at a height of about eight kilometers (five miles). The explosion was equivalent in force to about a thousand of the atomic bombs dropped on Hiroshima and Nagasaki. An area of forest of about 2,500 square kilometers (around 960 square miles) was flattened and/or scorched. Only perhaps three people were killed, thanks to the very sparse human population of the Tunguska region. If that asteroid had exploded somewhere else—e.g., over Broadway and 50th Street in New York City—the death toll would have been millions. An asteroid of that size is expected to hit the Earth every few centuries.

Whereas you readers may or may not experience a nuclear winter or a Tunguska-size asteroid collision in your lifetime, you are already experiencing three sets of dangers that are slowly killing or undermining the lives of most people. One of those dangers is global climate change, caused by human burning of fossil fuels, and harming us in many ways: e.g., through higher average global temperatures, increased climate variability, increased storm frequency, decreased average rainfall, decreased agricultural productivity, ocean acidification, coral reef death, decreased coastal protection against tsunamis, melting of polar ice caps and lower-latitude glaciers, rising sea level, and spread of tropical disease vectors to temperate latitudes.

The dangers of climate change itself are exacerbated by the dangers of some scientists' suggested response by geoengineering, whose proponents wish to test ways of chemically modifying the atmosphere or oceans so as to combat climate change. But geoengineering's potential for instead causing us irreversible harm is extremely high, because laboratory tests of safety are poor guides to atmospheric outcomes. (Think of chlorofluorocarbons, hailed in the 1930s as non-poisonous refrigerator and air-conditioner gases, and benign in laboratory tests—until the 1970s, when was it discovered that once in the atmosphere, they deplete the ozone layer.)

The second slow danger to all of us is inequality within and between human societies. In today's era of globalization, the coexistence of poor societies and rich societies now painfully aware of each other is a recipe for support of international terrorism, war, emergence and spread of diseases, and unregulatable population movements.

The remaining slow danger to all of us is the extinction of much of our heritage of wild animal and plant and microbe species, which evolved over the course of hundreds of millions of years. Among the ways that extinctions harm us are through the declines and disappearance of aquatic species, which provide the main source of protein for nearly half of the world's people, the degradation of farmland (earthworms really are important for generating soil!), the declines of tree species and whole forests upon which we depend for so many uses, and losses of ecosystem services through which living creatures ranging in size from big to microscopic maintain our water supplies and atmosphere at no cost to us.

Those are practical, selfish reasons for being concerned about biodiversity, its accelerating decline, and extinctions. I haven't even mentioned yet the moral reasons for concern. Read on!

Preface

Human beings are the only animals that speak about morals, about "right" and "wrong." Morals differ among communities and individuals as well as over time. In this book, we deal with a widespread sense that it is right to preserve the only living beings known to exist, as well as the civilization of *Homo sapiens*. Over the course of our professional lives, we have witnessed the massive assault on nature that is pushing so many species to the brink of extinction, and we have been obsessed with trying to understand the magnitude, the impacts, the drivers, and the remedies for the horrific recent losses of ecosystems, species, and populations resulting from human activities. Many have written on this topic, and evidence of individual species extinction has been the subject of numerous newspaper articles. This book is different in that it is primarily concerned with what is happening at the level of *populations* of species. Another important distinction of this book is its focus on the process of extinction, which most simply can be seen as the progressive loss of the populations of a species until the last one is gone.

The three of us have been friends and colleagues now for decades. We have published scientific papers, book chapters, and books together. We have spent many hours discussing our interest in all kinds of topics, but most of the time, our discussions have been focused on the current extinction crises. In 2015, Gerardo, Paul, and Anne Ehrlich, Paul's accomplished wife, published *The Annihilation of Nature: Human Extinction of Birds and Mammals*, which was devoted to the species extinction crisis. Since then, we (together with Rodolfo) started to think about a sequel devoted to population extinctions. During the spring and summer of 2019, the three of us had lunch every Tuesday in the Nexus Cafe at Stanford University. Both Paul and Rodolfo are professors in Stanford's Department of Biology, and Gerardo, a professor at the National Autonomous University of Mexico, was doing a sabbatical in Paul's lab at the time. We called our meetings the Catalpa Dialogues because we sat in a garden under the shade of several magnificent catalpa trees, many times in the company of our friend and ecologist colleague, Hal Mooney.

Often in our conversations, we discussed the complexity of the extinction crisis, which involves both species and populations. Until relatively recently, the customary way to view the issue was to focus on the accelerating disappearance of species. Since the 1970s, however, Paul has argued that the extinction problem required a focus on population exterminations as well. Indeed, a species becomes extinct when the last individual of its last population is lost. Understanding the population aspects of the extinction crisis has fundamental implications for ecology, evolution, conservation biology, and, of course, the future of civilization. It is important to note that it is *populations* of species that deliver critical life-support services to humanity at local, regional, and global levels. In general, the greater the numbers of populations, the more benefits humanity receives. Some of these benefits—*ecosystem services*, as scientists call them—include supplying freshwater, animal, and plant food, and the proper combination of atmospheric gases that allows civilization to exist.

Populations of species also pollinate crops and wild plants and enhance our world by providing aesthetic services such as bird song and mountain vistas.

The magnitude of the extinction crisis is mind-blowing, as enormous as the catastrophic collision of the meteorite that likely wiped out the dinosaurs. That is why we need to deploy our best effort to address humanity's greatest challenge. In the last five centuries, some 1,000 species of vertebrates—mammals, birds, reptiles, amphibians, and fishes—have been recorded as becoming extinct. Countless more have surely disappeared before they were ever scientifically described or even publicly noticed. Our own research has shown that the vertebrates lost in the last century alone would have taken some 10,000 years to vanish under normal conditions of extinction and *speciation* (the evolutionary process whereby new types of living things develop from existing ones). But the ever-growing human population, with its increasing demands for consumption of goods and use of services, has exploited a one-time energy bonanza from fossil fuels, causing those extinctions in an evolutionary blink of the eye. An iconic example is the African elephant. As early as the 1900s, their population was estimated at 10 million individuals but has now been reduced to less than a quarter of a million—an astonishing reduction of about 95%. Many thousands of other non-vertebrate species are threatened with extinction because they are rapidly losing their populations. We have coined some terms—such as *defaunation, biological annihilation*, and *the mutilation of the tree of life*—to try to describe this sad and civilization-threatening situation. Since we published *The Annihilation of Nature*, there has been growing concern about and awareness of species extinctions, but until the loss of populations is widely understood, the seriousness of the extinction crisis will continue to be underestimated.

Our book is organized into twelve chapters describing the losses, their causes and consequences, and the possible solutions to species population extinctions. The window of opportunity to save most biodiversity is still open, but it is rapidly closing. That is why we titled the book *Before They Vanish: Saving Nature's Populations and Ourselves*: to emphasize the dramatic action that we and other conservation biologists feel is urgently needed. The time to start shrinking the human enterprise is now, or nature will shrink it for us. In chapter 1, "From Origins to Extinction," we briefly describe the roots of diversity of life on Earth and the differences between population and species extinctions. Chapter 2, "The Sixth Mass Extinction," puts the current extinction crisis in a geological context. It describes the previous five mass extinction episodes that have occurred over the last 550 million years and explains why scientists believe we have entered the sixth one—this time caused by humans, in contrast with the previous ones, which were caused by natural catastrophes.

The following five chapters, "Lost and Vanishing Mammals" (chapter 3), "Bird Songs Long Gone and Declining" (chapter 4), "The Silent Crisis: Other Vertebrates" (chapter 5), "Ignored Victims: Invertebrates" (chapter 6), "Vanishing Green: Plants—Our Emerald Treasure" (chapter 7), and "Microbes: A Hidden World" (chapter 8), provide accounts of the natural history and ecological roles of organisms representative of these groups, and examine relevant statistics underlying the tsunami of current biological extinction. Chapter 9 introduces the notion of defaunation—the decline in abundance leading to the loss of animal populations even in places where the natural ecosystems are still standing.

Chapter 10, "Drivers of Extinction," discusses the factors underlying the current to address the biodiversity crisis, along with possible ways to help. Chapter 11, "Nature's Decline: The Costs," is devoted to summarizing the direct and indirect costs associated with the extinction tsunami. Finally, chapter 12, "The Cure: A

Bittersweet Pill," deals with some of the actions being taken as well as ones urgently needed to tackle the present and impending negative impacts of the extinction crisis. Although this is a popular book, we have been very careful to include references where we cite data or in any other paragraphs where they are relevant, trying to ensure we do not improperly use any specific information.

This book is part of our wider efforts to share information and our personal insights and concerns, as well as those of our colleagues, with decision-makers and the public. We wish to make everyone aware of the very serious threat posed by population extinction and the urgency of the need to address this challenge successfully and expeditiously. The current rapid loss of populations and the subsequent extinction of species is a self-inflicted human disaster that threatens the stability and well-being of all societies and poses an existential threat to civilization, indeed maybe even to the survival of humanity itself.

We are indebted to Jonathan Cobb, who has once again lived up to his reputation as one the best freelance science editors. His contributions to our book have been huge. We thank Tiffany Gasbarrini, our editor at Johns Hopkins University Press, for her encouragement, advice, support, and for sticking with us for a long time with great patience. We are extremely grateful for discussions on topics we deal with in the book; thank you to Anthony Barnosky, Dan Blumstein, James H. Brown, Regina Ceballos, Pablo Ceballos, Gretchen Daily, Partha Dasgupta, Joan Diamond, Arturo Dirzo, Anne Ehrlich, Andres Garcia, Jose Gonzalez Maya, Larry Goulder, Nick Haddad, Elizabeth Hadly, Nate Hagens, Cheryl Holdren, John Holdren, Tom Lovejoy, Lourdes Martínez-Estévez, Pablo Marquet, Simon Levin, Tom Al MacCuish, A. Hal Mooney, Jeff Morgan, Dennis Murphy, Peterson (Pete) Myers, Trevor Nielson, Jesus Pacheco, Peter Raven, José Sarukhán Kermez, Yanet Sepulveda, Jorge Soberón, David Tilman, Alon Tal, Erik Joaquín Torres Romero, and Ed Zacchary. We are extremely indebted to Anthony

Barnosky, Ronald Bjorkland, Nick Haddad, Barnabas Daru, Jen Martiny, Dennis Murphy, Peter Raven, Pete Myers, and Taylor Ricketts for kindly reviewing chapters and providing insightful comments and ideas. We would like to specially thank Dennis Murphy for his invaluable discussion of the ways the endangered species act relates to population losses. Yolanda Domínguez helped us with the gigantic work of reviewing the references and the scientific names. Our appreciation to Anne H. Ehrlich, Guadalupe Mondragón, and Guillermina Gómez for their continuous support and encouragement. Gerardo's work on extinction has been supported in part by Marcela Velasco and Hector Slim of the Fundacion Telmex/Telcel, the Climate Emergency Fund (Aileen Getty and Trevor Nielson), and the Universidad Nacional Autónoma de México. Science artist Darryl Wheye created figures 2 to 14 and extensively researched the captions. We thank the Stanford Libraries Digital Production Group, especially Chris Hacker and lead photographer Wayne Vanderkuil, for providing excellent digital files of his illustrations. We also thank Mattias Lanas for providing plate 18, which illustrates the defaunation of African savannas. Finally, we are grateful to our friends Aileen Hessel, Mary Ann Hurlimann, Sandra Khan, and Paul Perret for great wine dinners and logistical support.

Left to right: Hal Mooney, Rodolfo Dirzo, Paul Ehrlich, and Gerardo Ceballos at one of their "Catalpa Dialogues" lunches at Stanford University, 2019.   Photo courtesy of Gerardo Ceballos.

Chapter 1

From Origins to Extinction

IN THE 1700s, RUSSIAN CZAR Peter the Great (1672–1725) and his
successor, the empress Catherine Alekseyevna, sponsored two historic expeditions to Siberia and Russia's northeastern Pacific region. Historians believe this large-scale and costly adventure was
motivated by Russia's desire to colonize North America. But Peter
also wanted to determine whether Asia and North America were
connected by land (or, if not, locate a northeast passage), as well as
map and catalogue this massive polar area.

As part of this effort, Danish-born explorer Vitus Bering was
commissioned to lead an expedition organized by the powerful
Russian Academy of Sciences[1] with support from the Russian empress. About 3,000 soldiers, scientists, and assistants participated
in what was dubbed the Great Northern Expedition (1733–1743),
leaving St. Petersburg in early 1733 and arriving many months later
at Okhotsk on the Siberian coast. There, the expedition separated
into four contingents to explore different regions of that vast unknown territory in eastern Siberia, especially the Kamchatka Peninsula and adjacent waters.

Then, after years of such exploration, Bering built two ships that
set sail on a rainy morning in June 1741 into the unexplored sea
between Siberia and Alaska. From the beginning, the expedition

was marked by problems and setbacks—nothing surprising if one considers that they sailed from the unmapped east coast of Siberia, which is washed by one of world's roughest seas.

The two ships were destroyed by a storm. Bering and his crew were stranded on an island that was later named after him. He and forty-five out of a total crew of seventy-seven perished from scurvy, starvation, and the dangerous weather. One of the survivors, however, German-born medical doctor and naturalist Georg W. Steller, wrote detailed notes on the geology, oceanography, and biology of the landscapes he explored throughout the expedition.[2] In the harsh environment of the northwest Pacific, he recorded sighting previously unknown species including the spectacled cormorant (*Urile perspicillatus*) and the gigantic Steller's sea cow (*Hydrodamalis gigas*), which bears his name.

Unfortunately, the Steller's sea cow, a nine-tonne (ten-ton) marine mammal related to the much smaller dugong and manatee of warmer waters, very rapidly went extinct. Limited in range to a very restricted area in the northwest Pacific, with no known extant populations elsewhere, they soon became victims of overhunting, and the last survivor was killed only twenty-seven years after Steller's initial sighting of the species.

But that is only part of the story. Other factors also contributed to the creature's demise. For example, one indirect cause was the local overexploitation of sea otters, almost completely wiping out their populations. But scientists needed more than two centuries to finally unravel the intricate connections of the marine region's food webs that killed off an entire species. Steller's sea cows primarily ate a type of large brown seaweed known as kelp. Kelp is also eaten by sea urchins, which in turn are prey for sea otters. When explorers and coastal residents put intensive hunting pressure on the sea otters for their fur, an overexploitation involving thousands of pelts, it triggered what ecologists call a "trophic cascade," in which the actions of predators change the abundance of the prey. In this case, decline of the overhunted sea otter population enabled the sea urchin population to increase, which in turn eliminated the kelp beds. This caused starvation among the sea cows, which, in synergy with them also being overhunted, contributed to their extinction. In other words, local population declines or extinctions in one species (otters) likely contributed to the disappearance of all the populations of another species, the Steller's sea cow.[3] This sort of linkage is common in almost every ecosystem and helps highlight how important it is to save populations.

Steller's sea cow and the spectacled cormorant are just two of the many vertebrate species that have sadly become extinct in a relatively short span of time. Similarly, billions of populations and thousands of species of animals and plants have become extinct in the last few centuries. Indeed, within the last sixty years, there has been a more dramatic change in the abundance of life forms on Earth than in the last 10,000 years.[4]

Species and Population Extinctions

In our previous book, *The Annihilation of Nature*,[5] we described the current species extinction crisis in some detail. But there is also a pressing need to document and raise awareness among the public and decision-makers alike of how population extinctions—the disappearance of component populations within species—are increasingly being seen to play a central role in the current extinction crisis.[6] Population extinction and its erosive effects on the distribution and abundance of our living companions are thus the main themes of this book. We also carefully document the impacts of that erosion not only on human civilization but also the natural world, and explore what we can do to reduce those impacts. We examine the causes and consequences of the loss of populations of animals, plants, and microorganisms, where a *population* is a constitutive part of a species, but unless there is only one extant population, not the entire species.

We hope this book will contribute to further understanding of the grave nature of the current "extinction crisis" and how the loss of populations, as a byproduct of human activities, also destroys the fabric of humanity's life-support systems. Population extinctions, of course, also erode the entire species' ability to survive because they reduce capacity to respond to issues such as local disasters that ultimately lead to species extinctions. In some cases, both population and species extinctions happen rapidly, with little warning, due to one or a few dominant causes (e.g., volcanic activity, asteroid collision, overhunting). In other cases, there are many interacting stressors operating over a prolonged period of population reductions and extinctions that eventually threaten survival of the entire species. Regardless of how the process unfolds, as populations, and then species, become extinct, the fabric binding life in the biosphere frays and weakens; natural services required by civilization decline, such as the provision of freshwater and amelioration of the climate; and ultimately the continuation of life as we know it is threatened.

Much of our own work and that of colleagues has focused on evaluating recent and current species extinction rates compared with the geologic "normal" or "background" pace of such extinctions that have largely prevailed outside of previous mass extinction events. Recent analyses, using very conservative assumptions, have shown that the average rate of vertebrate species loss over the last century has been in the vicinity of 100 times higher than the background rate, clearly suggesting that a sixth mass species extinction is already underway.[7] Estimates for other groups of plants and animals also suggest that current extinction rates are between 100 and 1,000 times higher than background extinction rates.[8]

These sobering conclusions are greatly reinforced by data on population extinctions. Many studies are finding enormous declines of populations in recent years. Our own work showed that 32% of 33,000 vertebrates, including the full range of abundance levels from very common to common, rare, and endangered species, have decreasing populations. Clearly, all kinds of species are being affected. A recent World Wildlife Fund report calculated a dramatic 69% decline between 1970 and 2018 for a large sample of terrestrial and aquatic vertebrates. Another more recent study, analyzing 70,000 species of animals, showed that 48% of them have declining populations.[9] All kinds of taxa are declining, including mammals, birds, reptiles, amphibians, fishes, insects, snails, clams, and other kinds of invertebrates, as well as plants, fungi, and microbes, as we will relate as far as the uneven statistics demonstrate.

We can think about the first five mass extinction events as critical junctures in the history of life on Earth that paved the way for the evolution of *Homo sapiens* by, among other things, wiping out potential predators and competitions. These five mass extinctions all occurred in the last 550 million years (chapter 2). They are known as the Ordovician–Silurian (440 million years ago), Devonian (356 million years ago), Permian–Triassic (250 million years ago), Triassic–Jurassic (210 million years ago), and Cretaceous–Tertiary (65 million years ago). Now *Homo sapiens* itself is poised to cause a sixth mass extinction event. The main trigger is human-driven reshaping of Earth and its living passengers to meet the needs and desires of a growing population of eight billion people. The sixth mass extinction crisis may equal or exceed in magnitude the fifth cataclysmic die-off that wiped out the non-avian dinosaurs and many other life forms about 66 million years ago. If the sixth extinction crisis is of that magnitude, the consequences for humanity of such enormous loss of populations and species will certainly be catastrophic. Before we dive into the population loss problem, let us review here, briefly, the origins, scientific classification, and variety or "biodiversity" of life on Earth.

Origin of Life

All living populations and species share a long evolutionary history that goes back to the origin of life. The oldest indirect evidence of life we currently have is fossil filaments and tubes from hydrothermal vents dating to some 4.28 billion years ago.[10] The earliest known remains of living beings are fossil bacteria that lived in shallow waters 3.8 billion years ago, which have been found in ancient stromatolites of Australia and Greenland.[11] These stromatolites consist of rock layers formed by cyanobacteria (blue-green algae) that secreted calcium oxides (lime) and entrapped sediments, producing mounded structures. Living representatives of these ancient bacteria now are found only in a few places with shallow waters, such as Shark Bay in Australia, the Cuatro Ciénegas Lagoon system in northwest Mexico, and other salty lagoons and lakes.

For the next 3.3 billion years, called the Precambrian eon, living organisms were microscopic in size and confined to the oceans. Animal life was largely limited to cnidarians, relatives of today's corals and jellyfish.[12] But over this time frame, changes were occurring to Earth on a massive scale: land masses formed, along with mountain ranges and ocean basins. At the same time, these dramatic changes were accompanied by barely perceptible shifts in the planet's structure and composition. Frequent and large-scale volcanic activity released gases that produced an atmosphere rich in carbon dioxide (CO_2) and nitrogen (N_2). This abundant atmospheric CO_2 fueled the metabolisms of the first living organisms, allowing them to flourish. Initially, the oxygen (O_2) produced as a waste product of photosynthesizing microorganisms (cyanobacteria) was largely used up in the weathering of minerals, especially iron. This "mass rusting" slowed a gaseous oxygen build up in the atmosphere, and after about 50 million years, or 2.3 billion years ago, led to the great oxygenation event[13] in which O_2 gradually displaced CO_2 as a prominent atmospheric gas.

Interestingly, O_2 was poisonous to these early life forms, and this radical shift in atmospheric composition led to an early mass extinction event. Relatively little is known about the magnitude and timing of this event because of the microscopic nature of those early living anaerobic inhabitants of the planet and a lack of fossils—and for these reasons, it is not ordinarily included in the list of five mass extinctions preceding the current one. Relatives of these organisms currently live in low-oxygen places such as deep in the oceans.

But even as a range of microorganisms died off due to the changing atmosphere, other types of microorganisms that could mobilize energy in an oxygen-rich atmosphere flourished and evolved—and eventually dominated the planet. A bonus for life in an oxygen-rich atmosphere is the formation of a layer of ozone (O_3, three oxygen atoms bound together) in the stratosphere, the second-closest layer of the atmosphere to Earth. This ozone shield absorbs most of the sun's lethal form of ultraviolet radiation (UVB). While life in the sea was largely protected from the effects of UVB by the first meter or two (five to six feet) of water, this newly formed ozone "shield" permitted plants and animals to invade the land for the first time. Although specific timings remain controversial, this pivotal event is dated to around some 500 million years ago.

In one of the most profound milestones in the evolution of life, Precambrian microscopic life experienced a massive adaptive radiation, evolving into many different forms and ecological roles 500–600 million years ago.[14] However, the exact triggers of this radiation remain uncertain.[15] During this period of biological expansion, now called the Cambrian revolution, a spectacular diversification of living organisms, including the first of many animals with mineralized skeletons, began to populate the oceans, which, up to this point, had supported only microscopic and soft-bodied life.[16] Fossil records show that in a surprisingly short time, the oceans became teeming with an astonishing array of organisms visible to the naked eye (macroscopic) such as trilobites with tough external skeletons, soft-bodied animals such as "worms" living in mineral tubes, and other organisms that have basic body plans similar to today's faunal communities: head, tail, eyes, segments, and appendages.[17] The factors that led to such massive and unusually rapid radiation of life are, we reemphasize, still poorly understood and a topic of much research and debate.

This explosion of life continued over the next 500 million years, with an almost unbridled diversification of forms. Evolution produced populations of a wide range of now familiar but extinct creatures such as dinosaurs, pterodactyls, and mammoths, and those whose descendants are still living: mammals, birds, reptiles, amphibians, fishes, clams, octopi, and insects, in addition to a huge and varied array of other animals, plants, and microorganisms. Despite the setbacks of the five mass extinction events over the last 500 million years, the diversity of the floral, faunal, and microbial communities at the present time is greater than at any other point in the history of the planet (chapters 5–8).

Classifying Life

With such a vast diversity of ever-evolving life forms, it is not surprising that human beings have struggled to classify them into categories—including that of populations—in order to discuss them, their attributes, and ecological roles; indeed, to *comprehend* them. This is what people do with any vast collection in which they are interested—colors, stamps, temperatures, airplanes, geographic features, and even other people.

In 1758, Swedish physician, botanist, and naturalist Carl Linnaeus (1707–1778) published the tenth edition of his immensely important treatise, *Systema Naturae*, in which he established the basis for the modern classification of animals, having previously started on plants.[18] The son of a Lutheran minister, Linnaeus, an amateur botanist, was interested in plants from early childhood and spent much of his youth studying the local Swedish flora and reading about botany. At the age of twenty-five, he made an extensive trip around Lapland and more limited tours through other parts of Europe, including England. As a professor and later rector of Uppsala University, he trained many students, some of whom traveled widely with him and contributed to and expanded on his work and the influence it has had on modern life sciences.

Linnaeus's greatest contribution to science was the development of a classification system of all living and extinct species (kinds of organisms) based on a hierarchical ordering. He proposed that all species be given a unique, two-part Latinized name in which the first part of this binomial identified a group to which the specimen belongs and the second, or trivial name, identified a subset within the group (species) that had characteristics distinct from other members of the larger group.

Universally accepted and still in use to this day, the Linnaean system of classification groups together organisms that share obvious physical traits and provides the basis for a Latinized scientific name that can be used to classify and identify an organism (whether animal, plant, fungus, or microbe) regardless of its common name in any language. For example, the scientific name for the gray wolf, *Canis lupus*, identifies an individual wolf as a member of a unique subgroup or species (*lupus*) but also a part of the larger, more inclusive group, the genus *Canis*. The genus *Canis* also includes the coyote (*Canis latrans*), and the domestic dog (*Canis familiaris*), and other dog-like animals. Similarly, for a botanical example, consider the scientific name of the California rose (*Rosa californica*), which designates one kind of rose found in California. The genus *Rosa* also contains the French rose (*Rosa gallica*) from that country, and many other members of this popular assemblage of plants in many parts of the world, all distinguished by their different species name but identified as being related by their common genus name.

Linnaeus's new classification system assumed increased relevance with the work and 1859 follow-on publication of Charles Darwin's *On the Origin of Species*.[19] Darwin's work on evolution heralded a scientific revolution that helped explain the driving forces creating the diversity of life forms. Darwin also revolutionized the way biologists and naturalists viewed how communities are shaped and the inherent value of biodiversity to civilization. The huge change ushered in by Darwin's theories should not be underestimated. Before Darwin, the (nonscientific) explanations for the natural world held that the wide array of different species had been instantaneously produced by supernatural forces. Due to Darwin's theories, this view gave way to acceptance that different life forms (species) had appeared gradually over time. Although multiple instances of so-called "rapid evolution" have been documented,[20] diversity overall evolved in different ways, most of which took immense amounts of time. Over thousands or millions of generations, organisms assumed different physical and behavioral traits and occupied different environments, and often modified those environments. The complex tension between the physical forces of environments and interactions among the organisms living in those environments continue to generate a palette of spectacular life forms, a rich panoply of life.

In the absence of a more rigorous definition[21] or biological metric to identify and classify individuals, the term "species" currently provides a useful medium to communicate the degree of organic diversity among individual specimens in a collection. The term species is used to group together those individuals (animal, plant, fungi, or microbes) that have identical or very similar physical, behavioral or other characteristics, that loosely are judged to be the same "kind." Simultaneously, other collections of individuals that have different characteristics are classified as other species, that is, other kinds. Drawing the line between different species can sometimes be difficult. In sexually reproducing organisms, it is traditional to use the inability to mate and/or produce fertile offspring as the line separating species.

Although the number of species in an area can provide one measure of biodiversity, the relative paucity of historical and ongoing field and taxonomic research in many areas of the world has hampered an accurate assessment of the totality of species diversity or the full number of populations, along with many details of how they are changing.[22] The clearly defined and heavily entrenched concept of recognizable species has in some ways obscured a perhaps more important unit of diversification: the population. The general term population is used to describe any collection of objects, but biologists use it more specifically to describe collections of organisms of the same species found in a particular location.

The Diversity of Life

How many species and populations are there in the world or at any one location? As mentioned earlier, the answer to this deceptively simple question is actually very complex, and for many areas of the world and most groups of organisms, it has not yet been established. Although scientists know, for example, pretty much how many species of birds and butterflies exist (many thousands of each) and where they occur, when it comes to mites, midges, mushrooms, and other less charismatic groups, it is a different story— as it is, of course, with bacteria and viruses (if we decide to consider them "alive"). The varying definitions of species further cloud the problems of counting them. For example, should organisms with similar morphology occupying different areas be considered two different species or two different populations of the same species, such as the brown bears (*Ursus arctos*) of Europe and the grizzly bears of North America? Current taxonomic opinion in this case suggests they are different populations of the same species, but the grizzly is also a member of a separate subspecies (*Ursus arctos horribilis*) from its Swedish conspecifics.

Use of modern molecular genetic techniques has yielded more precision on the differentiation of organisms. However, there is no universal agreement yet on the kinds of differences (e.g., in the gene structure or in the mechanisms that control the expression of those genes) that would justify classifying organisms as "different." Indeed, molecular studies mostly show, as one would expect from the generally gradual process of speciation, a continuum of differentiation. Creating clear differentiation standards thus seems as difficult as establishing a single metric for separating "tall" from "short."

Despite the logistical challenges in obtaining species-level data, some estimates of global species counts have been made. Thousands of new species are named and described each year, and not just small and more obscure organisms. Between 1993 and 2008, around 400 new mammals were added to the catalogue, about 40% of them large and with very distinctive characteristics.[23] Based on the rate of discovery of new species in the last few decades, the current number of formally recognized and classified species, approximately 2 million, is surely a gross under-estimation of the probable total number, as is likely to become increasingly evident once naturalists and taxonomists get going on the mites, mushrooms, and bacteria.

Estimates of the number of living species have varied widely. Collecting beetles from tree species in Panama, American entomologist Terry Erwin,[24] for example, used the density relationship between beetles and other arthropods to derive an estimate of the species richness; he calculated there were at least 30 million different tropical arthropod species, far exceeding the estimated

1.5–2.0 million total for all species of organisms. Although some subsequent studies[25] have contested this high number, other researchers have suggested Erwin himself might have underestimated tropical species diversity!

The most comprehensive recent studies, ones that include only eukaryotes (a higher-level classification grouping containing organisms whose cells have a distinct nucleus) estimate almost 9 million species, of which more than 2 million are marine.[26] The estimates of the number of species, when prokaryotes (the higher-level classification grouping containing single-celled organisms lacking a well-defined nucleus, namely bacteria and archaea) are included, increases substantially to 1 to 6 billion species, with bacteria representing the largest number by far.[27] A recent study using mathematical "scaling" laws (such as the relationship of number of species to area of habitat) estimated the species richness of microbes to be 1 trillion species.[28] That incomprehensible number did not include viruses, of which there are millions of "types," because viruses cannot reproduce independently and hence are not always considered to be "alive." As molecular-level methods to evaluate genetic variation improve, estimates of the number of eukaryotic species are most likely to increase. Considering that many species comprise many (hundreds or more) populations, the diversity of life at this level is mind-boggling, comprising billions of populations.

Based on current and historical trends of species and population losses resulting from anthropogenic forces, many of the world's species, especially the animals and plants with which we interact, or which provide critical services, will disappear before we can fully understand and appreciate their ecological importance or before the human species itself becomes extinct.

Chapter 2

The Sixth Mass Extinction

PREHISTORIC FOSSIL RECORDS indicate that as *Homo sapiens* spread out of their original home in Africa, extinctions of large animals followed—to the extent that human-implicated extinctions eventually deleted about half of the large-bodied species with which we evolved, as well as many bird species on continents and islands across the world. A well-documented example of this displacement can be found if we study the spread of the Polynesian people several thousand years ago throughout the Pacific.[1] With them, perhaps unwittingly, the Polynesians carried the Pacific rat, an efficient omnivorous predator. The rats preyed heavily upon adult birds, hatchlings, and eggs, devastating populations of numerous species. Many of those species of birds had evolved in the absence of mammalian predators and lacked defensive behaviors to avoid them.[2] However, people themselves were also an invasive species that put an end to more than a thousand bird species; most notoriously, they ate to extinction the diverse and huge moas of New Zealand.

Our knowledge of the dodo (*Raphus cucullatus*), memorialized in the phrase "dead as a dodo," was based on written descriptions and illustrations from the seventeenth century depicting it as a flightless bird with small wings, chubby body, and a massive curved beak. It was related to pigeons, but don't think of the pigeons that swarm around London's Trafalgar Square or similar urban sites around the world. Think instead of the giant crowned pigeons (*Goura* sp.) of New Guinea that live primarily on the ground and are more than 0.60 meters (two feet) long. These plump birds are endangered by people cutting down their forested habitat and hunting them for food. The dodo was found exclusively on the island of Mauritius, located in the Indian Ocean some 1,931 kilometers (1,200 miles) from the southeast coast of Africa. It has the dubious distinction of being the first species recorded to become extinct in historical times, an early victim of what has become known as the sixth mass extinction.

In 1598, a Dutch vice-admiral described the dodo in his journal after visiting the island. His written and other accounts from those times describe the big (up to twenty kilograms / forty-four pounds or more) flightless bird as being "tame and stupid."[3] Until recently, it was believed that the dodo disappeared because it was intensively sought as a food supply by both the island's inhabitants and the sailors that would stop by to replenish water and food supplies. It has been said that people could kill the dodos by walking up to them and hitting them with stones and sticks.

More recently, however, scientists have uncovered new evidence suggesting that in addition to being directly killed by people, dodos were also driven to extinction by rats, such as the ones introduced on the Pacific islands, as well as other animals that preyed upon their eggs and competed for food.[4] The dodo became extinct most likely because of a combination of habitat loss, hunting, and introduced animals. The last dodos were seen in the 1660s; a computer estimate of their final extinction date is 1690.[5] A few skeletons, a dried leg, and a head deposited in the collections of Oxford University's Museum of Natural History are all that remain of the dodo today.

In contrast to the early extinction of the dodo, what is believed to be one of the most recent recorded vertebrate extinctions

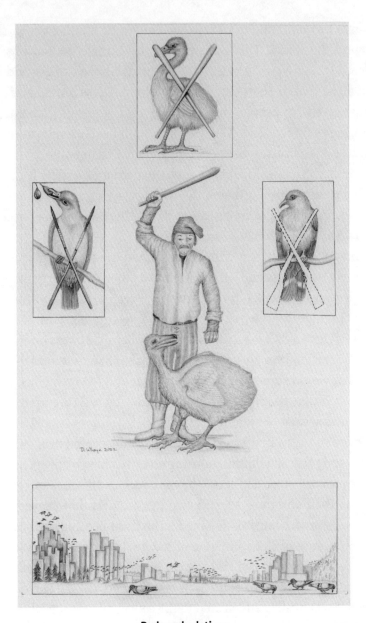

Dodo and relatives

The dodo was found exclusively on the island of Mauritius, which is located in the Indian Ocean some 2,000 kilometers (1,200 miles) from the southeast coast of Africa. It has the dubious distinction of being the first species recorded to become extinct in historic times, an early victim of what has become known as the sixth mass extinction.

occurred in the Bramble Cay, at the very northern tip of Australia's great barrier reefs. It is a tiny 3.6-hectare (approximately nine-acre) sandy coral atoll about half covered by grasses, located in the Torres Strait islands off the Cape York Peninsula in Queensland. Bramble Cay is some fifty-five kilometers (approximately thirty-four miles) south of the Fly River delta in the southern shores of Papua New Guinea. This obscure cay has become important in the conservation literature because in 2016, Australian scientists reported the extinction of the only population of a rodent known as the Bramble Cay rat (*Melomys rubicola*), in what appears to be the first reported mammal species extinction caused by today's climate change.[6] It is possible that another population of this species still exists in the nearby but little-explored Fly River delta. In that case, the Bramble Cay rat extirpation would just be one of the best-advertised population extinctions of a mammal.

The Bramble Cay rat population, which occupied the grass-covered area, was so large when it was discovered in the mid-1800s that sailors used them for bow-and-arrow target practice—just as the now-extinct Passenger pigeon was used for target practice in New York shooting galleries around the same time. By 1978, the numbers were estimated to be several hundred, but a proper census revealed only around 100 animals in 1998. From then on, the population declined until the last specimens were seen in 2009. The tiny island had by then been reduced in size by sea level rise, and it had been completely flooded on several occasions. The Bramble Cay rat decline had long been noted, but humanity contributed more than just the cause of its decline; it also took no steps to preserve the species, probably because it was a "rat" and lacked the iconic status of, say, a panda.[7] Early on, the Bramble Cay rat could likely have been established on safer islands in the Fly delta.

The dodo and the Bramble Cay rat are but two of hundreds of examples of species lost in historical times due to human activities, as we will see in the next chapters. The loss of a species, however, has been a common event in the billions of years of life on Earth long before humans appeared on the scene. This is hardly surprising. Earth has been around slightly more than 4 billion years.[8] To put that into perspective, if that timespan is scaled to a twenty-four-hour day, people capable of documenting anything (with writing) only emerged a few seconds before midnight. During the rest of the 86,400 seconds in that twenty-four-hour Earth-existence "clock," trillions of different kinds of living things had been evolving and going extinct continually, mostly lasting the equivalent of a few minutes or less of that "day." The details of those prehistoric extinctions are lost in the depths of geological time, although the fates of a tiny few individual species can be inferred from traces of them embalmed in strata of fossil-bearing rock. And those same strata have told scientists a great deal about the fates of entire floras and faunas in the distant past.

Scientists know more about how many species, and which ones of those among big plants and animals, have gone extinct since the year 1500, when the first recorded written descriptions are available. Early accounts are becoming better known to the general public, with extinct animals such as the thylacine (*Thylacinus cynocephalus*), the Passenger pigeon (*Ectopistes migratorius*), and the Golden toad (*Incilius periglenes*) starting to take their place alongside the dodo. More and more, people seem interested in learning what animals and plants have become extinct in historical times before they personally had a chance to see them alive. What birdwatcher wouldn't thrill to ticking off a Passenger pigeon from their list, or better yet a dodo?

Mass Extinctions

By analyzing the fossil record, scientists have learned how rates of extinction have varied over geological time.[9] Under usual conditions, extinctions occur at what we call "background" rates. These are the predominant, "normal" rates occurring through most of geological time, distinguished from very high rates of extinctions during periods defined as "mass extinction" episodes. The difference is analogous to your "normal" body temperature (98.6°F or around 37°C) that distinguishes that background level from brief episodes of high "fever." Another useful metaphor is that of individual human births, lifespans, and deaths. Many, many births and deaths occur, like the origin and extinction of species in geologic time. We don't think it's unusual when people die. But we do think it's highly unusual when a lot of people die in a short time—such as during pandemics. The contrast with background rates is the same—and they tell a worrying story.

In the last 550 million years of geological history, there have been five known periods of extremely rapid rates of species disappearances globally—so-called "mass extinctions."[10] We say "known" because for the roughly 3 billion preceding years, all life was microscopic, so likely mass extinctions cannot be identified from a nearly-nonexistent fossil record. In addition to the five known mass extinctions, it is becoming increasingly clear that for the sixth time, Earth is now entering a period when the pace of extinctions has risen rapidly above the geological "background" rate, and this time the cause can be traced to the expansion of the human enterprise and the form it has taken. But so what?

Maybe the best way to answer the "So what?" question is with an example. The Mesozoic was a truly remarkable era in the history of life on Earth that began 252 million years ago. During this time, Earth saw the first appearance of birds, mammals (our distant ancestors), and flowering plants. Despite it being a crucial act in the evolutionary play of life's diversification, it is most popularly known because of the Hollywoodesque "stars" that dominated the Mesozoic scene: giant reptiles. Those were the actors that reigned throughout that era and the Jurassic period whose name Steven Spielberg's 1993 movie *Jurassic Park* made famous.

British geologist John Phillips bestowed the name "Mesozoic" (Greek for "intermediate animal") on this era, perhaps in reference to the first appearance of small mammals in the fossil record among giant reptile bones. This geological era corresponds to a timespan of approximately 250 million years. Earth in the Mesozoic can be thought of as an ecological theater characterized by acts and scenes of peace and violence in the daily life of the most diverse and dominant cast of large animals until then—the dinosaurs.

Time traveling on land, you couldn't miss vegetarian dinosaurs, some of them deservedly called *Titanosaurs* and other names in recognition of their titanic size. They included animals ranging in size from 63.5 to 90.7 tonnes (70–100 tons) such as *Argentinosaurus*, found in South America, which usually peacefully foraged on exuberant evergreen vegetation or seasonally dry forests of the Mesozoic's Jurassic and Cretaceous periods. Dwarfed by these giants, but still enormous, you could see 5.4–9.07-tonne (six to ten ton) mega-predators such as the famous attorney-eating *Tyrannosaurus*, engaged in what must have been violent dramas of hunting (and sometimes scavenging), eating hundreds of kilograms of meat at a time. Their prey would have been selected from populations of a variety of spectacular beasts, including the charismatic *Triceratops* and the majestic and popular-with-the-children "Steggy"— *Stegosaurus*. Other violent dramas of hunting and scavenging would have been a common sight and involved smaller, more agile predators such as the now movie-famed *Velociraptor*, whose diet included pterosaurs[11] and likely other small reptiles, amphibians, insects, small dinosaurs, and even some early mammals.

The fossil evidence, fragmented as it is, illustrates the exquisite adaptations of these animals to the myriad biological interactions, such as those of plants and grazers, predators and prey, and competition for prey and for carrion, in which they participated. It also shows the physical conditions prevalent in the different regions of the Mesozoic Earth to which they were forced to adapt. Indeed, this was a time of great spatial heterogeneity fueled by a formidable land mass dynamism. Indeed, 250 million years ago, Earth's land was lumped in a single, mega-continent called Pangea, but by around 70 million years ago, it had fragmented into a mega-archipelago. This was accomplished by the relentless splitting apart of continental plates, whose slow movement wouldn't have been noticed from year to year, but was nonetheless cumulatively fast enough to radically change the surface of the Earth over millions of years.

Based on what we know of current natural biological communities, it is fascinating to imagine what existed in the Jurassic period in terms of an *entangled bank,* as Charles Darwin described the living landscape of his day.[12] One can envision complex food webs and networks of interacting dinosaur species coevolving among themselves and with other animal lineages on land and in the seas, as well as plants, fungi, and other microbes, under the dramatic and frequent changes in the physical conditions on Earth occurring within that gigantic timespan. It would have been a truly vast ecological theater in which a plethora of evolutionary plays took place, engendering a long-lasting dinosaur-dominated biodiversity.

The supremacy of the dinosaur players throughout the Mesozoic period came to its end some 66 million years ago. The speed and trajectory of an Earth-bound celestial body and the rotational movement of Earth conspired, as it were, to make what we now call the Yucatan Peninsula, in Mexico, the target of a meteorite mega-collision. The collision was a cataclysm that brought about dramatic changes to the abiotic (that is, the geophysical, non-living) conditions of the globe.[13] This led to a major mass extinction (the fifth over the last 550 million years of Earth's life) that swept away a significant portion of the planet's biodiversity in one of the most well-known "defaunation" events. This collision, which has become known as the Chicxulub Impact (after a Mexican village near the crater edge), unleashed many times the power of a nuclear explosion, vaporizing everything within miles of the impact, triggering huge tsunamis, and casting debris, gases, and heat into the atmosphere that rapidly spread around the planet and literally cooked everything that wasn't safely underground, protected by deep-enough water, or in fortuitous topography. Malcolm McKenna has called it, appropriately, the Cretaceous barbecue.[14]

However, any life forms that didn't succumb immediately faced having to do their best to survive on a literal scorched Earth. This mass extinction worked across the scale of organism sizes, but mostly removed large and medium-sized animals, and thereby changed forever the course of evolution. Beyond the global extinction of all dinosaurs (with notable exception of the birds), it is speculated that around 70% of the entire world's species went extinct in this geologic "instant." In some groups, the percentage was even higher. For example, 96% of the foraminifera species, those predominantly marine unicellular organisms, became extinct, probably because of the high temperatures of the ocean.[15] Thus, this mass extinction event was in fact a major trimming of the tree of life involving not only the big branch of the dinosaur lineage, but many others as well.

Even when a trigger event seems clear, some mass extinctions may nonetheless result from multiple causes[16] that can be difficult to sort out. For instance, some of the famous extinctions associated with the Chicxulub event 66 million years ago may not have come from the asteroid collision itself, but from the volcanism resulting from the collision, which produced gigantic lava flows in south Asia (the remnants of this activity are now known as the Deccan Traps).[17] However, one can still feel the enormity of this dramatic change when visiting the charming town of Chicxulub in Yucatan, located not far from the meteorite's impact crater.

Being at Chicxulub evokes a sense of all that was but exists no more, along with all that might have been as well as all that might never have been. For example, from a planetary perspective, the exact zone of the meteorite's impact—and its consequences—can be considered an accident, a mere throw of the cosmic dice. Indeed, had the meteorite collided with Earth a few hours later, it would have hit the deep ocean, and its subsequent chain of events might have been very different, setting the scene for a world in which we humans may never have come into being. This geological accident was also, in effect, one of Earth's major evolutionary "experiments": a mass extinctions phenomenon from which some important lessons can be learned, particularly regarding the types of victims and survivors of rapid environmental change and the time it takes for biodiversity recovery in the aftermath of a catastrophe.

Lessons of the Aftermath

In extinction events, a common pattern tends to be detectable among the animals. There is a contingent of more vulnerable victims, the "losers" we might call them, frequently represented by large-bodied, long-lived species with large home ranges, slow reproduction (few babies per reproductive event), and slow population growth rates. In contrast, there is a contingent of less vulnerable species, the "winners," largely comprised of small-bodied, short-lived species with small home ranges, rapid reproduction (many babies per reproductive event), and rapid population growth rates. Of course, some species fall somewhere in between these two extremes of vulnerability. This implies that in extinction episodes, the demise of some lineages or species is accompanied soon after by the flourishing of others. Indeed, the extinction of the dinosaurs created a very different ecological theater dominated by a new cast of birds and mammals, a novel *dramatis animalia*. It featured comparatively very small creatures that had previously lived in the shadows of the dinosaurs. In general, such a pattern of differential animal loss typically leaves a defaunation signature of "animal community downsizing," whereby smaller-sized species and lineages become the surviving—and potentially flourishing—fauna.

This evolutionary pattern that followed the dinosaurs' demise has also been detected in the aftermath of several other global biodiversity catastrophes. Recent research, for example, convincingly argues that before a mass extinction at the end of the Devonian period (419–359 million years ago) of the Paleozoic Era, when fishes reigned in the oceans, the seas were populated by gigantic "sea monsters" such as the ten-meter (thirty-three-foot) shark-like and armored *Dunkleosteus terrelli*. Such species almost completely disappeared or decreased in abundance during the extinction event, while smaller kinds thrived and underwent an explosive "evolutionary radiation"—which meant that creatures less than a meter (one yard) long dominated the seas.[18] This pattern of differential defaunation is aptly described by the (slightly modified) saying that extinction events literally "had bigger fish to fry."

The second important lesson to be gained from analysis of such mass extinctions is the time needed for post-extinction recovery of biodiversity. After the dinosaurs' demise, Earth's biodiversity recovery process took millions of years.[19] Consider species diversity, for example, using Foraminifera (the tiny single-celled animals with hard shells that have been around for more than half a billion years), an abundant and reliable bio-indicator.[20] Using forams as a guide, post-Chicxulub species diversity recovery is estimated to have taken around 10 million years. On land, the truncated tree of life that remained after the meteorite crash included a small-sized bestiary of species poised to become winners over millions of years undergoing a trajectory of expansion, diversification, and colonization of the entire planet.[21] Although some fossil-rich areas, such as Corral Bluffs, near Denver, Colorado, indicate that some mammals were present in that area as early as some 700,000 years after Chicxulub, globally it took millions of years to undergo body-size reconfigurations and return to an ecological theater with a well-represented cast of large animals.

We are frequently asked, "Why should we care about the extinction of populations and species of animals and plants, or about the replacement of local native populations with those of invasive species, or about the downsizing of animal communities?" After all, extinction is a normal and essential part of the process of evolution, a process that produced *Homo sapiens*. A large extraterrestrial body hit Yucatan, and yet here we are. Similarly, biological invasions are also a natural part of that process; in the Western Hemisphere, human beings are an invasive species, arriving hundreds of thousands of years after we evolved in sub-Saharan Africa. So, what if we're destroying biodiversity now at an especially high rate? It's happened at least five times before from other causes, and we're here, fat and happy.

Let's look at some of the other mass extinction events in which myriad populations disappeared and were replaced by new life forms. One occurred about 440 million years ago at the end of the Ordovician period of geological time, when some 85% of the existing species were wiped out. It caused a gap in the fossil record used by geologists to divide the Ordovician period from the Silurian period. Indeed, the division of periods was based on discontinuities, or changes in the fossil contents of rocks, before it was realized that they represented episodes of mass extinctions. The likely culprit of this late Ordovician event was a reorganization of marine nutrient cycling during a short, sharp ice age that chilled and toxified the oceans, tied to uplift and weathering of the Appalachian Mountains[22] and volcanic events.[23] Climate change, both cooling- and heating-related, was almost certainly involved.

Some 75 million years later, or 365 million years ago, another disaster for life created the fossil boundary between the previously mentioned Devonian period and its successor, the Carboniferous period. A die-off was possibly caused by the flourishing of plants that had invaded the land and led to flows of nutrients into the oceans, dramatically changing the environments of sea animals by periodically exhausting oxygen in the water.[24] Rapid changes in sea level, associated with size fluctuations in the polar ice cap, may also have contributed.

The next mass extinction occurred around 250 million years ago, ending the Permian period, and has been called "the great dying." There is little doubt that it was a biodiversity holocaust, an opening of the drain wide, not a turning down of the faucet. More than 95% of marine species vanished from the fossil records, accompanied by some 70% of land-dwellers—together, a gigantic pulse of population extinctions. It very likely was the result of a catastrophic volcanic explosion, releasing huge amounts of lava (Siberian flood basalts) and blasting huge amounts of the greenhouse gas carbon dioxide (CO_2) into the atmosphere. The episode apparently featured extensive areas of fires ignited by the lava, resultant deforestation, and massive erosion. The extinction event may have lasted some 100,000 years.[25] It was followed by a period in which conditions were generally harsh, and ecosystems underwent dramatic shifts as new organisms evolved.[26]

There was another episode at the end of the Triassic period (251–201 million years ago), when perhaps four-fifths of the species that had evolved since the Permian disaster disappeared. The cause here again looks like volcanism and resulting global warming[27] along with ocean acidification. One result of this mass extinction was clearing the way for the rise of the dinosaurs, just as was the case for the next mass extinction event, at the end of the Cretaceous period, which cleared dinosaurs away (except for the ancestors of birds) and made room for the rise of mammals. In addition, all the pterodactyls were exterminated, as were the dominant shallow-water marine predators, the mesosaurs and plesiosaurs, and the ammonites, the predacious mollusks that ruled the oceans, faded away as well.[28]

So, it is clear that these five mass extinction episodes were a nasty business for the dominant life forms of their time. They, however, had no direct impact on humanity (because humans hadn't yet evolved), although the one that ended the Cretaceous period was beneficial for us because it eliminated the dinosaurs and made ecological room for a flowering of the previously inconspicuous mammals. The accident of each of the last five mass extinctions, then, adjusted the trajectory of life in ways that made it possible for *Homo sapiens* to evolve. In that sense they were lucky breaks for us. But for those organisms involved, who were alive at the time, they were a catastrophe.

That's the real take-home message for us. For humanity today, the ongoing sixth mass extinction is one vast crisis—an existential threat to civilization, many human populations, or even to our species. True, recovery from previous mass extinctions has always occurred, but it took hundreds of thousands to millions of years, and leading species of one period seldom made it to the next. So, the silly argument that those recoveries "show" that there is no need to try to protect biodiversity is pretty much meaningless for humanity. We'll be long gone—after much suffering—before biodiversity recovers without us as members of Earth's species.[29] A victory for what comes next, perhaps. But definitely not for us.

Humanity and the Sixth Mass Extinction

Extinction rates over geological time are defined as species extinctions per million species-years (abbreviated as E/MSY) or number of extinct species per 10,000 species per century.[30] In 2011, paleontologist Anthony Barnosky and his colleagues published a breakthrough study showing that the "normal" ("background") extinction rate for mammals in the past 2 million years was roughly two species extinction every century for every 10,000 species in existence. So, if there were 10,000 species in existence at a given time, one would expect two species to disappear each hundred years, but if there were only 5,000 one would expect only a single extinction, and so on. Yet in the last 100 years, extinction rates show unambiguously that species are disappearing much more rapidly than they have for most of geological time. In 2015, a study that clearly demonstrated this shift was published, changing the paradigm for understanding the extent of the current extinction crisis. The study estimated the extinction rates of vertebrates (i.e., mammals, birds, reptiles, amphibians, and fishes) over the last 500 years[31] using data from roughly 39,000 vertebrate species. These "current" extinction rates could then be compared with the "background" or "normal" extinction rate estimated by Barnosky and colleagues.[32]

That paradigm-changing research, carried out by Gerardo Ceballos and Paul Ehrlich, two of the authors of this book, was based on two conservative estimates and assumptions: (1) the normal rate of extinctions in the fossil record was rounded up, making the background rate likely higher than the true background rate, and (2) the current rate was kept low because it omitted many possible extinctions where the disappearances had not been thoroughly documented. Thus, despite the comparative evidence for a higher extinction rate showing a dramatic increase, indicating the beginning of a sixth mass extinction episode was underestimated, this higher level was nonetheless likely to be an underestimate of the true scale of the extinctions currently underway. Our conservative estimate indicated that the vertebrate species that became extinct in the last 100 years would have taken around 10,000 years to disappear under "normal" conditions. This roughly 100-fold increase in the species extinction rate shows the magnitude of the current extinction crisis.

Some of the most important evidence of the human role in causing this current extinction episode comes from data on *population* extinctions. You may recall our argument in chapter 1 that they are the usual precursors of species extinctions. Population extinction is the *process* by which species diversity is destroyed; it is the "virus" causing the feverish "spike" in the rate of biodiversity destruction. Many dramatic losses of populations we describe against a background of escalating species extinctions provide the frontline evidence that biodiversity loss is an existential crisis for humanity.

Populations supply a vast array of natural services, not the least of which include providing humanity with food, water, and medicines, without which civilization cannot survive. So it is fair to say that in wiping out populations, humanity is busily sawing off the limb on which it is perched. Similarly, many people—ourselves included—see gigantic ethical issues in wiping out our only known living companions in the universe. Concerted effort on a scale running from local to global will be needed to protect as much biodiversity as possible.

Sitting in the Louvre's Rembrandt Room, with so many manifestations of unique, exquisite, invaluable, artistic creations, one of the authors of this book, Gerardo Ceballos, found it difficult not to recall what William Beebe, the famous twentieth-century naturalist, said about *Homo sapiens* becoming a gigantic force pushing so many populations and species to extinction: "The beauty and genius of a work of art may be reconceived, though its first material expression be destroyed; a vanished harmony may yet again inspire the composer, but when the last individual of a race of living things breathes no more, another heaven and another earth must pass before such a one can be again."[33]

A similar reflection came to another of this book's authors, Rodolfo Dirzo, during his visits to Mexico City's spectacular Museo Nacional de Antropología. In these visits, he strolls in awe from the exhibits that display the demise of the Pleistocene megafauna and across the sections of the beautiful artwork of the many Mesoamerican cultures. The thought that an unethical disdain of such artistic manifestations of culture could lead to their loss painfully resembles the disdain we are exercising as a force of biological extinction—in a similar way to the Chicxulub meteorite that "vaporized" all dinosaurs and so many other species of plants, animals, and microorganisms 66 million years ago. We have entered a sixth mass extinction, and this time *we* are the cause. It is difficult for all three of us not to wonder why we cannot also become the solution to such an unthinkable destruction of biodiversity. With this book, we hope to contribute to the reflection that saving our living companions will, paradoxically, save us from a ghastly future.[34]

Chapter 3

Lost and Vanishing Mammals

ONE OF THE MOST EXTRAORDINARY and visually powerful accounts of the massive impact of human activities in decimating populations of wild mammals in eastern African was offered by the book *The End of the Game* by Peter Beard, published in 1965.[1] He recorded the extermination of thousands upon thousands of elephants, rhinos, and other large mammals in the savannas and bushlands of Kenya, Tanzania, and Uganda in the 1960s. Together, the text and its astounding photographs offer sobering evidence of the capacity of *Homo sapiens* to destroy nature. Beard wrote, "The deeper the white man went into Africa, the faster the life flowed out of it, off the plains and out of the bush . . . vanishing in acres of trophies and hides and carcasses."[2] One of the most remarkable photographs in the book is a two-page black and white image of a herd of hundreds or maybe thousands of elephants . . . now only a cruel reminder of times forever gone. Another brutal photograph shows a huge landscape littered with countless carcasses and bones of elephants.

The End of the Game was a rare book when it was published because it offered one of very few accounts describing the beginning of the most recent massive destruction of mammal populations inflicted by human beings. Although it is centered on eastern Africa, the rampant assault on wildlife quickly spread as ripples to cover the entire planet. Similar losses have occurred in the last decades in the Americas, Asia, Australia, Europe, the poles—in other words, nearly everywhere. Statistics on the scale of the biological annihilation that has occurred since 1970 are mind-blowing, revealing that a high proportion of mammals, large and small, marine and terrestrial, are now in danger of extinction. For example, one estimate indicated that since 1970, 70% of the individuals in 3,000 analyzed animal populations have disappeared, and another more recent study on 70,000 populations estimated that they have experienced the loss of 48% of all individuals.[3]

Since the year 1500, roughly 116 species of mammals have become extinct.[4] A famous example of human contribution to animal extinction is the thylacine, also known as the marsupial tiger or the Tasmanian wolf. The thylacine was a superficially dog-like animal originally widespread in New Guinea, Tasmania, and mainland Australia, although it was never abundant in number despite this large habitat. We know its ancient distribution largely thanks to Aboriginal artists who portrayed it in their magnificent rock art. When the British began to unload their criminals in Australia in the eighteenth century, the thylacine was already relatively scarce or even nearly extinct on the mainland. Its decline was perhaps attributable to competition from the increasing Aboriginal populations with improved hunting technologies[5] or negative interactions with the dingo, a close relative of the domestic dog that was larger and had a proportionally larger brain and that invaded Australia from Asia, or both.[6]

The thylacine held on in Tasmania for a while, but then was determinedly helped on its way out by a bounty in response to demands from grazers who (probably falsely) said the animals were a threat to their sheep. Other factors may have been habitat destruction, co-declines of prey species, competition with wild domestic dogs, a distemper-like disease, and possible long-term inbreeding.

The last captive in a zoo died in 1936, but undocumented reports of its persistence in the wild in Tasmania, mainland Australia, and even New Guinea still periodically emerge.

The long list of mammal species already gone, most only perpetuated in old naturalists' chronicles, natural history museums, and scientific collections, include the Steller's sea cow (chapter 4), the Bramble Cay rat (chapter 2), the Sardinian pika (*Prolagus sardus*) from the islands of Corsica and Sardinia, the Guam fruit bat (*Pteropus tokudae*) from the Pacific island of that name, the Bluebuck (*Hippotragus leucophaeus*) from South Africa, the Falkland Island fox (*Dusicyon australis*) discovered by Charles Darwin, and the Caribbean monk seal (*Neomonachus tropicalis*) from the Caribbean Sea.[7]

Mammal species have become extinct in all continents and seas of the planet. However, most extinctions have occurred in Australia, where some thirty-four species, such as the Lesser bilby (*Macrotis leucura*), the Southern barred bandicoot (*Perameles notina*), and the Toolache wallaby (*Macropus greyi*), became extinct in the last centuries. Most of the Australian extinct mammals are classic victims of the rampage of invasive species such as foxes, cats, and rats that ate them to extinction.[8] In general, extinct mammals have succumbed to human activities resulting in habitat loss and fragmentation, overhunting, introduced species, diseases from domestic animals and, more recently, climate change.[9] This chapter will visit some of these species on the verge of extinction, revealing more about these creatures with whom we share a planet and highlighting the tremendous loss to both present and future, humanity and other species, if we let them disappear forever.

Vanishing Mammals

In addition to the mammalian species already extinct are those in clear danger of extinction. An important problem in determining the number of endangered species of mammals or any other group is the availability of data, which are costly and time-consuming to amass. The International Union for Conservation of Nature (IUCN) provides the most complete compilation of the status of all at-risk plants and animals. There are 1,340 species of mammals considered at risk of extinction on a recent IUCN list (2022), including 233 critically endangered, 550 endangered, and 557 threatened. All these species are at risk because they are experiencing population extinctions. As would be expected, however, the drivers of population losses in marine and terrestrial mammals are quite different.

Large-mammal populations especially are under enormous pressure in poor countries where major threats, besides hunting by people, are human population growth–related land-use change, especially by clearing for agriculture, and reduction of supplies of plant foods by livestock competitors. Their loss ricochets throughout functioning ecosystems, upward by reducing the food available to large predators such as lions and jaguars and downward, among other things, by greatly influencing populations of plants. Indeed, the extermination of Western Hemisphere large herbivores such as mammoths, at least partly by the human beings when they invaded around ten millennia in the past dramatically altered the entire flora of the hemisphere, causing extinction cascades and changes in plant community composition and vegetation structure as well as ecosystem function.[10]

Fierce Predators

Larger animals with longer generation times are less able to adjust the characteristics of their populations evolutionarily to meet environmental challenges. In addition, differences in behavioral patterns can largely determine how rapidly populations of species disappear. For example, consider the fate of that icon of climate-caused population extinction, the polar bear (*Ursus maritimus*). It is the world's only sea-going bear, a species that evolved from grizzly (brown) bear–like relatives.[11] Polar bears are also the only totally carnivorous bears. Living in an environment of ice and icy water, they have very high caloric requirements,[12] and to satisfy that need, they are heavily dependent on one food source, fat-loaded seals captured on sea ice. The massive recent shrinkage of the area of sea ice, a function of climate change, has likely accelerated their decline.

Although current population estimates are uncertain because of widely varying conditions in the polar bear's circumpolar range, there is little doubt that further global warming makes its future bleak.[13] Added to this problem is northward expansion of the brown bear populations and the uncertain results of their resumption of hybridization with the polar bears.[14] Furthermore, it has been found that at least on the Beaufort Sea, polar bears facing diminishing sea ice are also spending more time ashore and suffering more from pathogens associated with terrestrial animals, further complicating their chances for long-term survival.

Like polar and grizzly bears, wolves and coyotes are very close relatives. But gray wolf populations have been driven to extinction over much of their once-huge distribution—which covered most of North America and Eurasia. Where they have not been extirpated, as they have been in most of the United States, their populations have usually been much diminished. Although some wolves are protected, others often are the victims of deliberate hunting as enemies of livestock, pets, and people. Very often, killing off wolves leads to changes in ecosystems that people regret. Wolves have huge impacts on landscapes because they usually regulate populations of large herbivores such as elk. They do this not only by eating them, but also by changing their patterns of grazing out of fear for wolf predation. Huge populations of elk at Yellowstone caused overgrazing, changes in the structure of vegetation, erosion of river margins, loss of beavers by lack of food trees, and losses in fishes, amphibians, and other animals associated with habitats created by beaver dams.[15] Decimation of top predators such as large cats and killer whales can also have cascading effects on ecosystems, as the wolf–elk ecosystem at Yellowstone has illustrated.

Top predators are not just beautiful and fascinating creatures that attract tourists, they also tend to control the functioning of entire ecological systems. If, for example, they are exterminated in an area, the middle-sized carnivores whose numbers they had curbed may undergo population explosions, with consequences that cascade through that area's ecosystem. Scientists have observed just such a "mesopredator release" in many different circumstances. In a classic case, populations of the top predator, the coyote, disappeared from a fragmented sage-scrub system in southern California.[16] This led to an explosion of smaller predators such as skunks, foxes, raccoons, house cats, opossums, and snakes, that the coyotes had previously controlled. That in turn produced a reduction in population sizes of scrub-breeding birds that these now-more-abundant smaller carnivores dined on.

If a new top predator is introduced to an area, similar chaos can result, as the spread of escaped Burmese pythons in the Everglades of Florida has demonstrated. It has resulted in the near extinction of populations of opossums, rabbits, raccoons, and many other mammals and birds. The pythons also decimated bobcat populations, to the great joy of some rodents. Effects are not always so clear-cut, but ecologists have begun to pay more attention to the extinction of populations of top predators, especially "charismatic" ones like jaguars and pumas. Efforts to conserve populations of such wide-ranging creatures not only can have desirable effects on the properties of natural systems, they can also have the advantage of protecting populations of many other organisms; that is, with enough populations, they can serve as "umbrella species," protecting large slices of nature.

Coyotes, on the other hand, thrive in the presence of human disturbance, and their populations are almost impossible to exterminate. The difference is that wolves take larger prey, generating more enmity among people, they hunt in packs, and tend to be forest denizens. Coyotes tend to take smaller prey, or when they eat venison, it is often from carrion. Coyotes are more at home in open or savanna areas and are much-hated by sheep ranchers because they eat lambs, and by many suburban people because they eat dogs and cats. But since the arrival of Europeans in North America, the habitat favorable to coyotes has spread, as have these clever relatives of dogs. There are probably more coyote populations today than there were hundreds of years ago. Wolves are making some small comebacks, but many of their remaining populations are in much greater jeopardy.

Iconic Pandas

Perhaps the most "iconic" mammal in the conservation world is not the tiger or some other skilled top predator. It is a strict vegetable eater, the giant panda. The complexities of relationships between human populations and animal populations are typified by the situation of this black-and-white icon of conservation. A relative of bears, the giant panda once occupied a wide swathe of China, Myanmar, Laos, Vietnam, and Thailand. By 1800, however, expanding *Homo sapiens* had wiped out most panda populations, and the gentle giants were restricted to about 262,000 square kilometers (100,000 square miles) in just five provinces in southwestern China. In 1961, when it was further endangered, it became the logo of the World Wildlife Fund, which quickly transformed it into an international symbol for conservation. The Chinese government gave pandas as gifts to zoos in various countries, and they were made world famous by the press coverage they received and the establishment of online cameras featuring their breeding. But China's population increase of nearly a billion people since 1900, and an accompanying increase in per capita resource use, has continued to threaten panda populations in nature. Its low point may have been in the late twentieth century, when all populations may have totaled fewer than 2,000 individuals, although census results are questionable.[17]

Conservation efforts such as the setting aside of reserves, restrictions on timber harvesting, suppression of poaching, and a successful captive breeding program appear to have halted the giant panda's decline, and some panda habitat is recovering. Today, the black-and-white bears only exist in the wild in over thirty scattered, isolated populations in six mountain ranges of Gansu, Sichuan, and Shaanxi Provinces of southwestern China. They are still threatened by climate disruption and habitat fractionation, especially road construction, livestock grazing, land clearance by farmers at higher altitudes, and, ironically, mass development for tourism. Pandas, however, are now such a matter of national pride for China that continued efforts to protect the various wild populations seem assured.[18] That is hopeful—pandas are an important "umbrella species," and their protection of other populations, together with creation of nature reserves for them, benefits a series of ecosystem services including supply of freshwater and forest products, carbon sequestration, and income from tourism.

Gorillas, Our Most Spectacular Relatives

Gerardo remembers that flying from Arusha in Tanzania to Kigali, the capital city of Rwanda, is an easy one-hour trip. For most tourists, Kigali is an exotic destination. It is very different from other cities in Africa and many other countries throughout Asia and Latin America because it is incredibly clean and there are no street dogs. Researching a little into the recent past of Rwanda will reveal other astonishing things, some remarkable, some shameful. Walking in the relatively quiet city does not give any clue of its recent history. In 1994, Rwanda experienced one of the worst (and certainly most rapid) genocides of the twentieth century.[19] At that time, the Tutsi majority, incited by the country's president, carried out a massive persecution of the Hutu minority. In a little under four months, almost a million people were murdered.

In that time of turmoil, Western conservationists evacuated the Virunga National Park, home of one of the rarest and most fascinating of our wild mammal close relatives: the mountain gorillas (*Gorilla beringei beringei*). Mountain gorillas, one of two species of gorillas, are found exclusively in the Virunga volcanoes, located on the border of Uganda, Zaire (Congo), and Rwanda. The volcanoes are covered by lush bamboo and montane forests at elevations ranging from 700 meters (2,200 feet) to 5,109 meters (16,762 feet). The mountains famously trap fog and are damp and quite humid; they can also be very cold at night.

Mountain gorillas are adapted to this environment, covered as they are with thick fur. In the early 1900s, an estimated 5,000 gorillas roamed these forests. But illegal hunting took a heavy toll, and by 1960, they were rare. Now, population shrinkage and inbreeding have left them with comparatively little genetic diversity.[20] The scientific research of George Schaller and Dian Fossey (who was murdered in the mountains by poachers in 1985) gave people elsewhere in the world their first insights into these magnificent creatures and their hidden world.[21]

Like many other wild mammals in all corners of the planet, mountain gorillas are poached using snares and guns. It is difficult to imagine the pain and confusion that such an intelligent animal experiences when caught in a snare. Many hours will pass until it finally dies, for no other purpose than to feed the implacable appetite of the illegal trophy and live animal trades. Once the guerrillas (led by Paul Kigame, nicknamed the African Napoleon) won the Rwandan civil war, the Western conservationists arrived back in the Virungas, fearing the worst for the gorillas.

Surprisingly, they found that several guards had stayed behind to take care of their beloved gorillas and managed to survive. Wearing rags because of the many months they spent in the mountains taking care of the gorillas, these heroes managed to convince the armed forces that swept through the forests to spare the gorillas. Although several gorillas were killed, about 350 survived. It is those dedicated, dirt-poor guards, such as ranger Digirinana Francois, who during the war removed snares and guarded the park, that we all should remember. Asked why he stayed during the genocide, he said simply, "I was in love with the park, with the gorillas."[22] When Gerardo visited Rwanda to search for mountain gorillas in 2015, its population had increased to more than 1,000.[23]

Visiting the park to look for gorillas is extremely well-organized. Indeed, ecotourism represents the third most important economic activity in Rwanda. From the main base, Gerardo and another eight tourists, including his wife Guadalupe and his daughter Regina, started a three-hour walk following a guide and an armed guard. The park is limited by a stone fence standing one meter (3.28 feet) tall. The contrast between the park and the vicinity could not be more striking: outside, there are only crops, but inside features a lush bamboo forest. Surprisingly, they saw a group of golden monkeys (*Cercopithecus kandti*), a very beautiful species endemic to the Virunga mountains.

After one hour of a relatively easy walk, the guide found a group of some twelve gorillas, including a massive silverback male and several females and young individuals. They observed these most impressive animals for more than an hour, then headed back to the park entrance. Observing the gorillas, their beauty and strength on the one hand and fragility to human threats on the other, was a long-lasting experience of hope for Gerardo and his family. The mountain gorilla experience was possible because of the successful conservation of the gorillas, an effort that took decades of hard and constant work. It is one example of how even in very complex situations, conservation success can be achieved by dedicated individuals, institutions, and governments.

Mountain gorillas are a subspecies of the Eastern gorilla (*Gorilla beringei graueri*), which live in the lowlands of the Congo River near Rwanda and Uganda. It is estimated that the total Eastern gorilla population is less than 5,000 individuals.[24] The Western gorilla (*Gorilla gorilla*) is broadly distributed in the Congo basin, but it is considered at risk because of habitat loss and the bushmeat trade.

Orangutans, the Persons of the Forest

It might not be surprising to learn that the name orangutan means "person of the forest" (*orang* = person; *hutan* = forest) in Malayan. In the twentieth century, there was only one recognized orangutan species. In 1996, genetic and morphological studies showed that the orangutan populations of Borneo (*Pongo pymaeus*) and Sumatra (*Pongo abelli*), which have been isolated from one another for thousands of years, were so distinct they were best considered two different species. But the story did not end there. In 2017, a small population of orangutans in the Tapanuli region in northwest Sumatra was also described as a new species (*Pongo tapanuliensis*). That unique orang population is critically endangered; it had already lost most of its habitat when it was described, and it currently comprises only about 800 individuals.

The drastic decline of orangutans in Sumatra and Borneo is a textbook example of population extinction because the two species have been lost in many areas of their geographic range. Orangutans build new sleeping nests each night, and nest surveys indicated that between 1999 and 2015, the number of Bornean orangutans dropped about 50%, from about 300,000 to only some 150,000.[25] The reasons surveyed villagers gave for the killings that occurred ranged from conflict—fear of attack (self-defense), damage to orchards, interruption of timber operations, and "killed by the company"—to killing for food.[26] Some of the orangutans were shot for the bushmeat trade, or so they could be used in traditional medicine ("for sale of gall bladder"), or to sell the young for the pet trade, or for sport. Overall, killing for food was the most common reason given.

The results of studies showing high rates of humans killing orangutans have been controversial.[27] It is part of the culture of conservation biology, and quite rightly so, to focus on habitat destruction—which, driven by the constant expansion of human numbers and activities, is the overall major cause of the ongoing sixth mass extinction. But in the case of the slow-breeding orangutan, which has been hunted for thousands of years, hunting and other direct killing by its dominant primate relative, rather than habitat loss, may now be a major cause of its populations plunging toward extinction. In fact, humans killing for food has become a major threat to wild primate populations globally. Fortunately, orangutans are protected in several nature reserves. In Sumatra, they live in the massive 2-million-hectare (almost 5 million acres) Leuser ecosystem, which includes the Gunung Leuser National Park. In Borneo, there are several reserves in both Malaysia and Indonesia.

Chimpanzees and Bonobos, Our Closest Relatives

Half a century ago, Mary-Claire King and Allan Wilson published a groundbreaking paper demonstrating that the structural genomes of chimps and people are extremely similar. They proposed that the prominent phenotypic (outward observable characteristics) differences between us and our closest ape relatives are created by differences in the expression of genes rather than differences in genes themselves.[28] Subsequent studies of chimpanzee and bonobo populations in the field and laboratory have underlined a key finding from the viewpoint of this book—extinction of populations of apes closely related to us makes understanding *Homo sapiens'* behavior more difficult. As Gruber and Norscia have put it, "Taking into account variation both within and between *Pan* (chimpanzee) populations provides a more nuanced view of the behavioral flexibility of our closest living relatives, who can be considered to sit along a continuum shaped by various ecological factors within their surrounding environments. Such an approach enables more valid reconstructions of the last common ancestor between *Pan* and *Homo*."[29]

Some of Paul's fondest memories are of his adventures with his wife Anne and daughter Lisa while doing research at Tanzania's Gombe Stream, Jane Goodall's famous site for studying chimpanzees. By the time the Ehrlichs arrived in 1972, Jane had made some very intelligent decisions—such as decreeing that people should not have physical contact with their closest living relatives. One reason was that she did not want the chimps to learn how much stronger they were than people. The other was that she wanted to protect the chimps from human diseases. That latter reason has recently been shown to have been extremely wise. Chimps are in fact very susceptible to rhinovirus C, a virus that evolved when people settled down to practice agriculture. We have evolved with it long enough that it merely causes the common cold in us, but it can be lethal to chimps, who lack a coevolutionary history with it. Sadly, the virus has become widespread in chimp populations and "represents a grave threat to chimpanzees all across Africa."[30]

As if habitat destruction and poaching were not enough, tourism, even when involving people who are generally chimp fans, can also endanger them.[31] Although conservation efforts to preserve populations of a single organism are complicated, there are instances that we would encourage. One such case is the attempts to protect bonobos (also called "pygmy chimpanzees")—humanity's closest living relatives. Bonobo populations have been undergoing rapid decline, with evidence cited in the 2021 IUCN Red List indicating they may have lost half of their number between 2000 and 2015.[32]

In addition to the ethical reasons for preserving these intelligent, relatively peaceful humanoids, there are selfish reasons as well: the potential they provide for informing people about the roots of our own physiology and behavior.[33] Further, bonobos have characteristics that could be beneficial if adopted by *Homo sapiens*.[34] Bonobos do not fight with nuclear weapons or even guns, but rather often settle disputes by genital rubbing. Just a small genetic divergence from chimpanzees seems to have made a key difference in behavior between chimp and bonobo females, where the latter play a major role in making bonobo societies relatively peaceful.[35]

In another important study, researchers compared both chimpanzees and bonobos on a wide range of cognitive challenges, such as tests and puzzles.[36] The findings suggested that environmental (including cultural) differences could alter mental skills relatively rapidly in evolutionary time, at the most, it has been a mere 2 million years since the two kinds of chimps had a common ancestor[37]—a conclusion useful in thinking about human behavior. Of course, bonobos, like other apes, also play important ecological roles such as seed dispersal in the tropical forests they inhabit,[38] but it's their closeness to us that, psychologically, provides the greatest impetus for protecting their remaining populations.

Bonobos are threatened by the usual anthropogenic pressures on non-human great apes: hunting for bushmeat (where the demand has steadily increased) or sport, disease,[39] and habitat destruction for purposes of agriculture, logging, mining, and infrastructure development.[40] In some areas, bonobos come into conflict with

human beings. These problems are increasing because of the burgeoning size of human populations and contact in degraded habitats, often involving such things as ape "raids" on crops. Direct
attacks on human beings do occur, and that of course can lead to
both "revenge" killing by people and loss of support for conservation efforts.[41]

Heroic attempts are being made in some areas to protect the remaining bonobos, and even reintroduce groups into now-empty
habitat. The NGO *Amis des Bonobos du Congo* ("Friends of Congo's
Bonobos") reintroduced a dozen once-captive bonobos into vacant
bonobo habitat. They even had trackers follow the bonobos to protect them and ensure that they were adapting to their newfound
freedom. The bonobos, however, "became increasingly confused by
the persistent presence of the trackers."[42] A male got separated
from the troop, felt threatened by the trackers, and called the others
back to help him. One tracker threw a large stick at the troop to disperse them, but instead, they attacked. In the resulting brawl, one
tracker was knocked unconscious. When he recovered consciousness and the bonobos saw he was alive, they made gestures he interpreted as asking for forgiveness. He had suffered serious wounds
to his face, requiring half a year of reconstructive surgery. The incident underlines the problems of protecting the bonobos in the
presence of growing human populations. The general opinion is
that despite some hopeful community-based approaches to bonobo
conservation,[43] it is unlikely that there will be any wild bonobo populations left at the end of the century.

Costa Rica's Squirrel Monkeys

The mountains and the remaining patches of natural vegetation
around San Vito, a small city in southern Costa Rica, are incredibly beautiful. The largest remnant of the cloud forests near the

city, a mere 200-hectare (almost 500 acres) patch, is found in the Las Cruces Botanical Garden. In the late 1990s and early 2000s, Paul and Gerardo together with colleagues carried out a study in the region evaluating the value of small forest patches for mammal biodiversity.[44] One day, Gerardo and two of his students visited the Gamboa family, owners of a farm (*finca* in Spanish), to get information on the mammals seen in the finca. Gerardo was intrigued when one of the Gamboa brothers mentioned that sometimes they had seen squirrel monkeys (*Saimiri oesterdii*), a species endemic to the lowland forests of a small region along the Pacific coast of Costa Rica and Panama. The cloud forests and the elevation of San Vito were not the suitable habitat recorded for the species. All known records were below 1,200 meters above sea level (MASL) (3,937 feet).

Before leaving the finca, Gerardo asked them to call him if they sighted any of the monkeys. Almost miraculously, a few days later, the Gamboa brothers called him very excited—a small troop of squirrel monkeys were crossing a pasture in the finca to reach the forest along the Java River. When Gerardo and his students arrived at the finca, they managed to photograph the last squirrel monkey of the troop as it crossed the pasture. It was an amazing discovery that led to new insights about the species. After collecting data for more than twenty years, Gerardo and colleagues had gathered enough information about the distribution and status of this very endangered species to finally publish it.[45]

The discovery of the squirrel monkey population in San Vito proved to be important because it was found in a large region where the species had not been recorded and in the end, accounted for almost 40% of its total distribution. This species is endangered with extinction mainly because of habitat fragmentation and loss, the pet trade, and because they become infected with diseases when in contact with domestic animals.[46]

Other Primates

Primates, which include monkeys, lemurs, apes, and our other close relatives as well as ourselves, are, as we have indicated, important to humanity in part because they are close relatives. For many of us, they are important for the same fundamentally ethical reasons that parents, cousins, grandchildren, and other relatives are important to us—shared genes, shared cultural features, shared evolutionary history. But there are also practical reasons to preserve our fellow primates, as suggested in our discussion of bonobos. They tell us a lot about ourselves and our behavior—although fishes, octopuses, rats, and wolves (just to name a few) also can and have improved our scientific understanding of behavior. Anyone who has spent time with chimps can't avoid learning about our evolutionary roots as well.

Various primates are also culturally important to some human groups who have integrated their view of them into their religious beliefs, as exemplified by the Hanuman langur, the daring, clever, loyal Hindu monkey god. Monkeys even play a role in some peoples' food systems. For example, in Sulawesi people tolerate crop-feeding macaques because of their role in local folklore and their help in harvesting crops. The macaques eat only the fruit of cashews and leave the nuts on the ground, where farmers collect them. Unfortunately, the major role primates play in some human food systems is as a highly valued, protein-rich food item.[47]

Another contribution primates make to humanity is the roles they play in the maintenance of tropical ecosystems (with few exceptions, the main primate that is happy outside of the tropics is *Homo sapiens*). Primates play multiple roles in ecosystems, as, for example, predators, prey, frugivores, and pollinators. Large primates can be very important seed dispersers in the forests, helping keep tree populations genetically healthy and aiding forest regeneration. Despite these critical roles in our lives, the existence of primate populations around the world, from squirrel monkeys in Costa Rica to orangutans in Borneo, is severely threatened by habitat destruction in the form of logging, industrial agriculture, mining, drilling, and dam and road building. Primates are also killed for bushmeat. More than half of primate species are now considered to be threatened with extinction, and some three-quarters have declining species populations.[48]

Elephants, the Giant Ghosts

Elephants play an important role in ecosystem function. We like to emphasize two very different contributions they make. They are crucial for maintaining the savanna ecosystem because of their feeding habits. Elephants eat bark and destroy some trees, thus preventing the savanna from becoming a thick forest.[49] Elephants also provide humans with aesthetic ecosystem services, as anybody who has taken a child to a zoo can attest.

The African savannas where our species originated a few hundred thousand years ago were full of some of the largest land mammals—and to a diminished extent still are, including real giants such as elephants, rhinos, hippos, and giraffes. However, in many regions, today's elephants are but ghostly remnants of the gigantic herds that existed until recently. Although people have impacted those landscapes for thousands of years, the number of individuals of those giant species surviving until relatively recently is mind-blowing. Early aerial photographs where one can see a thousands-strong elephant herd now seem to be taken from a science fiction movie.

It is even more difficult to credit that those majestic views are now history. The decline of elephants in the last few centuries has been nothing but tragic. The African elephant population in the 1800s is thought to have been around 26 million individuals. Demand for bushmeat and ivory encouraged hunting that reduced elephant populations to roughly 10 million by 1913,[50] and to an estimated 1.3 million by 1979.[51] By 2021, the populations had dropped even further, to around 400,000.

The now-illegal ivory trade has been the leading cause of this steep decline. There is a long history of ivory use for piano keys and billiard balls, for example, and now, ivory is mainly being used for ornamental luxury goods, especially in China. It is estimated that some 36.3 tonnes (forty tons) of elephant tusk are illegally traded annually, although policy changes in China are somewhat reducing the incentives of poachers.[52] Nevertheless, well-organized gangs control the illegal trade from the African bush to high-rise skyscrapers in places such as Shanghai and Hong Kong.[53] Currently, an elephant is illegally killed every thirty minutes, totaling some 17,000 annually. That means that around 8% of the total population is disappearing every year. If these massacres are not halted, wild African elephants are unlikely to survive outside very few places by 2040. Strong international pressure forced China to impose (in a monumental effort) a ban on ivory imports in December 2017, but the current ivory trade situation is unclear. The unusual Central African population of "forest elephants," for example, suffered a "devastating decline" and its population size has reduced by about 80%.[54]

So far in this discussion, we've focused on the statistical impact of the decline in elephant populations, not on the elephantine (as opposed to "human") impact. Do we have any indication of what it has meant to millions of smart, sensitive social animals killed by human hunters over the last few centuries, or dying because *Homo sapiens* has occupied their home ranges? Can we imagine the impact on the survivors of the violent loss of companions? What does it mean to an elephant who has most of its closest friends murdered in minutes by those seeking their tusks? Should we even consider what it means to a young elephant to become an orphan? What does a female elephant suffer when her offspring that she carried for

Elephant and vultures
This illustration depicts poachers killing both targeted elephants and vultures. Poison is the means, watermelon the vehicle. An elephant will die outright. The vultures will die scavenging the poison-laced carcass. Eliminating the birds is advantageous to the poachers, as circling vultures can reveal elephant killing sites to local authorities.

twenty-two months and who she tended as a child for as much as a decade is shot and killed? What are the consequences for a population to be deprived of older individuals that may be invaluable sources of lessons in childcare or information on the history of the group, including, say, the locations of critical water sources in time of extreme drought?

We don't know the answers to these questions for sure, but having watched a group of elephants play with their kids in a lake or seen a pair of adults simultaneously squirting water down each other's throats, we suspect that elephants (and many other animals) can suffer emotional shocks not so unlike those we can suffer. If you want to think harder about such questions, read Carl Safina's brilliant book *Beyond Words*.[55]

Asian elephants are found from India to the Malay Peninsula and Borneo. Like its African relatives, this magnificent social species is threatened by habitat loss and fragmentation. Ivory trade is not as important a factor of decline for this species, but illegal trade to keep them in captivity to be domesticated for tourism and work is a problem for some populations.[56] It is estimated that there are currently around 40,000 to 50,000 Asian elephants, a drastic decline from the 1 million estimated to be extant in the early 1900s.[57]

Rhinos, Tropical Hornbearers

"Rhinos have an ancient, eternal beauty. With their massive bodies, clad in thick folds of prehistoric armour topped with a magnificent scimitar horn, they fascinate like few other creatures . . . they are the largest land animal in the world after the elephant." This is an eloquent description by Lawrence Anthony of the northern white rhino.[58] Anthony was a South African conservationist who tried to save the last northern white rhinos that lived in the Democratic Republic of Congo, in a region dominated by a guerrilla group called the Lord's Resistance Army, one of the most ruthless armies in the world. The remarkable story of his efforts, told in his book *The Last Rhino*, is another example of the heroic struggle of many conservationists to save endangered species.

It is difficult to explain why one has the urge to try to save disappearing populations and species many times, even at the risk of one's own life. At one point of his efforts, Lawrence did manage to convince the leaders of the Lord's Resistance Army to work with him to save the last lone rhino survivors. Ultimately, however, Lawrence was not able to save the Congo rhinos for multiple reasons beyond his control. He died from a heart attack in 2012, and remarkably, his death was not unnoticed by those other endangered giants, the wild elephants in his private wildlife reserve in South Africa: "Two herds walked half a day to his home in a funeral-like procession to mourn his passing even though there was no apparent way they could have known he died. They stayed for two days."[59]

Like elephants, rhinos are large herbivores. Elephants can weigh more than 5.4 tonnes (six tons); rhinos can approach 1.81 tonnes (two tons) and are another of the most threatened elements of biodiversity. At one time, an estimated 500,000 of today's five rhino species roamed large areas of Africa and Asia. No more. The most abundant survivors are the African white rhinos, which are grazers, grass-eaters. That gives them their name from *"weit"* — Afrikaans for wide—describing their lawnmower-style mouths. They were on the brink of extinction with 100 or fewer in the wild a century or so ago, but there now are close to 20,000, the largest populations being in South Africa.

Black rhinos are browsers; their hooked mouths are great for stripping leaves off bushes. Black rhinos numbered only some 65,000 in Africa around 1970, and they dropped to fewer than 2,500 by 1995. A genetically distinct population called the West African black rhino (*Diceros bicornis longipes*) was declared extinct in 2011.[60] That population aside, intense conservation efforts have allowed the number of black rhinos to double from their 1995 numbers to today, although this is still a dangerously low level. Rhinos are prime targets for poachers because their horns (the chemical equivalent of human fingernail clippings) have huge monetary value for use as ceremonial dagger handles in parts of the Middle East and in ground form, nonsensically, as an aphrodisiac (chew your fingernails, guys). For a while, we thought that Viagra (which actually works) could save the rhinos—but sadly, we've been told that dealers are now adding that chemical to ground horn, making the horn appear to be effective.

White rhinos are herd animals and not especially aggressive. Black rhinos tend to be loners, and isolated males do not take kindly

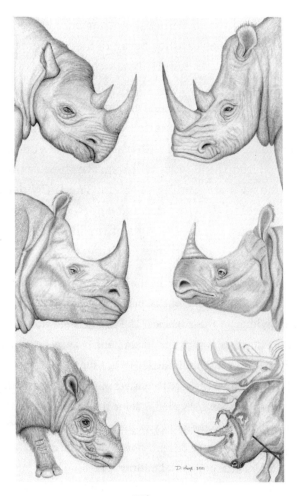

Rhinos

The top two species in this illustration, the black rhino (*top left*) and the white rhino (*top right*), are found in Africa. Both species have been heavily hunted for their horns, which are used in traditional medicine in China and Vietnam. The black rhino is more endangered, with fewer than 3,000 individuals surviving. The northern subspecies of the white rhino is extinct, while the southern one (found mostly in South Africa, with small relocated populations in Kenya, Namibia, and Zimbabwe) is estimated to have around 20,000 individuals.

The other three living rhino species are found in Asia. The greater one-horned or Indian rhino (*middle left*) is found in India and Nepal. It is on the rebound from near extinction, but remains at risk to poaching and habitat loss. The Javan or lesser one-horned rhino (*middle right*) is critically endangered and can be found in only two sites in Java's Ujung Kulon National Park, where poaching and habitat loss jeopardize the thirty or so remaining individuals.

The Sumatran rhino (*bottom left*) is also critically endangered. The last twenty to thirty individuals are now restricted to the Leuser ecosystem in northern Sumatra. The final illustration (*bottom right*) appeared on a wall in France's Chauvet Cave 32,000 years ago and is thought to depict a woolly rhino.

to approaches by human beings. Another behavioral difference partly accounts for black rhinos being in more trouble than their white rhino relatives. As loners, black rhinos only mate when a male runs into a female. As black rhino population density declines, the chances of males and females encountering each other drops, as do the chances of the population persisting. White rhinos, although they share the problem with black rhinos of gullible humans thinking powdered rhino horn will improve their sex lives, don't have as serious a mate-finding problem. The IUCN declared the Western black rhino (*Diceros bicornis longipes*) extinct in 2011, with the last individuals dying as victims of the illegal horn trade. Gerardo contemplated their last days, writing

> I can imagine the great distress of the last two rhinos, perhaps a female and her calf, wandering for weeks, perhaps months, without finding another individual of their species. The two of them alone, day and night, in the savanna. Africa and its immensity surely seemed even more immense for those individuals. Over time they learned to avoid humans, who pursued them so relentlessly. Alert to any strange smell or noise, they managed to evade for some time a death sentence handed down thousands of kilometers away, perhaps in an office in Shanghai or Beijing, where the price of their horns was determined. A chain of corrupt diplomats, police and military personnel, businessmen, and traffickers would be involved in transporting the horns to their destination, where their price would reach thousands of dollars paid by unscrupulous buyers. One day the savanna echoed with a burst from an AK-47 military assault rifle. The female collapsed in the dust. The crazed calf ran following its survival instinct, only to be felled by another burst of fire almost immediately afterwards. The mercenaries broke the silence in which they had remained during the stalking with laughter. Happy, they ran to remove the horn from the female. The calf, still small, did not even have a horn but fresh tender meat. Quickly as they had appeared, the mercenaries disappeared, leaving behind the

corpses of the last rhinos in that corner of the planet rotting in the sun. No one will ever be able to see a member of that population of black rhinos again. The Earth is now a sadder and more desolate place.[61]

There are three rhino species in Asia that get less publicity than their African cousins, but they are also in trouble. The big, armored-appearing Indian (or one-horned) rhino is down from some half-million to fewer than 3,000 individuals, primarily in two genetically distinct populations, one in Assam and one in Nepal. This is an instance where the genetic structure of an endangered species has been examined, and its possible consequences for conservation discussed.[62] One interesting—but sadly unanswered—question is whether hybridizing individuals from the two Indian rhino populations would lead to reproductive problems that might harm conservation efforts.

Until about 2000, a small population of the Javan rhinoceros, a once-widespread close relative of the Indian, existed in Vietnam. The last individual of that population of this somewhat smaller, also armored-appearing beast was killed by poachers in 2010. After that population extinction, only a small population of far fewer than 100 Javan rhinos is thought to persist in Ujung Kulon National Park on the western end of Java. These few individuals are under great threat because their powdered horn is worth more than $22,000 per kilogram ($10,000 per pound) on the Chinese medicine black market.

The Javan rhino is one of the least known of the big mammals. None are alive in captivity, only four have ever been exhibited in zoos (all dead by 1907), and researchers have rarely seen the elusive creatures alive. One possibly apocryphal story is of a scientist spending a year studying them without a sighting—working only on footprints and droppings. Near the end of the year, one walked through his camp, but he was out busy searching elsewhere.

The last of the extant rhinoceros' tribe is the smallest and hairiest—the Sumatran rhino, once widespread in Southeast Asia. It is a disappearing denizen of montane tropical rainforests, and its situation is truly "hairy." As late as 1980, one Sumatran population was estimated to be comprised of about 500 rhinos. Poaching rapidly pushed that population to extinction. Now, fewer than eighty individuals may survive in tiny, fragmented populations on Sumatra and Borneo, but the only one with a chance of persisting is in western Sumatra. Efforts at captive breeding in reserves are now being undertaken, and after much effort, a female of this elusive creature was trapped on Borneo in November 2018. Ironically, it is a relative of the extinct woolly rhinoceros that was a dominant herbivore over much of Europe and northern Asia during the Pleistocene, fading away at the end of the last ice age, about 10,000 years ago.

Pangolins, a Mammal with Scales

Pangolins are mostly cat-sized or smaller, anteater-like creatures covered in large scales. There are just nine known species, all of which are found only in Africa and southeastern Asia. There are lucrative black markets for the animal in China and Vietnam, where its scales are pulverized for use in traditional Chinese medicine and its meat is served in restaurants as a delicacy (chapter 10). The magnitude of the illegal trade is unbelievable. From 2016 to 2020, an investigation showed that more than 207 tonnes (228 tons) of pangolin scales were confiscated, representing hundreds of thousands of individuals.[63] Sadly, several of the nine species of pangolins are endangered or at the brink of extinction because of the illegal trade coupled with habitat loss.[64]

Pangolins represent a very unique group of mammals. Although little is known about their biology, pangolins feed almost exclusively on termites and ants. Each individual can eat up to 200,000 ants and termites every day. This makes them very important in keeping the populations of those species from growing unchecked, thus helping to maintain the functioning of their native ecosystems.

Bison and Prairie Dogs

North American bison were one of the most abundant big mammals in the world. It is estimated that some 30 to 60 million roamed the grasslands in North America from southern Canada to northern Mexico when the first Europeans colonized the continent in the 1500s.[65] When the colonizers expanded to the western United States, bison began to be killed in large numbers, but the construction of the Transcontinental or Pacific Railroad in the 1860s unleashed the extermination of most bison populations. The large herds were rapidly hunted to extinction, and by 1870, only small herds persisted scattered across their range.

One last large herd, however, estimated at 4 million strong, known as the Southern Herd, had managed to survive in Kansas, Colorado, New Mexico, Oklahoma, and Texas. Ultimately, this herd was exterminated in a mere five or six years under the brunt of extensive railroad-supported market hunting. The slaughter, which began in 1871, was brutal. From 1872 to 1874, the Santa Fe railroad company transported more than 450,000 bison furs, and the total of bison killed was estimated to be around 3,100,000 individuals!![66] The US government actively supported the extermination of the bison herds, especially to destroy the last strongholds of the native people by removing their main food supply.[67]

Alarmed by the disappearance of the bison, some conservationists managed to create the Yellowstone National Park in 1872, which later protected the last bison herd on the planet.[68] Fortunately, bison do not require huge populations to breed, so that even though a few small populations survived, they could gradually come back from the brink—not to the point of resuming their exosystemic roles, but to a point where they potentially could if any attempt is made to refurbish American prairie ecosystems. Bison grazing patterns is just what the doctor ordered for that ecological restoration task.

Black-tailed prairie dogs (*Cynomys ludovicianus*) were a common sight in the grasslands where the North American bison roamed. There are five species of rodents we call "prairie dogs," but the black-tailed prairie dog was by far the most abundant and widespread, being found from southern Canada to northern Mexico, occupying the extensive grasslands that dominated the landscape in that vast region east of the Rocky Mountains. Prairie dogs live in colonies, and the size of some of those original colonies is difficult to imagine. Edward Mearns estimated that a single colony in Texas extended for more than 400 kilometers (almost 250 miles) in the late 1800s.[69] Habitat destruction to convert the grasslands to crops and poisoning the prairie dogs because of the common but incorrect idea that they compete with cattle has caused a reduction of 98% of their range and populations.

Prairie dogs are considered a keystone species (a species that is influential beyond its abundance within natural animal communities) because they are fundamental to maintaining the grassland ecosystem where they live. They change the landscape by preventing the invasion of the grasslands by scrubland. They live in burrows and feed on plants. Digging their burrows and destroying high-standing vegetation influences the physical and biological characteristics of their habitat, such as soil fertility, water infiltration, and plant community structure, such as the relative abundance of certain plant species.

Losing the prairie dogs is associated with the loss of many other species that depend on the prairie dog grasslands, such as the black-footed ferret (*Mustela nigripes*). Additionally, losses in ecosystem services such as nutrient cycling, runoff and water infiltration, and supplying forage for cattle and wildlife, is estimated at millions of dollars annually in regions such as Chihuahua, Mexico, where it has been evaluated.[70]

Great Whales and Other Marine Mammals

Human beings have not yet pushed any great whale species to extinction, but seven out of their thirteen-species group are classified as endangered or vulnerable. Overall, large whales have been shockingly reduced by overexploitation even decades after most whaling ceased. The original anthropogenic threat to their existence was predation to obtain meat and oil. Although some whales were hunted in prehistoric times, the devastation of great whales such as sperm, blue, right, and bowhead became a global business in the seventeenth, eighteenth, and nineteenth centuries as industrial-style whaling accompanied the industrial revolution itself and a desire for fuel for oil lighting.

For example, the North Atlantic right whale (so-called because it swam slowly and was rich in fat—therefore the "right" whale to pursue) was reduced from roughly 10,000 individuals prior to exploitation to some 300 individuals when given protection. The great whales today are not only far reduced in numbers from preexploitation levels, but their important contributions to marine ecosystem functioning, such as the huge sequestration (i.e., capture) of CO_2 greenhouse gas in their massive and long-lived bodies, and the vertical circulation of nutrients as they dive to feed and come up to the surface to breathe, are likewise much reduced.[71]

There is an interesting connection between what has happened to whale populations and what is likely to happen to polar bear populations. During past interglacial periods, warming also threatened polar bear populations by reducing the extent of sea ice. It has been hypothesized that during those periods, survival of polar bear populations may have been enhanced by the availability of stranded whale carcasses as supplemental food.[72] Polar bears today frequently do feed from whale carcasses—they are concentrated sources of calories on arctic shores—but the anthropogenic whale holocaust may well end up depriving polar bears of what could be a badly needed supplement in the ice-impoverished arctic of a warming world.

Some smaller whales are also in trouble—species that were not highly valued during the great whale holocaust. One species Paul met when he lived with the Inuit on Southampton Island in 1952 is the beautiful beluga or white whale, which runs up to 6.7 meters (twenty-two feet) in length. This creamy white marine predator of arctic and subarctic seas was a prized catch for the Eskimos or Inuit people, to whom whale skin, "*muktuk*," was such a delicacy that Paul recalls the Aivilikmiut expression for "tasty" is "*ah nah muktuk*." Belugas can enter freshwater in big rivers and consume fish and crustaceans there, especially along the edges of arctic ice flows (they lack dorsal fins, making it easier for them to swim under ice in shallow waters). They move together in pods (the term for social groups of whales), heading south as ice forms in the fall and north following the edge as it melts in spring, and communicate with each other by a wide variety of clicks, grunts, chirps, whistles, and other sounds produced in their flexible bulbous foreheads. That makes them look cute and gives them many unique facial expressions.

Belugas are vulnerable to reduction of their prime food source by commercial fishermen and to toxic chemicals, habitat degradation (including reduction of sea ice), energy exploration noise pollution, and harassment.[73] They are hunted by polar bears and orcas, as well as by native peoples in some populations for subsistence. Orcas have become a larger hazard for belugas in recent years due to the decline in sea ice, which has given them access to beluga breeding grounds in shallow estuaries. There are many identified populations of belugas now, some substantial, but some consisting of only a few dozen individuals. Globally it is estimated there are about 135,000 adult belugas in existence. How long they may persist in the face of melting ice, orca movements, and other risks is an open question.

The narwhal is the beluga's closest living relative. Its most distinguishing characteristic is its long spiral tooth (tusk), which is full of nerves, has a sensory function, and may play a role in male dominance battles. Narwhal ecology and behavior are in many ways like those of belugas, and narwhals are often found with belugas primarily in the waters off northern Canada, Greenland, and western Siberia. Narwhals, for example, seem to show responses like belugas to the northward spread of orcas as the arctic warms.[74] They appear to be slightly less abundant than belugas, not surprisingly because of their more restricted range. Of the estimated 125,000 adult narwhals today, hundreds are killed by Inuit hunters annually in Canada for food and tusks (sold as souvenirs and ivory carvings). Controls are in place for narwhal ivory exports from Canada and Greenland, and the trade seems presently sustainable, although how long that will last may depend on the rapidly changing environments and politics of the north polar regions. Like polar bears and belugas, narwhals' fates are tightly tied to climate disruption, and the climate's impacts on orca distributions, commercial fisheries, tourism, and energy exploration in newly opening polar waters.

Vaquita Porpoise

Whales are not the only cetaceans to have suffered gravely at the hands of *Homo sapiens*. The vaquita is a child-sized porpoise, a denizen of the Gulf of California. There is only one remaining population in the northern part of the Gulf. It is plummeting toward extinction because it is an incidental victim of fishing nets (gill-nets), many of them illegal. These air-breathers get tangled up and then drown. The Mexican government has placed restrictions on the use of gillnets where the vaquita lives, but they still get caught by poachers who, ironically, are after another endangered species that occurs with it, the giant totoaba fish. That fish is also in decline from increased salinity of the northern Gulf due to diversion of freshwater from the Colorado River, as well as from overfishing. Sadly, the Chinese put special value on totoaba swim bladders (medicinal) and flesh (a delicacy). The bladders can sell for more than $100,000 each—and they have even been referred to as "aquatic cocaine."[75] Poachers and smugglers are active despite strong law enforcement, and a secondary victim is the vaquita, whose head unfortunately fits perfectly in the net mesh designed to catch totoaba.

So grave is their plight that the issue is whether to attempt a captive breeding program to save the vaquita, or to hope that Mexico can crack down hard enough on the poaching to give the vaquita a chance. Captive breeding of porpoises (unlike dolphins) is a tricky business[76] and might easily fail at a cost of killing some of the roughly twelve to fifteen individuals now surviving in nature. Complex efforts to move some of the vaquitas into safety even involved the use of trained US navy dolphins to herd them.[77] There is also the moral hazard that attempts to captive breed might lead the Mexican government to slack off on its anti-poaching efforts.

In September 2018, what was assumed to be a mother and calf pair were photographed, and acoustic records indicated a small population staying together. By 2019, however, scientists remained uncertain of the status of the remaining vaquita population. In 2023, a mere ten to thirteen vaquitas, including two young individuals, were estimated to be the only survivors in the wild.[78]

Flying Mammals

One of the most frightening examples of mammalian population extinctions going on right now in North America is the decimation of bats. Bats play critical roles in ecosystems and are especially major players in pest control. For instance, one population of Mexican free-tailed bats in Texas is estimated to consume 226.7 tonnes (250 tons) of insects each summer night. This protection of the summer cotton crop provides an ecosystem service valued at several million dollars every year. Overall, bat predation on agricultural pests in the United States is estimated to be worth almost $4 billion annually.[79]

North American bats are subject to two major new threats. One is a nasty fungus disease known as white-nose syndrome that has been attacking wintering populations of bats that hibernate in the Northeast. The fungus, which thrives in the cold, coats the muzzles and other parts of hibernators, awaking them. They try to forage, but in the food-short winter, they burn their fat reserves and starve to death. In 2015, the journal *Science* reported, "researchers identified massive population declines, ranging from 60% to 98% for all six North American bats studied, and extensive local population extinctions, the most severe being for the northern long-eared bat, which has disappeared from almost three quarters of its former hibernation sites."[80] The most likely hypothesis was that the pathogen was transported across the Atlantic on a spelunker's boots. We can only hope that North American populations will soon begin to become more resistant.

The second major new threat to North American bats is more directly tied to human activity. Sadly, it is caused by devices that are trying to help shield bats and the rest of Earth's organisms from catastrophic climate disruption—wind turbines. One study estimated that turbines could have killed over 600,000 bats in 2012 alone.[81] Here, help could be found by minor adjustments in turbine blade speed, slightly raising the wind speeds at which they start.[82] With wind turbines becoming ever more common, these are important steps to take, especially because bats are also threatened, in addition to white-nose syndrome, by habitat loss and climate disruption.

Beavers, Engineers of the Meadows

One rodent that tends to call attention to its presence in no uncertain terms is the North American beaver (*Castor canadensis*). Or perhaps we should say it calls attention to its absence. California, for instance, used to have huge areas of wetlands that have now disappeared. In part, that is because most of the United States (including most of California) once had myriad populations of beavers made up of several hundred million individuals. Through their act of damming streams, these "ecosystem engineers" had huge impacts on landscapes, often to human benefit—creating as many as 250 million beaver ponds and building some of the soils that made the United States an agricultural power. In the process of building secure habitats for themselves and their lodges, rich in the trees they favored as food, beaver populations created wetlands that purified water, controlled its flow, and provided habitats for a wide range of other animals, including endangered species of migrating songbirds as well as, naturally, frogs and toads and fishes beloved of fly fishers and gourmets.

That landscape pattern almost ended in the eighteenth and nineteenth centuries because Europeans decided that beaver fur made great hats, and hunters and trappers obligingly wiped out population after population to feed the demand for pelts, making huge profits but coming close to exterminating the species. Fortunately, public concern, and the beavers' own resilience, saved the situation. Once the slaughter for their pelts ended, beaver populations rebounded, and reintroductions restored some beaver populations in areas such as the eastern United States where they had been largely extirpated.[83]

Restoration of the overall pre-European invasion situation is, of course, impossible, but efforts to restore beaver benefits in some areas continue. Beavers and people are now cooperating to build beaver dam analogs. Cooperating people start a barrier in the general style of a beaver dam, driving poles into stream beds and then weaving sticks among them. Then, local (or sometimes introduced) beavers pitch right in. They gnaw down nearby trees and add them to the beaver dam analogs, and also pack in mud. The dams are aimed at restoring badly eroded streams, now running through narrow clefts. The dams force water onto the floodplain, forming marshes and pools, as well as recharging aquifers. They provide irrigation for the alders and willows that colonize the streamsides and poolsides. In the process, beaver dam analogs supply habitats for populations of many species that previously could not maintain themselves in an area.

However, not everyone is friendly to beavers or their dams. The eager beavers often mess with irrigation ditches, girdle valuable timber or eat its saplings, and thoughtlessly build dams that flood highways, railroads, fields of crops, and other infrastructure, actions that often don't endear them to their human neighbors. But beaver dams could be considered a sign of a switch in human attitudes toward an eventual situation where *Homo sapiens* learn to encourage biodiversity, live with it, and benefit from it.

Corollary

What we have described here and in the following chapters of the book is a small sample of the nearly 40,000 species of mammals and other vertebrates that are losing countless populations because of human activities. Most of the mammals in trouble are not as iconic as bonobos or as ecologically important over broad areas as bats, but are interesting and inherently valuable nonetheless.

Climate change may be especially threatening to rare, highly specialized species, whose loss is no less a tragedy or danger to humanity for being less well-known.

All those populations and species represent an evolutionary legacy that we, all humans, have inherited from billions of years of evolution. By losing them, we are changing the evolution of life; we are eroding Earth's capabilities to maintain current life in general, and human life in particular. Humanity must undertake an effort without precedent to save all these disappearing organisms. Ironically, by doing so, we would likely contribute to saving ourselves.

Chapter 4

Bird Songs Long Gone and Declining

IN THE 1800S, ESKIMO CURLEWS were one of the most abundant birds in the Americas. They also made one of the longest bird migrations, from the Arctic to Patagonia in the very bottom of South America. In the late nineteenth century, up to 2 million birds were killed every year along their migration routes. As a result, despite being as numerous as they were, their populations have dwindled rapidly until they became extinct. In his 1952 novel *Last of the Curlews*, Fred Bodsworth wrote a sad epitaph for this species: "The odd survivor still flies the long and perilous migration from the wintering grounds of Argentine's Patagonia, to seek a mate of its kind on the sodden tundra plains which slope to the Arctic Sea. But the Arctic is vast. Usually, they seek in vain. The last of a dying race, they now fly alone."[1]

Many other bird species have suffered the same fate as the Eskimo curlew, some of them abundant and some rare. The impact of human activities goes a long way back. The "future of those birds—and innumerable other creatures," as we have been pointing out, "is in our hands. Our horrendous ethical and esthetic sin is one that could be assuaged if we as species dedicated ourselves to preserving—instead of destroying—nature."[2]

Unfortunately, the Eskimo curlew is just one of some 130 species of birds that have become extinct since 1500.[3] However, it has recently been demonstrated that the rate of species losses is increasing.[4] Under the "normal" extinction rates occurring in the last few million years, only one bird species on average should go extinct in a century. Sadly, the current extinction rate indicates that the birds lost in the last 500 years should have taken 18,000 years to be lost if those earlier "normal" rates had pertained.[5] Habitat loss and fragmentation, legal and illegal trade, introduced predators and diseases, and climate change are the main causes of this dramatic increase in extinctions and population declines.

Island birds have been especially harshly hit by extinctions, beginning with the dodo and including a plethora of other species such as the Great Auk (*Pinguinus impennis*), last known from Iceland, the Canary Islands' oystercatcher (*Haematopus meadewaldoi*), and the Hawkins's rail (*Diaphorapteryx hawkinsi*) from the Chatham Islands.[6]

The Hawaii archipelago has lost many birds such as a crow, honeycreepers, and including species such as Kauaʻiʻōʻō, (*Moho braccatus*) from Kauaʻi island. Guadalupe Island off Baja California, Mexico, once known as the Mexican Galápagos, lost some eight species, such as the caracara (*Polyborus luctuosus*), of which the last eleven known individuals were killed by a professional hunter to supply the European collector's market.[7] And on the island of Guam, in the Pacific, the introduced brown tree snake (*Boiga irregularis*) caused the extinction of at least ten bird species in less than four decades.[8]

On the continents, many interesting and extraordinary species have been extirpated, mostly because of habitat loss, hunting, and illegal trade. The Imperial woodpecker (*Campephilus imperialis*) was decimated by logging of the old-growth forest where it lived and was probably hunted to extinction. The last reliable record is from 1956. It is said that when searching for the last individuals, a local reported that Imperial woodpeckers were "a great piece of meat."[9] The Atitlán grebe (*Podilymbus gigas*), known from famous Lake Atitlán, in Guatemala, vanished in 1989, mainly because of habitat loss, hybridization with other grebe (*Podilymbus podiceps*), competition for food with introduced trout, and interference of civil war with conservation efforts.[10] The Carolina parakeet (*Conuropsis carolinensis*) was exterminated because it was considered an agricultural pest in eastern North America. Sadly, the last individual died in captivity in the Cincinnati Zoo in 1918.[11]

The first Western Hemisphere vertebrate to go extinct in modern times was the Labrador duck (*Camptorhynchus labradorius*). It was a sea duck species closely related to Steller's eider. Its specialized bill suggests that it fed on mollusks in sediment, perhaps mostly snails and clams. Its population, never large, declined rapidly after 1850, but the cause of its decline remains somewhat obscure. It was notoriously bad-tasting to our species, so overharvesting was unlikely. Perhaps the best explanation was its restriction to coastal waters off northeastern North America, an area transformed by human development in the period of its decline.

Perhaps the most historic avian extinction is that of the Passenger pigeon (*Ectopistes migratorius*). It didn't go extinct hundreds of years ago on some obscure Indian Ocean Island like the dodo, nor before the American Revolution in a nearly uninhabited area of the North Pacific like the sea cow. It went extinct in the good old United States of America, with the last captive bird dying in 1914. These pigeons once occurred in flocks of millions or even billions, where all the individuals bred in a single giant colony. It is thought that by forming such giant colonies (each of which could be considered a population), the pigeons overwhelmed their natural enemies. They bred and fled before hawks and owls could make a significant dent in their population, and before their enemies could themselves reproductively build big populations on a diet of pigeon.

But the situation changed in the nineteenth century, when *Homo sapiens* became a major predator who used traps and firearms. With railroads penetrating the forests of the Passenger pigeon's Midwestern strongholds, it became profitable to harvest the fat and tasty young (squabs) for US East Coast markets, and trapped adults were even forced to suffer being used as live targets in shooting galleries. Relatively quickly, the pigeon's populations were greatly reduced in size by market hunting, so that there were no longer the vast numbers available to swamp predators. Population after population of the pigeons dwindled away as the billions of birds rapidly became thousands, then fewer, and the sky-darkening flocks and gigantic breeding colonies disappeared.

But market hunting did not do away with the Passenger pigeon; by the end of the nineteenth century, this hunting had become uneconomic, and even though the population had plummeted, there were still thousands of surviving individuals. But these thousands couldn't make big enough breeding assemblages to swamp predators and find forests offering abundant acorns. In 1914, the last known Passenger pigeon individual, a female named Martha (after Martha Washington, the wife of the first president of the United States) died in the Cincinnati Zoo.

The Passenger pigeon is the most extreme example in modern times of the rapid disappearance of a superabundant species, one comprised of many populations. But like the bison, these birds may have been so common in the nineteenth century because the original *Homo sapiens* populations of eastern North America lacked the guns and railroads that were central in causing their demise. Additionally, Native Americans may have harvested just enough birds to keep their populations in the millions rather than billions. In any case, two ecosystem services provided by the pigeons were lost: supply of an abundance of "bushmeat" and reductions in the threat of Lyme disease by competing with deer mouse populations for food, especially acorns. Lyme disease has existed in North America since before the arrival of humans. Deer mice are now the major reservoir for the microbe that causes Lyme, and they have thrived in the absence of Passenger pigeons, as, sadly, has the nasty disease.[12]

Few examples of population extinctions leading to species extinction are as well-recognized as the progressive and rapid loss of the Passenger pigeons. Sometimes, forest fires are a boon to preserving forest-dwelling creatures. A classic example is the severely endangered red-cockaded woodpecker of the southeastern United States. This charming bird was once common, adapted to the broad expanses of long-leaf pine stands that once covered much of the area. The birds thrived in open stands of old pines, pines old enough to be infected by a heart fungus that eases the birds' tasks of excavating a nest hole (populations of another American woodpecker, the red-naped sapsucker, or *Sphyrapicus nuchalis*, also benefit from a heartwood fungus that eases nest excavation).[13]

The red-cockaded woodpecker's open-forest habitat is maintained by periodic fires during the growing season, removing dead pines (snags) and leaving the fire-adapted large trees. Unfortunately, the long history of fire suppression in the United States has negative consequences on these pyrophile (fire-adapted) species. Fire suppression and harvesting of pines before they grow old, along with frank deforestation, have pushed many populations of woodpeckers and numerous other species close to extinction.

A similar fire-dependent bird is Kirtland's warbler (*Setophaga kirtlandii*), which breeds in a part of Michigan where natural fires produced their preferred habitat, stands of young jack pines extensive enough to contain grassy clearings. The area of such habitat had been increasingly restricted by human activities, making the warbler severely endangered by the late mid-twentieth century, but since then, human intervention to artificially expand such habitat

Sapsucker

Within an ecosystem, the removal of a species on which others generally depend—that is, a keystone species—may lead to the disappearance of the dependent species. In a Colorado subalpine ecosystem, a keystone species complex includes sapsuckers, willows, aspens, and a heartwood fungus that infects some of the aspens. In this illustration, the wells that sapsuckers drilled to forage on the sap of shrubby willows were visited by an array of sap stealers, and the nest holes that the sapsuckers excavated in infected aspen were later used by two species of swallows. The red-naped sapsucker also drills nest holes in fungus-infected aspens. The fungus, evident as circular black protuberances on the aspen trunks, indicates heartwood infection, which softens the wood enough for drilling. Tree swallows and violet-green swallows, which are unable to drill nesting cavities (even in infected aspen), are obligate secondary cavity nesters. They take possession of the nest holes of other swallows or use naturally occurring cavities. Here, the two swallows flying overhead make use of old sapsucker nest sites.

has led to a resurgence of the warbler populations.[14] A second threat was parasitism by growing populations of cowbirds that parasitize Kirtland's nests, but cowbird control has also been a feature of Kirtland's conservation efforts. On top of habitat destruction and parasitization as threats, Kirtland's populations are also threatened by climate disruption and drought in its Bahamian wintering grounds that reduces the vigor of males returning to breeding sites.[15]

Declining Songs

The current loss of myriad populations of the 11,000 species of wild birds because of human activities, including habitat destruction, illegal trade, overharvesting, and climate disruption and its consequences, is very dramatic, and the current impact is gigantic. Of the total bird weight on the planet, 30% is accounted for all wild birds, while an astounding 70% is poultry (chickens, turkeys, and geese) descending from just a handful of wild ancestors. However, it is hard to think what these numbers mean in terms of number of individual birds. But one can get a better understanding from bird statistics. The 10 billion individuals estimated in a 1970 bird census in North America had dropped to 7 billion in 2018, a 30% decline.[16]

In a different scale, across the planet, human activities kill about a billion birds annually.[17] Pet and feral cats are believed to kill several million, human hunters kill more than 100 million, cars and trucks kill perhaps 60 million, and wind turbines are responsible for the deaths of some 400,000. The champion here, though, is glass, both transparent and reflective. An estimated billion plus birds smash into glass panes and die trying to reach the habitat seen through the glass or reflected in it. Skyscrapers, rich in glass and lighted at night, compete with cats as bird-killing machines.

The basic fact is that *most* bird populations should be considered endangered. Climate disruption and most other existential threats to human civilization will impact every bird population.[18] Silent springs are a genuine prospect. Some populations are, of course, more threatened than others. Human beings and their accompanying rats ate most of the endemic birds as they invaded the Pacific a thousand years ago, but the survivors include many island-nesting populations whose breeding grounds will soon be underwater. All birds that do those spectacular long-distance flights, like the Eskimo curlew, tend to face high human-caused mortality along most of their routes. Birds dependent on certain kinds of forest, like today's populations of Clark's nutcrackers are on conifers, may dwindle away as those forests are destroyed by climate disruption and accompanying wildfires. We could go on and on—grassland bird populations wiped out by agriculture, albatross populations in trouble because adults feed their young plastic trash, shorebird populations threatened on their migration routes but also by tourist development of beaches—but we're sure you understand the lesson, underscored.

One of the great challenges of saving populations involves preserving birds that are important members of marine food chains. The Pacific was the last area of Earth invaded by *Homo sapiens*—in historical times, just a few thousand years ago. The fossil record and biogeography tell us that once there were hundreds of Ducie Island equivalents. Not only did those islands support hordes of breeding seabirds, but each hosted one or more species of endemic flightless rails as well as a variety of songbirds. But thanks to the wonderful navigation abilities of the Polynesians, those biological riches are long gone.

An example of the early impact of those humans is Henderson Island in the Pitcairn group, a *"makatea"* (chunk of raised coral reef) with cliffs about twelve meters (forty feet) high that is covered with an almost undisturbed forest growing on razor-sharp coral.[19] It has been declared a World Heritage Site. The island boasts four endemic land birds, the Henderson fruit dove (*Ptilinopus insularis*),

Stephen's lorikeet (*Vini stepheni*), the red-eyed crake (*Zapornia atra*), and the Henderson reed-warbler (*Acrocephalus taiti*). In 2011, a multimillion-dollar operation was mounted by the Royal Society for the Protection of Birds (RSPB) and the Pitcairn government to annihilate the island's rats. A ship brought two helicopters to Henderson to distribute poisoned bait, and it was thought the operation was a success. But a rat was briefly filmed by a member of a National Geographic Society expedition in March 2012, and a rapid response team of the RSPB confirmed their presence in May. The surviving rats quickly exploded to a population of as many as 100,000—roughly the pre-poisoning size. A second attempt to eradicate the rats could be very beneficial for the island's endemic species but has not been carried out.[20]

Palmyra, a remote Pacific atoll, represents a natural experiment that illustrates the cascade of ecological consequences triggered by the absence of birds resulting from human disturbance—the introduction of coconut palms into otherwise native forest in this case. Comparing islets dominated by native forest versus coconut palm–dominated islets, Rodolfo and a team of researchers[21] discovered that roosting birds occupy the native forest but avoid the coconut islands. Birds (more specifically, their guano) in native forests serve to fertilize soils, increasing coastal nutrients. This in turn favors the abundance of phytoplankton, which attracts manta rays—a magnificent spectacle to watch. This complex interaction food web breaks down in the absence of birds, signaling the impoverishment of ecological processes—and beauty—of bird-defaunated ecosystems.

Caged Birds

The legal and illegal bird trade is a multimillion-dollar enterprise involving thousands of people all over the world. Unfortunately, birds with beautiful colors or melodious songs are among the most sought after by the trade. Millions of wild birds are sold every year in legal and illegal transactions, affecting avian populations all over the world. The rarer the species, the higher the price. Parrots, macaws, cockatoos, and caged songbirds represent the largest group of captive wild animals in the United States. Surprisingly, they are the most popular pets after dogs, cats, and fishes.[22] Heavily traded species such as the green macaw (*Ara militaris*), red macaw (*A. macao*), hyacinth macaw (*Anodorhynchus hyacinthinus*), Philippine cockatoo (*Cacatua haematuropygia*), yellow-crested cockatoo (*C. sulphurea*), Palm cockatoo (*Probosciger aterrimus*), and yellow-headed parrot (*Amazona ochrocephala*) all fetch high prices in the black market.[23]

The populations of two Australian birds, once secure from extirpation, now are down to one or maybe two populations balancing on the brink. Southern Australia's beautiful orange-bellied parrot has been reduced to roughly fifty individuals, and northern Queensland's gorgeous Golden-shouldered parrot is moving rapidly in the same direction. Both species are threatened with extermination, but their differences highlight the problems of tabulating population extinction.

In the isolated and spectacular southwestern corner of Tasmania, where the southern ocean breaks along the cliffs and islets of a rugged coast, there is the last group of orange-bellied parrots breeding in the wild, some fifty individuals. They clearly comprise a single population. The birds migrate from their breeding grounds around the tiny settlement of Melaleuca in this small area of Tasmania north of southern mainland Australia. On their wintering grounds, the parrots are subject to the usual anthropogenic threats of habitat fragmentation and degradation by grazing, crop agriculture, urbanization, industrial development, introduced predators and competitors, and so on. Ironically, the breeding grounds in Tasmania are now less desirable than in the past because of the extirpation of the aborigines. The fire regimes of Tasmania's original inhabitants created landscapes more favorable to the parrots' favorite food plants.

Comprehensive conservation efforts for the parrot are now underway, including the supplementation of natural tree holes with nest boxes at the main Tasmanian breeding site, and some 250 individuals are now in captive breeding programs. Sadly, the last wild population of the birds has recently been threatened by a parrot plague called "beak and feather virus." It is not clear whether the infection came originally from captive-reared birds released into the Melaleuca area or from contact with other parrot species in nature. The latter seems slightly more likely, but the disease clearly adds to the threat of extinction.

Whether the conservation efforts will prove adequate as Australia suffers more climate disruption and human population growth remains to be seen. The orange-bellied parrot is one of only two parrots that do long-distance migrations, the other being the closely related swift parrot, a member of the same genus. The swift parrot also breeds in Tasmania and risks the dangerous crossing of Bass Strait to winter on the mainland. And it, too, is highly endangered. But for Paul, seeing fully half of the remaining wild population of orange-bellied parrots, and getting photos of the stunning little bird, was both thrilling and depressing.

The other gorgeous Australian parrot that is also in deep trouble lives at the northern extreme of the continent. The Golden-shouldered parrot is a denizen of Cape York, the northeastern tip of Australia, where it is now restricted to the eastern edge of its historical range on the peninsula, especially near Musgrave. It nests in cavities excavated in termite mounds in tropical savanna woodland, when the mounds are soft after the rainy season, and both members of a pair feed the three to six young. The species is now considered divided into just two populations, both threatened primarily by widespread habitat change from human disruption of ancient fire regimes and heavy grazing pressure.

For example, when managed fires are not hot enough, woodland develops at the expense of the grasses that provide seed for the parrots at critical seasons. Cockatoo grass (*Alloteropsis semialata*) is particularly important throughout the tropical savannas because it provides seed for both birds and small mammals, often at times when other foods are scarce. This grass is also highly susceptible to overgrazing by pigs and cattle. Further, it has been shown that Pied butcherbirds, thought to be the chief predator of the parrot, are far more able to ambush both adults and young in the woodland environment that provides them with greater cover. Green tree ants are known to attack nestlings, and because the nests of these ants are constructed by sewing leaves together, they reach greater densities in woodland where the foliage is appropriate for nest construction. Feral pigs, one of the most destructive species *Homo sapiens* has spread around the world, are an additional problem, digging up vegetation and even damaging termite nests where those habitats suitable for the parrot reproduction may be in short supply.

Sold for a Song

Songbird populations are rapidly disappearing due to heavy legal and illegal trade. The magnitude of the trade and the conditions of the traded birds have been described as follows:

> Pramuka Market in Jakarta is said to be the largest wildlife market in Asia, and possibly the world. The open-sided, multi-storey concrete structure is several storeys high, with walkways and stalls packed with cages—large and small. Some are filled with dozens of birds, piled on top of one another. Some of the highest-value birds have ornate cages of their own. But it is the sheer volume that is striking. Cages hang like a chaotic array of Christmas tree baubles. The sellers hand-feed birds with long, skinny spoons that they poke through narrow gaps between cage bars.[24]

Paul was involved in a one-afternoon survey of the market in 1996, which estimated it held between 80,000 and 150,000 individuals of some 150 species. Their sales value was in the vicinity of $1 million. For the right price, even a California condor could be purchased. The fright of the bird flu early in this century restricted the market's operations, but it has apparently since recovered.[25]

Thus, as can be seen, the situation has become quite bad and is only set to get worse for many species of wild birds. Populations of songbirds have been decimated in many regions of Asia, Europe, the Americas, and Africa. Many, such as the Javan green magpie (*Cissa thalassina*) from Indonesia, with fewer than fifty surviving individuals, are teetering at the brink of total extinction.[26] We cannot lose more species and see our skies become more barren as humanity dithers about what should be done. As we will discuss at the end of this book, there are still many steps that can be taken—but we must act now.

Chapter 5

The Silent Crisis

Other Vertebrates

ALTHOUGH MUCH LESS IS KNOWN about population and species losses in vertebrate groups such as reptiles, amphibians, and fishes than in mammals and birds, recent studies have shown that those groups are experiencing similar negative effects from human activities, including habitat loss, species legal and illegal trade, pollution, toxification, introduced species and diseases, and climate disruption, and the interactions among these factors. Some fifty species of reptiles, more than 200 species of amphibians, and around 200 species of freshwater fishes have become extinct in the last five centuries, and thousands more species of those groups are facing extinction.[1] Furthermore, recent estimates indicate that 60% of a sample of species of fish and around 40% of a sample of amphibians and reptiles have experienced local population losses attributed to climate change.[2]

There are many illustrative accounts about how species in those groups are on the brink. Long-lived tortoises in California's Mojave Desert, for example, are endangered by myriad natural enemies and by human activities including, perhaps surprisingly, development of solar power. And, counterintuitively, it looks like global warming may be leading to massive extinction of lizard populations. Frogs and salamanders are under assault by fungal diseases likely imported from Asia by the pet trade. The Chinese sturgeon, a "living fossil" fish, is similarly well on its way to becoming known only as a real fossil due to construction of large power stations on the Yangtze and Jinsha Rivers. Populations of these vertebrates, and many others, may be in even more trouble overall than those of the more attention-grabbing mammals and birds.

An Amphibian Holocaust: The Chytridiomycosis Fungus and Other Causes

The current extinction crises threaten all kinds of organisms, from vertebrates and invertebrates to plants, fungi, and microorganisms. But amphibians—frogs, salamanders, and limbless, worm-like caecilians—are probably different from other animals in that the whole group may have become globally more endangered in recent decades, especially threatened by the introduction of infectious fungal diseases.[3] In the late 1970s and 1980s, amateur observers and scientists began to find many dead frogs and toads, sometimes in clusters, in numerous regions across the globe. They suspected that some of those die-offs were related to a new disease.

They were proven correct in 1988, when Lee Berger, an Australian veterinarian, discovered a new fungus, *Batrachochytrium dendrobatidis*, that caused what became known as chytridiomycosis in amphibians. The fungus responsible for the disease severely damages the skin of amphibians, affecting their balance of water and salt and eventually causing their death. The chytridiomycosis fungus disease has caused a devastating amphibian holocaust. No other vertebrate group has lost more species and likely populations than amphibians, especially frogs and toads.

Around 80% of all extinct amphibians, more than 200 species of frogs and toads, are suspected to have become extinct by the chytridiomycosis fungus in the last four decades alone. Those extinct species, whose common names usually refer to their colors or regions where they were found, include the Golden toad (*Incilius periglenes*) from Costa Rica, the Sierra de Omoa streamside frog (*Craugastor omoaensis*) from Honduras, Sri Lanka's Bubble-nest frog (*Pseudophilautus adsperus*), the Loa River frog (*Telmatobius dankoi*) from Chile, and the mountain mist frog (*Ranoidea nyakalensis*) from Australia. This has been one of the most devastating disease introductions ever recorded.

A second fungus that causes chytridiomycosis and death in salamanders, *B. salamandrivorans*, was described in 2013.[4] Native to Asia, it has been introduced to Europe and could be accidentally introduced to the Americas by the wildlife pet trade.[5] If introduced in the Americas, which are home to the largest number of salamanders on the planet, it has the potential to cause thousands of populations and species extinctions (chapter 11).

Two very interesting and sad cases that exemplify the road to the extinction of amphibians are those of the gastric-brooding frog and the Golden toad. The story of gastric-brooding frogs is fascinating. Only two species are known to science; the Northern (*Rheobatrachus vitellinus*) and Southern (*R. silus*) gastric-brooding frogs were discovered in 1972 and 1984, respectively.[6] Their discovery would have been most likely unnoticed beyond a few specialists. However, David Liem, who described the Northern gastric-brooding frog, discovered that they had a unique way to reproduce. They lived in a very small area of Queensland's human-dominated rainforest in Australia, where they became extinct in the 1980s, probably because of the introduced chytridiomycosis fungus, logging, fires, and disturbances to the water quality.

Their extinction due to human pressures represents an instance of loss of opportunity for humanity. Their reproduction systems were unique; the females swallowed the newly fertilized eggs and brooded the tadpoles in their stomachs, which were converted into

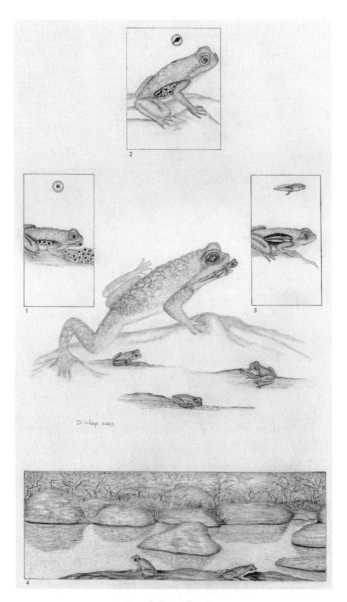

Gastric-brooding frog

The story of gastric-brooding frogs is fascinating. Only two species are known to science: the Northern (*Rheobatrachus vitellinus*) and the Southern (*R. silus*) gastric-brooding frogs, which were discovered in 1972 and 1984 respectively. However, soon after being scientifically discovered, they became extinct. Their extinction due to human pressures represents an instance of loss of opportunity for humanity.

wombs.[7] The frogs offered revolutionary potential for studying human diseases such as acid reflux and related cancers because their stomach acid had to be turned off to protect the brood. But now they—and all the potential medical breakthroughs that could have saved countless lives—are lost to us. Although that chance may have been lost, many other species and genera offer functional ecological traits fundamental for the provision of ecosystem goods and services, although these too are rare as the gastric-brooding frog.

One of the most dramatic stories of the loss of an amphibian is that of the Golden toad. It was a beautiful creature five to seven centimeters (around two to three inches) that was exclusively found in a very restricted area of about ten square kilometers (around four square miles) at around 1,500 MASL (meters above sea level) (almost 5,000 feet) in the cloud forests in the Monteverde Mountains in Costa Rica.[8] Golden toads, an obscure, little-known species that probably spent most of their lives underground, were only described in 1966.[9] In 1987, Martha Crump, a herpetologist, stumbled on a small pond with more than 100 reproducing Golden toads. She describes that encountered in her book *In Search of the Golden Toad*. The males were bright yellow, which made them an impressive sight. Five years later, she found very few, and in 1989, she saw just one surviving lonely male.[10] Sadly, the species became extinct only twenty-three years after being scientifically described. Droughts probably induced by climate change and the chytridiomycosis fungus disease are the likely causes of its extinction.

Vanishing Amphibians

Although good data on most species are scanty, it is estimated that half of the more than 6,300 species of amphibians have declining populations, and many are facing extinction.[11] Habitat loss is a major cause of amphibian declines. Most concerning are the devastating deforestation rates of tropical forests on all continents. Most of Madagascar's frogs and toads, such as the beautiful and colorful Williams' Boophis frog (*Boophis williamsi*) and the Rainbow burrowing frog (*Scaphiophryne gottlebei*), for example, are made vulnerable by the extensive transformation by human activities of its natural vegetation. Not surprisingly, habitat loss by water use and pollution is affecting many populations, such as those of the Longdong stream salamander (*Batrachuperus londongensis*) and the Chinese giant salamander (*Andrias davidianus*). The latter is up to almost two meters (nearly six feet) long, making it the largest salamander on the planet.

The chytridiomycosis fungal disease is affecting populations of most species of frogs, toads, and salamanders globally. In addition, introduced predators such as trout devour the eggs and tadpoles of many species. And climate change, already the likely culprit contributing to the extinction of the Golden toad in Costa Rica,[12] and indeed accelerating changes in temperature and humidity, is also an important factor contributing to the loss of many populations of amphibians.

Another remarkable story of a frog facing extinction is the one of the Loa River frog (*Telmatobius dankoi*). The Loa River is a small stream in northern Chile's Atacama Desert, one of the driest regions on the planet. Where the river crosses the city of Calama, it was the habitat of the Loa River frog, a small frog scientifically described in 1999 and found only in that portion of the 440-kilometer (273-mile) long river. When Chilean scientists learned in 2019 that excessive water extraction for agriculture, mining, and urban development was drying up that section of the river, they set up a rescue mission with other conservationists and the Chilean government—and they were just in time. When they arrived to rescue the frogs in late June 2019, they found only fourteen trapped in a "muddy puddle," as they described it, slowly dying. A few days later would have been too late to save the species.

After collecting them, the rescuers took the little amphibians on a fifteen-hour flight to the National Zoo in Santiago, Chile's capital city. However, the effort to save the species had just begun. Experts had to figure out how to recreate the conditions (water characteristics, diet, and the like) needed for the frogs to thrive. But life has many interesting twists, especially when aided by intelligent hard work. A little more than a year later, the zoo announced that they had successfully bred the frogs and had 200 tadpoles. More work lies ahead to continuing breeding the frogs, restoring their habitat, and reintroducing them. But the program seems promising, bringing hope for this tiny population of a small frog and its unique universe.

Relatively close to the site of the Loa River frog's habitat but in Bolivia, "Romeo," the last known Sehuencas water frog (*Telmatobius yuracare*), was in 2009 dubbed the loneliest frog in the world. This species, like many other frog species, had been devastated by chytridiomycosis fungus disease.[13] He had stopped calling to try to attract a mate and was kept in captivity in K'ayra Center at the Museum of Natural History in Cochabamba. Miraculously, in 2018, scientists discovered five more individuals of the species in a remote Bolivian cloud forest. One became "Juliet," and Romeo began to sing a courtship song again—suggesting at least a small chance a population of the species may be reestablished. But a difficult captive breeding program seems to lie ahead.[14]

Reptiles: Scaly Vertebrates

From having ruled Earth in the Mesozoic, reptiles have declined in prominence but still comprise a fascinating and important portion of Earth's biodiversity. Indeed, while attorneys are now safe from being an impromptu snack for *Tyrannosaurus rex*, they—and the rest of us—should be cautious when vacationing in northern Australia. There dwell dinosaur wannabes—salt-water crocodiles, of which males may be six meters (twenty feet) in length and weigh 1.36 tonnes (1.5 tons). Our ancestors millions of years ago, such as *Australopithecus*, were often prey to large animals including reptiles. But we've pretty much managed to turn the tables on big animals and become their predators. In fact, we have caused the extinction of some eighty-three species of reptiles, most of them found on islands.[15]

The giant tortoises of the Indian Ocean islands were some of the first recorded extinctions that have occurred since 1500. The Réunion giant tortoise (*Cylindraspis indica*) and another five closely related species, some weighing up to 400 kilograms (881 pounds), were used as food and hunted to extinction by the Europeans exploring the vast ocean from the fifteenth to seventeenth centuries.[16] All other extinct reptiles, including lizards such as the Jamaican giant galliwasp (*Celestus occiduus*) and snakes like the Round Island burrowing boa (*Bolyeria multocarinata*), succumbed, mainly to introduced rats, mongoose, and other predator species and, as usual, habitat destruction by human beings.

The conservation situation of endangered reptiles is very worrisome. A recent study has shown that 21% (1,829 species) of the known 10,196 reptile species are at risk of extinction.[17] Factors causing the decline of reptiles are, as one would expect, the same that affect other groups of animals, such as habitat loss by encroachment of agriculture, cattle grazing, and infrastructure, legal and illegal trade, introduced predators, and toxification.[18] Interestingly, because most species are found in regions with warm weather, climate change is already a major factor causing large population declines across all groups of reptiles.[19]

Thousands of species have suffered heavy population losses by legal and illegal trade.[20] Part of what makes it so difficult to stop is the sheer profit involved: animal illegal trade generates billions of dollars annually. Lizards, snakes, and turtles are sold in legal and black markets, pet stores, and increasingly online. Most of the trade is for pets, followed by meat, skins, and traditional medicine. As with birds, rare and newly described reptiles are in high demand and fetch high prices. However, these species are naturally more prone to extinction because of their small populations and limited geographic ranges. For example, the rare and likely endangered Mexican Sinaloan long-tailed rattlesnake (*Crotalus stejnegeri*) is in high demand by the pet trade.

As predators, human beings have been very successful at killing crocodilians and sea turtles. Of the twenty-six crocodilian species found across the world's tropical and subtropical regions, the Philippine (*Crocodylus midorensis*), Orinoco (*C. intermedius*), and Siamese (*C. siamensis*) crocodiles plus seven other species are at risk of extinction.[21] The gharial (*Gavialis gangeticus*), found in northern India and Nepal, is one of the rarest crocodilians. Captive breeding and reintroduction programs have restored some of its populations, but the species is still considered critically endangered.

Anyone who has done much scuba diving in tropical waters has likely had the thrill of seeing a giant turtle swimming gracefully, its shell decorated by a distinctive scattering of barnacles. Yet six of the seven species of sea turtles are on their way out—hunted for flesh or shell, their eggs laid in masses in pits on tropical beaches dug up by people to dine on, construction on their breeding areas, plastic pollution, and so on. Many smaller freshwater turtles are in great trouble, but none so little-known and extraordinarily rare as the Yangtze softshell river turtle (*Rafetus swinhoei*). This species is the world's largest in freshwater: around one meter (more than three feet) long and weighing around ninety kilograms (approximately 200 pounds). It was found only in eastern and southern China and Vietnam.[22] Three living individuals are currently known, one captive one in China and one or two in Vietnam. The only female of the last captive breeding pair died at the Suzhou Zoo in China in 2019.

In 2020, a wild female was discovered in Vietnam, but sadly, it was found dead in early 2023. The event was described as follows:

A dark oblong creature drifted from the depths of Hanoi's 1,400-hectare (3,459 acres) Đồng Mô Lake and began to float, motionless, on the surface last weekend. Local residents and conservationists locked eyes on the gloomy sight, sharing fears about what this might mean for the future of a revered species in Vietnam. On Monday, state media reports suggested it was not only a Giant Yangtze Softshell Turtle (*Rafetus swinhoei*), but it was likely the last known female of the species, leaving only two known males remaining—one in Suzhou Zoo in China and another in Hanoi's Xuân Khanh Lake.[23]

This severely endangered turtle shares something with rhinos, tigers, and pangolins: one of its main threats besides artisanal hunting for consumption is its use in traditional medicine. Captive breeding has been attempted to preserve it and a close relative, the Asian giant softshell turtle, but so far without success.

Of the roughly 500 kinds of turtles in the world, about 60% are experiencing declining numbers. Best known to most Americans is the eastern box turtle, probably because of a reason its populations are gradually moving toward extinction: the pet trade. Box turtles tend to be very "philopatric"—they usually remain in the same habitat, close to where they were born, and populations once discovered by collectors can be overharvested easily. The "box" name comes from parts of the shell that are hinged plates which, when head and legs are retracted, can close and form a "box" to protect the animal from predators.

At the opposite end of the spectrum in terms of common familiarity and pet popularity are the gigantic tortoises found on Aldabra and the Galápagos Islands. The difference between turtles and tortoises is quite simple: turtles live in water, and tortoises live on land. The giant tortoises of Aldabra, an isolated Indian Ocean atoll, grow shells to just over a meter (four feet) in length and have total weights of well over 250 kilograms (550 pounds), close in size to the superficially similar Galápagos tortoises. Giant tortoises once evolved, as far as the fossil record shows, on all continents except Antarctica and Australia, and had reached many islands, especially in the Indo-Pacific. But extinctions have expunged their populations from most of their distributions, largely, it is suspected, because they were easy sources of food for a spreading voracious predator, *Homo sapiens*.

One of Darwin's contributions to the understanding that organisms evolved and were not specially created was his pointing out that island organisms tended not to be related (similar in structural features) to each other but to organisms on the closest land. Aldabra giant tortoises are thus most closely related to smaller tortoises in Madagascar, not to the superficially similar Galápagos giant tortoises. The latter, in turn, are closely related to smaller tortoises on the South American mainland.

Lizards and snakes are two more diverse groups of reptiles. Some of the first modern extinctions were island lizards and snakes, which succumbed to introduced predators such as rats, pigs, and mongooses. The Mauritius skink (*Leiopisma mauritania*) and the Round Island burrowing boa (*Bolyeria multocarinata*) are two examples of naive island species that never developed mechanisms to cope with predators. Many more island and continental lizards and snakes are now threatened with extinction thanks to the usual culprits—such as habitat destruction and introduced species. Strange as it may seem, because reptiles are more diverse and abundant in warmer places, climate change is disproportionately affecting species of reptiles, especially lizards and snakes.[24] In northern Mexico, for example, nine endemic lizards to the Chihuahuan Desert with very restricted distribution are threatened by climate change. At high temperatures, they tend to reduce their daily activities, spending more time hidden to cope with the heat. This affects their feeding and reduces their reproduction.[25]

Many attempts are being made to conserve reptiles, including breeding them in zoos and setting aside protected areas. But efforts should not be restricted to these measures alone. Indeed, the most recently updated comprehensive study on reptile conservation concluded that "although some reptiles—including most species of crocodiles and turtles—require urgent, targeted action to prevent extinctions, efforts to protect other tetrapods, such as habitat preservation and control of trade and invasive species, will probably also benefit many reptiles."[26]

All Kind of Fishes

At risk of sounding repetitive, freshwater fishes are also some of the most affected animals facing population and species losses. They are declining at alarming rates, and their inconspicuous decimation has been called the silent crisis. Although the scientific data available is scanty, it is known that more than 160 species have become extinct in recent times and countless more populations are gone.

For example, the salmon-like deepwater cisco (*Coregonus johannae*) is one of fifty-seven species and subspecies of North American fishes that were wiped out between 1900 and 2010.[27] The deepwater cisco had been a member of a now-decimated group of six endemic (occurring nowhere else) ciscoes that once comprised part of a complex food web in the Great Lakes—the basis of what was known as the "chub fishery." The deepwater cisco was abundant in Lake Huron and Lake Michigan in the nineteenth century but was heavily exploited beginning in the 1930s and went extinct around 1952. It was followed down the drain by its congener, the shortnose cisco (*C. reighardi*), in 1985. Populations of other members of the complex have subsequently gone extinct in some of the Great Lakes. Overexploitation of these endemic species seems to be the original cause of their decline, but another factor was the invasion, almost certainly aided by human activities, of the alewife (a herring relative), which became superabundant in the lakes and outcompeted several native fishes, including ciscoes.

According to the International Union for Conservation of Nature (IUCN), more than 3,200 species of fish are at risk of extinction.[28] Mexico is a textbook example of a country blessed by a massive freshwater fish population that came to suffer species extinction losses.[29] Around fifty species have become extinct or extirpated in the last century because of habitat degradation, overexploitation, introduced species, and pollution.

The sad tale of the Catarina pupfish (*Megupsilon aporus*) is one that is becoming too familiar. It was a tiny 3.8-centimeter (1.5-inch) fish discovered in a single desert spring and only described in 1972.[30] It had an unusual structure related to its small size, as well as some behaviors different from that of most of its closest killifish relatives, from which it was estimated molecularly to have diverged evolutionarily some 7 million years ago. When its spring started to dry out from overexploitation for agriculture, several attempts were made to establish aquarium populations of it and its co-inhabitant, the Potosi pupfish (*Cyprinodon alvarezi*). Alas, the Catarina pupfish proved difficult to maintain, and all of its aquarium populations died out, with the last in 2015, making it extinct. The Potosi pupfish also survived many years in captivity,[31] but its last population died off from diseases in 2014.[32] Of Mexico's 500 freshwater species, around 150 are endangered and facing extinction.

Populations of many other fishes are doubtless declining throughout the world for the same reasons. An example is the Chinese sturgeon. Originally found across Japan, Korea, and China, habitat loss and overfishing have converted it by population extinctions to an exclusively Chinese sturgeon, now restricted only to the Yangtze. This big fish, which can reach up to almost five meters (sixteen feet) and weigh 0.45 tonnes (0.5 tons), lives part of its life in the sea but spawns in freshwater. This species is the record migrating sturgeon, once traveling over 3,000 kilometers (around 2,000 miles) up the Yangtze to breed. China's dam-building spree has cut it off from most spawning sites, and it is much disturbed by noise and can be easily run down by fast ships. Massive releases of captive-reared juveniles have done little to enhance the tiny remaining population. Down to a mere fifty individuals or so, it was greatly threatened in 2016 by flooding, which washed non-native sturgeon of several species from fish farming operations into the Yangtze.

Another heartbreaking example of population losses in fishes is the Chinook salmon (*Oncorhynchus tshawytscha*). Formerly, millions of salmon would hatch in northern California's cold, freshwater mountain streams, migrate to the Pacific, stay there for three or more years, and then migrate back to their natal stream to spawn before dying. The 640-kilometer (400-mile) Sacramento River, the largest in California, is the conduit. With time, however, access to 98% of riparian and floodplain habitat along the Sacramento needed by young migrating salmon would be lost, pollutants would accumulate, commercial and recreational fishing would require management, and disease, predation by non-native species, and drought would all take a toll. Meanwhile, obstacles along the way would increasingly constrain the route.

One obstacle, the 182-meter (600-foot) Shasta Dam, has blocked the Sacramento River since 1945. Its fish ladder, whose step-like pools help fish navigate past the dam, serves as a corridor. Just over six kilometers (ten miles) downstream, another obstacle, the forty-eight meter (157-foot) Keswick Dam, has blocked the Sacramento since 1950, although it also has a fish ladder. Further downstream by 177 kilometers (110 miles), the 234-meter (770-foot) Oroville Dam, the world's tallest dam, has blocked the Feather River that feeds into the Sacramento since 1968.

There are seventeen runs of Chinook each year, each named for the season when most adults return to freshwater. The winter-run Chinook that spawn in the summer, are at greatest risk. In 1993, only 186 adult winter-run Chinook returned from the Pacific to spawn. The following year, the endemic Sacramento River winter-run Chinook were officially listed as Endangered at the federal level. Managing drought conditions is increasingly critical for winter-run Chinook. They depend both on cold water releases from Shasta Reservoir to keep eggs and hatchlings alive and on hatchery production to boost numbers. The Livingston Stone National Fish Hatchery at the base of Shasta Dam produces approximately 200,000 juveniles annually and is expected to increase production.

There are cases of biologically interesting populations and species of freshwater fishes at risk of extinction in all corners of the planet. One evolutionary case that we find fascinating is the fish diversity in the seven Great Lakes of Eastern Africa, the largest of which are Lake Victoria, Lake Tanganyika, Lake Malawi, and Lake Turkana. Those lakes are a textbook example of what is known as adaptative radiation, where over several million years, a few species evolve into an incredibly diverse fish fauna comprising more than 600 endemic species.[33] These species of the family Cichlidae, endemic to those lakes, have diverged in unique ways, with interesting variations in the shape of the head, the mouth, and the teeth, all related to the feeding mode. Species evolved to specialize in feeding on different resources such as algae, plankton, insects, mollusks, and fishes. Specialization in some of these feeding habits is extraordinary, with two species feeding on the scales of other fishes and another on the eyes of other fishes.[34] Unfortunately, the fishes of the African Great Lakes are endangered by exotic fish introductions, overfishing, water use, pollution, and siltation, which have caused severe declines in native fish populations and even species extinction (see chapter 10).[35]

Marine Fishes

The oceans are some of the most degraded ecosystems in the world. Large-scale impacts of human activities such as acidification, pollution, toxification, overfishing, ship traffic, and noise, have

Chinook salmon losses in California

Formerly, millions of Chinook (king) salmon would hatch in Northern California's cold fresh-water mountain streams, migrate to the Pacific, stay there for three or more years, and then migrate back to their natal stream to spawn before dying. The nearly 650-kilometer (400-mile) Sacramento River, the largest in California, is the conduit. With time, however, access to 98% of riparian and floodplain habitat along the Sacramento needed by young migrating salmon would be lost, pollutants would accumulate, and commercial and recreational fishing would require management. Drought, along with disease and predation by non-native species, would also take a toll (as described in chapter 5).

impacted all the oceans. The IUCN has estimated that 100–200 million sharks are killed by commercial fishing each year, and they consider about a third of shark species to be threatened with extinction. In a sample of 230 different shark populations, an average 83% reduction of breeding size has been observed compared with historical levels. Populations of species such as the Atlantic halibut (*Hippoglossus hippoglossus*) and Beluga sturgeon (*Huso huso*), now only with about 100 mature individuals, along with the European eel (*Anguilla anguilla*) and the Chinese sturgeon (*Acipenser sinensis*), are some of the most endangered marine fish species. The southern bluefin tuna (*Thunnus maccoyii*), considered critically endangered, has an estimated population of 500 individuals, and the Canadian population of Atlantic cod (*Gadus morhua*) has declined 99% in only thirty years.

These figures are sobering as well as heartbreaking. An article in *BioScience* concludes that "failure to prevent population collapses, and to take the conservation biology of marine fishes seriously, will ensure that many severely depleted species remain ecological and numerical shadows in the ecosystems that they once dominated."[36] They are also likely to become extinct in the next decades. The loss of fish species not only means the loss of valuable food sources, but it also frays the fragile ocean ecosystems that, ultimately, are the foundation for all life on Earth. "Out of sight" most definitely cannot be "out of mind," for their sakes— and ours.

Chapter 6

Ignored Victims

Invertebrates

EVEN AS MANY EDUCATED PEOPLE and the more reliable news media outlets are becoming aware of the sixth mass extinction currently underway, as evidenced by the disappearance of considerable populations of mammals and birds (especially charismatic ones), a great decline in Earth's insect and invertebrate fauna has been little-recognized until very recently. But recent reports have been changing this, with talk even of an "insect apocalypse" in the offing.[1] Knowledgeable scientists are appalled by the evidence of these species declines, and more and more are demanding action, but so far, little has happened.[2]

From the viewpoint of the general public, however, the stories of disappearing populations of animals without backbones, and their critical impact on humanity[3] have been largely untold or ignored. In fact, there have even been hints of relief at being rid of what can be seen as "nuisances" who spread disease, eat our crops, and bite our bodies. But insects and other invertebrates are crucial to humanity as pollinators and as food sources for higher level species who are in our food chain—and more directly as an important source of nourishment for some 2 billion people who complement their diet with them.

In this chapter, we try to stress the irreplaceable role played by insects and other invertebrates by sharing leading examples of the fates of these populations. In the following paragraphs, we describe in some detail invertebrate extinctions often little recognized by society even though some of these invertebrate groups are important for humanity as sources of food, medicines, and pigments, and providers of environmental services like pollination, seed dispersal, and pest control.

Locusts: A Biblical Plague in North America

The biblical book of Exodus described massive hordes of locusts that threatened ancient Egyptian agriculture. In the time of the pharaohs, Egyptians thought them so important that they carved representations of locusts onto their tombs. Some parts of Africa still suffer from locust outbreaks, made more devastating by climate disruption because droughts provoke the lack of native food sources,[4] but it is little-known today that a century and a half ago, North American agriculture suffered similar disasters.

Indeed, the possible complexity of population and species extinctions in invertebrates is brilliantly illustrated by an insect, the Rocky Mountain locust (*Melanoplus spretus*). The "locust" in this case was a migratory grasshopper whose populations may have contained trillions of individuals when it swarmed over the plains of North America, devastating agriculture there as late as the end of the 1800s.[5] In the spring of 1874, for example, mobs of locusts darkened the skies over the prairies. Millions upon millions of locusts devoured almost all the growing crops: corn, wheat, barley, beans, potatoes, tree fruit, and on and on. The crops were totally consumed, right down to the ground in most cases.

Also destroyed were the livelihoods of hundreds of thousands of farmers, with economic losses approaching half of the value of US farm production that year.[6] Government action was demanded, and relatively ineffective countermeasures were initiated—using collecting devices to capture millions of the insects and subsequently burning them, for example. Capturing millions sounds impressive, but when there are billions flying around, one can appreciate the scale of the problem. One positive result was that public and political interest converged, permitting the development of a scientific approach to insect control and laying the foundation of the discipline of economic entomology.

When not in an outbreak phase, the locust was largely restricted to breeding grounds in Rocky Mountain valleys, where females laid their egg pods in the soil. In the late nineteenth century, agricultural development heavily altered those grounds as farmers raised livestock and planted forage crops. The alterations apparently reduced the locusts' breeding ability, which transformed them from a force of nature and serious agricultural pest to extinct in an evolutionary blink of an eye. As with the Passenger pigeon, huge numbers proved no insurance against population extinctions that lead inexorably to species extinction.

The Rocky Mountain locust is an early case of the documentation of population and species extinction of an invertebrate. It was thoroughly documented because of its enormous economic impact.

Butterflies and Moths

Lepidoptera (butterflies and moths) is the order of insects most attended to by naturalists. But one can safely assume that trends in butterflies, which account for less than 1% of the known species of insects, are nonetheless roughly representative of that gigantic slice of terrestrial biodiversity. For instance, a long-term decline

Rocky Mountain locust

Top panel: Until 1902, when Rocky Mountain locusts (*Melanoplus spretus*) were last seen, recently hatched grasshoppers would migrate from the bottomlands of the eastern slopes of the Rockies once their wings were flightworthy. They would form billowing swarms that could be mistaken for clouds or smoke and migrate to the western and central parts of the United States and Canada. *Middle panel*: A sample of primitive devices designed to control infestations on farms using human- or horse-powered apparatuses to capture and bag locusts for eventual burial (*top left and middle*). Other devices were designed to capture, poison, or burn locusts in kerosene, coal oil, coal tar, and so forth (*top right and bottom left*). Still others were designed to capture the insects on a conveyer belt and crush them (*bottom right*). *Bottom panel*: Efforts to control infestations, however, were no match for the swarms, and the devastation was generally unstoppable. Locusts even ate the paint off posts and garden tools.

of butterfly populations in the American West seems related to climate disruption, among other factors.[7] In lowland central California, a heavily agricultural region, it appears to also be related to increased use of neonicotinoid (a type of neurotoxin) pesticides.[8] Where some data are available on overall abundance, moths, perhaps fifteen to twenty times more species-diverse than butterflies (which are basically a group of moths that have evolved to function in daytime), show similar trends to butterflies in Britain,[9] indicating many populations declines and extinctions.

The major drivers of moth extinction seem to be habitat destruction, especially tied to agricultural intensification and climate disruption. Available data suggest that light and chemical pollution and invasive species are also important in the potentially catastrophic loss of the moth biota. For those of us over sixty and interested in nature, such data seem only to confirm a lifetime of observations. We can remember summer nights when the moths in our car headlights and around street and stadium lights were so abundant they resembled snowstorms.[10] Driving in many places across all the continents a half century ago often resulted in car radiators full of butterflies or fenders caked in sludge of insects hit at high speeds. Occasionally, they were abundant enough to make removing some wise from the viewpoint of engine heating. No more.

It is sad that only persons over a certain age will remember the blazing stars of night skies, which still can be enjoyed in isolated high-altitude locations. But the expanding human enterprise has banished deep darkness over much of Earth, to the detriment of more than insects and human cultural services. Many birds, mammals, and other organisms, especially those that migrate, now suffer disorientation, migration disruption, and mortality because of human lighting pollution.

Paul developed an interest in insects in a fairly common way, because of the beauty of butterflies. He remembers that when he was about ten and attending a Boy Scout summer camp, he became fascinated by the pipevine swallowtail (*Battus philenor*) nectaring in large numbers on lakeside flowers near Lebanon, Pennsylvania. He didn't get his hands on a butterfly net until a few years later at another summer camp, this one near Ely, Vermont, where a counselor introduced him to butterfly collecting. He was hooked immediately, and shortly thereafter, at age fifteen, he was introduced to Charles "Mich" Michener, then curator of butterflies and moths at the American Museum of Natural History in New York. Mich later became the world's expert on the evolution of bees and their behavior and was Paul's major graduate school professor. Mich not only encouraged Paul's interest in butterflies and their morphology, taxonomy, and evolution, but he directed him toward the Lepidopterists' Society, just then being formed.

The Society provided Paul with the addresses of people with whom to exchange butterfly specimens. His first package came from a nice man named Bill Hammer in Berkeley, California, an exchange for some common eastern species Paul sent him. Paul has never forgotten the smell of PDB (moth crystals) as he emptied the box, starting with a cascade of triangular envelopes labeled "*Euphydryas editha bayensis*," now known as the "Bay checkerspot." Little did he imagine that later he would develop his observations of the Bay checkerspot and its relatives into an experimental system that would occupy his attention and the attention of treasured colleagues for more than seventy years. The dried corpses of butterflies in those envelopes in 1948 would be instrumental in developing the scientific interest in population extinctions, the concept of metapopulations, and the entire scientific subdiscipline of coevolution.

The Jasper Ridge Checkerspots

When Paul arrived at Stanford in 1959, he learned there were Bay checkerspots on "Jasper Ridge," a portion of Stanford's huge campus (originally the stock farm of the corrupt robber baron Leland

Stanford). He had been looking for a butterfly to use as a system for studying how insect populations changed size and how they evolved. The checkerspot looked ideal, and he began a program of marking the wings of the butterflies with distinctive numbers so that when he captured them again, he could determine how they had moved around. It turned out that rather than a single population, there were three separate populations on Jasper Ridge, that is, a metapopulation. Marked individuals rarely moved from one population to another.

The mark-release-recapture (MRR) experiments also provided estimates of population size over time. The middle population of the three went extinct in the mid-1960s, was reestablished by migrants from other Ridge populations and then went extinct again in the mid-1970s. The other two populations subsequently exploded in size but then they too both declined to extinction. The Bay checkerspot was gone from Jasper Ridge by 1997.[11]

Paul had never imagined that the roughly 485-hectare (1,200-acre) area of what is now Stanford's Jasper Ridge Biological Preserve would be inadequate to preserve a small insect. It's a good example of how scientific assumptions need testing. Most studies of insect population changes tend to focus on the behavior of large populations—agricultural pests such as the locusts or disease vectors such as mosquitoes. But with the checkerspot work by Paul's group and the late Ilkka Hanski working on a different checkerspot (*Melitaea cinxia*), some of the focus has shifted to the dynamics of small insect populations and their possible extinction. The checkerspots are confined to some seven hectares (around seventeen acres) out of the 485-hectare (~1195 acres) reserve, areas featuring plants adapted to nutrient-poor, toxin-rich serpentine soil. It is also instructive that were it not for that tiny area of checkerspot habitat, Jasper Ridge would long since have become a housing development—prevented only by Paul and a biology faculty colleague successfully battling the university's financial boss to protect the checkerspots.[12]

Edith's checkerspot butterfly surviving both wet and dry years

Variation in land surfaces in a given area, such as differences in elevation and slope orientation toward the sun—known as topographic heterogeneity—can make it possible for populations of Edith's checkerspot butterflies (*Euphydryas editha*) to survive both wet and dry years. In this drawing, a female Edith's checkerspot is ovipositing (laying eggs) at the base of its foodplant, California plantain (*Plantago erecta*), which is growing on a south-facing slope. The four vignettes across the top of the drawing show differing densities of California plantain in wet and dry years on south- and north-facing slopes. The differing densities are linked to differing densities of adult Edith's checkerspot that will emerge. *Vignettes, left to right*: On south-facing slopes in dry years, California plantain is sparse, reducing the number of checkerspot larvae that survive to adulthood (1), while in wet years the foodplant is dense, increasing checkerspot survivorship (2). On north-facing slopes in dry years, California plantain is dense, increasing the number of larvae that survive to adulthood (3), while in wet years, the foodplant is sparse, reducing checkerspot survivorship (4). Thus, checkerspot populations occupying habitat that includes both north- and south-facing slopes have a safeguard against extremes in annual rainfall.

The extirpation of Bay checkerspot populations on Jasper Ridge was a gradual process, in which changing climate played a dominant role, but other factors were likely involved, including nitrogen deposition from Bay area automobile exhausts.[13] The added nitrogen allowed invasions of grasses that crowded out the forbs on which the checkerspot caterpillars fed, a process that occurs unless grass-feeding cattle are grazing the habitat and carrying away nitrogen as they are shipped to slaughter. Extreme weather events can, of course, also endanger or even wipe out populations. That includes populations of that most-resilient large group of organisms to which the checkerspots belong: plant-eating insects.

Plants have not taken the assault by insects lying down. They have evolved a wide array of defense mechanisms to ward off plant-eaters ranging from people and butterflies down to bacteria and viruses. To deal with insects and most others, many plants have laced their flesh with a great number of poisonous chemicals ranging from cyanide to alkaloids. Many of these compounds—aspirin, belladonna, pepper, opium, and so on—humanity has cleverly adopted for its own use.[14]

Humanity puts great effort into controlling populations of insects that like to dine on our crops—but weather often keeps them under control effortlessly. Indeed, one of the reasons that temperate-zone agriculture is more productive than tropical agriculture is the pest-control service of that extreme weather event called "winter." Pest outbreaks in temperate zones tend to be limited by the arrival of cold weather; in warmer climates, the pests can be active longer and do more damage.

A "natural experiment" at one of Paul's field sites illustrated this dramatically. He had been studying the impact a tiny blue butterfly (Glaucopsyche lygdamus—the silvery blue) had on the big lupine plant (Lupinus amplus) on which the blue caterpillars fed. The butterflies laid their eggs in late spring and early summer on unopened flowers of the lupine inflorescences (flowering stalks). In 1968, Paul and the late botanist Dennis Breedlove[15] removed the eggs from roughly every other inflorescence and were able to show that the tiny herbivore caterpillars destroyed almost half of the seed set of the lupines. That meant the butterflies were operating as powerful selective agents on the reproductive patterns of the lupines.

This experiment underlined the evolutionary impact a population of small plant-eaters can have on a population of large plants. Then in 1969, an early spring was climaxed by a late three-day snowstorm with heavy freezes starting on June 24. Although some populations of herbs, such as those in the understory of the spruce-fir forest, were undamaged, most lupine inflorescences were destroyed by the event. Deprived of their egg-laying sites, the blue butterfly population crashed to apparent extinction,[16] although in later years, the site was gradually reestablished by migrant blues. Other subalpine butterfly populations in the area, especially the common alpine (*Erebia epipsodea*), a blue whose larvae fed on lupine leaves rather than inflorescences (*Plebejus icarioides*), and the large marble (*Euchloe ausonides*), as well as a nasty pest, the Rocky Mountain bite fly (*Symphoromyia fulvipes*), all experienced population declines in the summer of 1969, as did many small mammal populations. But the special vulnerability of the *G. lygdamus* population led to its extinction.

Long-term data sets on the dynamics of well-defined insect populations over substantial stretches of time are generally lacking. Jasper Ridge Bay checkerspot records are unusual in that respect. But long-term data in the form of carefully labeled museum specimens do provide some insight into the decline of North American populations of a butterfly that can be declared to be "iconic"—the monarch (*Danaus plexippus*). When Paul started his PhD research, the first step was to document the anatomy of a butterfly.[17] The biggest common species available in quantity from a scientific

Saint Francis's satyr

The endangered St. Francis's satyr butterfly escaped notice for a surprisingly long time. It's a weak flyer that lives as an adult only three to four days on average, when females must mate, locate preferred host plants, and lay their eggs. But it was finally discovered in 1983, confined to a single, small wetland/grassland area in North Carolina's Fort Bragg (now Fort Liberty) army base. Its ten square kilometer (3.38 square mile) range is fragmented into small (under two hectares/five acres) segments that lie primarily within artillery ranges or beaver habitat, which ensure the disturbances upon which its host plant, Mitchell's sedge (*Carex mitchelliana*), depends. That disturbed habitat includes abandoned beaver pools that promote meadow development, open streams with beavers, and woodland patches whose trees have been cleared by small, low-intensity, shell-ignited fires.

supply house was the monarch, with 100 available bottled in preservative for a nominal price (dried specimens now sell for as much as $60).

In those days, Paul was unconcerned about this use of monarchs—today, the situation is different, as populations of this iconic migratory species have declined precipitously. In 2018–2019, for instance, there was an especially dramatic drop in the populations of this migratory species overwintering in California,[18] whereas the larger eastern populations overwintering in the mountains of Mexico happily rebounded somewhat from a long decline. Careful study of museum collections of monarchs, despite some sampling problems, suggest a general population increase from 1900 to 1950 in both the butterflies and the milkweed plants on which their caterpillars feed. That has been followed by a decline from 1950 to today,[19] with fluctuations in the eastern and western populations controlled at least in part by often contrasting weather conditions in the two regions.[20]

Scientists are working to understand the overall decline and fascinating changes in the shape and size of the monarch's wings over time, but much mystery remains. Climate disruption looks like a major threat to some populations of the monarch, just as it was to the Jasper Ridge populations of the Bay checkerspot. For instance, the monarch population that migrates from breeding grounds in northeastern North America to specific wooded groves in the Mexican volcano belt has suffered from climate-related decline in its favorite roosting trees.[21] But it also seems to suffer from climate disruption at other parts of its long and diverse migratory route.[22] The effects of climate on the monarch's summer breeding range may be more complex[23]—a northward expansion following its milkweed host plant, but an expansion into generally inferior habitat. Other factors such as the impacts of farm consolidation on milkweed habitat, and of herbicides and insecticides on milkweeds and monarchs may play a role.[24] Overall, though, climate disruption looks like the most dangerous of the many threats this iconic animal faces.[25]

Complex Ecological Interactions

Weirdly, one of the other possible factors in monarch decline also has a connection to Paul. In 1948, he collected some monarch larvae in northern New Jersey and raised them in his home. Two of the eighteen caterpillars produced not monarchs, but parasitic tachinid flies. Paul sent them to a fly taxonomist at the US National Museum, who identified them as *Compsilura concinnata*. Later, he found out that the tachinid was a European native that had been introduced into North America as a potential biological control of the gypsy moth (*Lymantria dispar*), a forest pest that was also a native of Europe. *Compsilura* turned out not to be a very effective parasite of the gypsy moth, but was an effective parasite of many other insects, including the beautiful silkworm moths of the northeast—such as the famous Luna moth.[26] Historical records suggest both a general decline of moth species in southern New England and a northward movement of some lepidopterans in the face of many changes, but the evidence is entirely correlational. The roles of such things as more moth-attracting streetlights and deer browsing of forest understory shrubs that are moth food plants have been hypothesized but not demonstrated. *Compsilura* seems likely a major player, however.[27]

The iconic and declining monarchs have increasingly been shown to be part of a complex web of interactions involving *Compsilura* and other tachinid flies,[28] a protozoan parasite,[29] host-plant chemistry,[30] and human alteration of ecosystems,[31] including the use of neonicotinoid pesticides.[32] It seems unlikely, however, that the monarch will follow the Rocky Mountain locust into extinction soon, even though its situation is somewhat similar.

Monarch butterflies were an abundant species dependent on a few key areas to support their life cycle—in this case, its overwintering habitat in Mexico.[33] Loss of habitat there is already reducing monarch populations that summer in the eastern United States, and if business as usual continues, will cause further extinctions. But non-migratory populations exist through much of the western hemisphere south of Mexico, and throughout much of the rest of the world's tropical regions, where it has dispersed.[34] Those populations appear to be in no more danger now than most other widespread insect species, which is fortunate because the risk of species extinction in these circumstances is reduced.

The monarch exemplifies an important pollinator that is also supplying an aesthetic service. The latter underlines the potential impact of population extinctions. If, say, destruction of monarch overwintering habitats in Mexico exterminates its eastern North American population, the monarch's persistence elsewhere on the planet will not compensate people in places like New Jersey for the loss of the beauty that monarchs once added to their gardens as they sipped nectar from *Buddleia* (commonly known as "butterfly bush") flowers. This is an example of aesthetic ecosystem service, a service that provides people with psychological rewards, not physical ones. Being recognized as such, though, also may raise motivation for doing something about such situations.

Extensive historical collections and long-term studies are not always essential, however, to further our understanding of butterfly conditions and distribution. General reports can pass along important information. For instance, the pioneering butterfly scientist William Henry Edwards described butterfly populations—beautiful species like the zebra swallowtail (*Graphium marcellus*)—swarming around his home near the Kanawha River in West Virginia.[35] Deforestation reduced those populations that fed on pawpaws (a common small understory tree), but it would be simple to assess the state of the swallowtail population now. There are many butterfly enthusiasts in the United States, and many of them con-tribute to an annual "season summary" published by the Lepidop-terists' Society. They report at random about which interesting species (usually rarer ones) they collected. When Paul was secre-tary of the Society around 1960, he tried (but failed) to get the organization to do what would be called citizen science today—carry out simple transect surveys of common species all over North America to get a systematic assay of the state of the butterfly fauna, including zebra swallowtails.

It may not be too late now for such surveys—rather, they may be needed more than ever. Currently, our detailed scientific knowl-edge of the overall butterfly situation is restricted to a few popula-tions like those of the checkerspot that disappeared from Jasper Ridge. But the sort of efforts Paul fought for were begun in Ohio in the 1990s, and after twenty-one years, have recorded a decline of roughly 33% in the butterfly abundance in the state.[36] The success of the iNaturalist program based on the web and the ubiquity of cell phone cameras, shows the great potential for citizen scientists to do systematic population censuses.[37] At a much less detailed level, Paul was able to extract evidence of extensive butterfly pop-ulation extinctions since 1940 from "square-bashing" censuses of British butterflies.[38] Records suggesting their decimation in Europe go much further back.[39]

Britain's depauperate (only around sixty species) butterfly fauna is the most intensely investigated national butterfly assemblage in the world. The results of the square bashing could almost serve as a textbook study of countryside biogeography—patterns of organ-ism distribution in human dominated areas. It has demonstrated the importance of land-use change, agricultural chemical inputs, plantations of exotic trees, landscape connectivity and corridors, habitat management in general, reintroductions of locally extinct species, and so on.[40] Many of these are important in the battle to preserve populations of organisms worldwide.

The loss of populations is common and can be dramatic. The Poweshiek skipperling (*Oarisma poweshiek*) has a large geographic range but there was rapid extinction of these populations in the late 1990s and early 2000s, concurrent with the development of neonicotinoid pesticides and the invasion of the Asian soybean aphid against which the pesticide is sprayed. This arc of points matches the northern range of the corn-soy planting region and hence concentrated pesticide use of the US Midwest. The only remaining populations, amounting to a few hundred individuals in 2023, were found in Manitoba and Michigan. Dr. Nick Haddad's research group currently has 500 caterpillars in a greenhouse, which is likely more than the numbers existing in the wild. Very scary!

Honey Bees, Champion Pollinators

Huge attention has rightly been paid in the popular media and professional press to the recent decline of honey bee populations because of their importance in crop pollination. Many plants either cannot reproduce or have their reproduction much curtailed if there is no living agent to carry pollen between individuals. Plant pollination by animals involves some of evolution's most fascinating results. For instance, various orchids, notably those in the genus *Ophrys*, have evolved to mimic, in appearance or smell, females of certain insects—often bees. The male insects attempt to copulate with the mimicking flowers, picking up a package of pollen from one flower, and then, with another attempt to copulate, transfer that pollen to another flower. Although it is not clear if these insects receive some reward—as is typical in insect pollination— these acts of pseudocopulation insure the pollination of these deceiving orchids!

Plants struggle when their pollinators decline in number. A classic example is described in a study of plant–pollinator interactions in a Midwestern US deciduous forest understory. The interactions there were first studied at the end of the nineteenth century and then again more than 120 years later, when it was discovered that the original plant–pollinator network had been disrupted by many factors, including extinction of half of the bee populations initially involved.[41]

The fate of pollinator populations takes on a special importance for our species because the quality of the human diet is heavily dependent on certain plant populations, especially animal-pollinated crops.[42] For instance, the health of multitudes of people living in poverty rests on inclusion of critical micronutrients in their diets, like vitamin A and iron, that are largely supplied by pollination-dependent plants.[43] The degree to which the future of pollination services depends on the diversity of different pollinators remains an open question[44] in view of all the uncertainties about the composition of future ecosystems on one hand and the fate of the nearly ubiquitous domesticated pollinator, the honey bee, on the other.

The critical importance of the pollination services that honey bee populations supply to agriculture has generated a large literature[45] on so-called colony collapse disorder (CCD). CCD was first reported in the early 21st century and is a syndrome in which worker bees abandon the colony, leaving behind queens and a few nurse bees and generally adequate supplies of honey and pollen. Those bee populations eventually went extinct when there were no workers to maintain the food supply. The possible causes of CCD and problems with populations of other bees have been much discussed; they include misused or novel pesticides,[46] especially neonicotinoids,[47] invasive mites, new or emerging diseases or parasites,[48] stress from management practices such as constant moving of hives, destruction of natural bee foraging areas, multiple such threats that reduce the hive population below a critical sustainable size,[49] economics of the honey trade,[50] and so on. Whatever the explanation and the future of CCD turns out to be, loss of pollinator populations is very bad news for the human feeding base.[51]

The loss of pollinators can have immediate economic effects as well. Coffee, for example, does not require pollination in order to reproduce, but the quality of the berries improves dramatically if the coffee plants are insect-pollinated. Coffee plantations in Costa Rica close to forests that provide habitat for local bees have substantially higher value plantation yields.[52] The lesson learned is that having forest fragments next to the coffee plantations is a better economic strategy than converting all the land into solid coffee plants.

Other Insects

Interest in insect extinctions, aside from those in "popular" groups such as butterflies and bees, has been slight compared with the popularity of iconic large animals, so there are relatively few stories to match accounts of rhinos or Passenger pigeons. But there are myriad other insects—many millions beyond the 20,000 or so kinds of butterflies and bees, and many or most appear to be sharing the high rates of population exterminations that beset the butterflies.[53] So, it is not surprising that there are few and less-detailed studies of insect populations in trouble.

Fruit flies, often mentioned by their Latinized generic name *Drosophila*, are widely used in lab and science experiments and have therefore taught humanity a significant portion of what is known about genetics, evolution, and development. Fortunately, *Drosophila* populations (and species) do not seem to be under much threat, although very little is known about natural populations of most *Drosophila* species. What is clear, especially in agriculture situations, is that flies are important in providing populations of crops such as, but not limited to, cacao (chocolate), mango, cashew, and avocado (and, by extension, serving people) with pollination services.[54]

For instance, there are rich fly communities whose conservation should be a matter of concern[55] because of their roles in ecosystems that supply important services such as pollination and waste disposal to humanity. Yet only a single fly species has been listed as endangered under the US Endangered Species Act. The Delhi Sands flower-loving fly (*Rhaphiomidas terminatus abdominalis*) was so listed in 1993. Ironically, this small insect now serves as an umbrella species. It is restricted to a small remnant of an inland sand dune area in southern California built up by Santa Ana winds blowing sand out of nearby mountains. Agriculture and other development have already destroyed something like 98% of this fly's habitat and populations. The area comprises a unique ecosystem, containing a series of plants and animals with limited distributions, but even given the listing, pressures on the remaining ecosystem persist.

Another remarkable example can be drawn from the world of beetles, the Coleoptera, which is the most species-rich order of insects. When a lady asked famous British evolutionist J. B. S. Haldane what he had learned about the creator from his study of creation, he replied, "He must have had an inordinate fondness for beetles."[56] Within the beetle fauna, not only because of their beauty, but because of the dramatically important role they play in ecosystems, members of the subfamily Scarabeinae, otherwise known as dung beetles, are often singled out. Their uncharismatic name reflects their invaluable service for ecosystems—and humans. They detect dung, dead bodies, or rotting fruits on the ground and cut pieces of these materials into ball-shaped sections, which are then rolled and buried into their below-ground nests. The importance of this ecosystem engineering is epitomized by the introduction of sheep into Australia, where there are no native dung beetles. The sheep carpeted the ground with dung, making it necessary to introduce Scarabeinae to fix the problem!

The dung beetles' susceptibility to anthropogenic impact was manifested in a tropical conservation biology classic study.[57] It showed a significant decline in dung beetle diversity and dung decomposition rate in relation to the size of patches in a Central Amazon fragmented forest landscape, changing a variety of ecosystem functions such as carbon in the soil and reducing sanitation services. Beetle population extinctions had a major effect on the functioning of forest patch ecosystems.

As we mentioned above, there has been much recent attention to studies showing great and rapid declines in Earth's insect fauna—an "ecological Armageddon"[58] in response to environmental deterioration and agricultural intensification.[59] For instance, some 80% of the flying insect biomass disappeared from protected areas in Germany between 1989 and 2016.[60] It is crystal clear from such studies and anecdotal evidence that insects, at least in some groups and areas, are in deep trouble[61] and maybe they are universally and globally as well.[62]

Spiders are also important in controlling insect pests, and they are also not immune to the dangers facing other invertebrates. Some spiders are losing populations, and others have become endangered. An example is the fen raft spider (*Dolomedes plantarius*), a large semiaquatic species. This spider hunts from vegetation at the water's edge. It hangs on to aquatic vegetation with its hind legs and uses tiny hairs on its front legs to detect vibrations of water-striders or similar prey. It can run on the surface of calm water to catch its prey, or dart down a plant stem underwater to nail something—including small fishes. The spiders tend to live in fens (British name for marshy or flooded land). Populations of the fen raft spider were once very widespread in Eurasia, but habitat destruction (primarily draining of wetlands) has wiped out many populations. There are only a few remaining populations in Great Britain, where it is listed as endangered. They are the first spider where the future of its populations in a changing world has been studied. Recently, efforts have been made to reestablish now-extinct populations and reinforce the few in existence by releasing individuals raised in captivity—and success has been reported.[63]

Crayfishes and Other Crustaceans

Mammals, birds, and butterflies are not, of course, the only kinds of animals undergoing population extinctions; it's just that we know more about groups that are popular with both amateur naturalists and scientists. But when scientists do get information about more obscure groups like crayfish, patterns of loss are similar. Crayfish are freshwater crustaceans related to lobsters in both appearance and (often) delicious flavor, and as in many invertebrate groups, we have sadly little information to evaluate the state of their populations. We know many are in trouble—one global survey of 590 species estimated about a third of them are at risk of extinction. We can be sure that many populations are disappearing under the impacts on their stream homes from mining, recreation, pollution, and habitat destruction in general. Indeed, the freshwater lakes of the planet appear to be gradually drying up and disappearing, and with virtually every one, numerous populations disappear.

The Benton's cave crayfish (*Cambarus aculabrum*) is known only from four caves in Arkansas and Missouri. This small cave-dwelling freshwater creature is almost blind, which fits its lightless habitat where it feeds on decomposing organic matter—one of nature's garbage collectors. It is endangered because of its small geographic range and tiny populations. There are over 500 species of freshwater crayfish worldwide, and they are important parts of many people's diets. Their interesting shapes and colors are beloved by many aquarists, and a huge aquarium trade market in crayfish has developed.[64] The family to which the Benton's cave crayfish belongs has hundreds of species and subspecies in the southeastern

United States—which is a center of diversity for the group (and where they are famous in local lore as "crawdads"). Twelve of them are listed as critically endangered, as is Benton's, and about half are considered endangered or vulnerable.

The primary threats to crayfish populations are the same old: habitat destruction and modification (especially changing water flows and chemical pollution) from human activities, overexploitation, and introduction of competing exotic species as fish bait.[65] This is exacerbated by the physiology of the crayfish themselves: if a water system becomes uninhabitable, it is very difficult for crayfish to escape to another system.

One unusual threat originated in the German aquarium trade, where the first example of a parthenogenetic crayfish—one where females reproduced without the participation of males—was discovered around the turn of this century. Aquarists noted that if they bought a single individual of the beautiful, marbled crayfish (marmorkrebs), they would sometimes find a swarm of babies in their aquarium. The marmorkrebs is genetically closely related to the slough crayfish, *Procambarus fallax*, a native of Florida, but no parthenogenetic strain has been found within the range of *P. fallax*. But it soon will be—as this rapid reproducer has turned out to be a dangerous invader, having spread widely not just in Europe but also to Africa and Asia. It is quite possible that it will start to exterminate populations of other crayfish globally. The aquarium trade is moving crayfish around extensively, and once an undesirable invader establishes itself, eradication becomes extremely difficult.[66]

Bivalves

Freshwater and marine ecosystems harbor many species of mollusks known as bivalves, such as oysters, mussels, clams, and scallops, which are sessile (immobile) invertebrates with a hinged shell. Many of the threats that face freshwater crayfish populations are shared by what may be the most unsung group of threatened populations in North America—freshwater mussels. These largely sedentary creatures play important roles in ecosystems, especially by feeding many other creatures and filtering and purifying water. But they are highly vulnerable of population extinction because, among other things, their populations are easily wiped out, especially by the nearly ubiquitous dams that *Homo sapiens* love to stick into any water with the gall to flow.

The reproductive strategy of mussels involves attaching their young larvae to the gills of specific fishes and dispersing that way, eventually dropping off the fish and (if they find a suitable bottom) settling down, attaching to a surface, and growing. Dam building has thus restricted the distribution of mussel populations when they block fish movements and tie mussel populations to fish populations that can be forced to extinction by the dams. For instance, of the almost eighty species of mussels in the US Midwest, over half are endangered, suggesting a major episode of population extinctions that have moved those creatures to the brink of total disappearance. In the United States as a whole, almost three-quarters of mussel species are losing populations at a rate that imperils their existence.[67]

Shellfish reefs, formed by millions of oysters or mussels, have long been huge contributors to the human food supply. As a single example, in 1864 in Great Britain, 120,000 people were employed as oyster dredgers, and 635,000 tonnes (around 700,000 tons) of oysters were devoured in London. In addition, oysters supply humanity with an array of ecosystem goods and services. They are an important food supplier, they filter pollutants from water, and they serve as ecosystem engineers by forming large structures that host a variety of other organisms from fish to algae.

But despite their huge importance to humanity, natural oyster populations are in decline; a recent peer-reviewed scientific study concluded that "the condition of oyster reef ecosystems is poor and the challenge in revitalizing native oyster reefs is great."[68] The reefs are, for example, disappearing from around Australian estuaries. Those formed by two species—flat oysters (*Ostrea angasi*) and the delicious Sydney rock oysters (*Saccostrea glomerata*)—have declined about 90% and 99%, respectively, with most of the natural reefs having been destroyed by overharvesting. The rock oyster is now, fortunately, extensively and successfully farmed, although there is a constant battle against the parasite causing "Queensland unknown disease" (*Marteilia sydneyi*).

We can't leave the example of oysters without mentioning the only bivalves we've ever dealt with personally one at a time. Those are the giant clams (*Tridacna gigas* and relatives) that may grow to just over a meter (four feet) across and a weight of almost 225 kilograms (500 pounds). They live buried in sand on coral reefs, with only the rim of their shells and their beautifully colored algae—zooxanthellae—inhabiting their mantles exposed. The zooxanthellae are often the same that help nourish the corals themselves, as we'll see. Paul was always careful not to stick his hands into the giant oysters' shells when diving on coral reefs, although the stories that they have actually trapped and drowned divers seem to be baseless.

Giant clams are sadly on their way out due to the usual anthropogenic threats. Ironically, they share with rhinos the threat from Chinese medicine—the powerful muscle that shuts the shells is rumored to have aphrodisiac properties. But they also have suffered substantial overexploitation as a human food item. Regional and global oyster populations are likely facing further declines as monocultures of non-native species, climate-change-related deterioration of ocean conditions, and epidemic diseases all seem to be worsening.

Snails

Gastropods (snails) are another large group of mollusks. The extinction of these little creatures is rarely documented, unlike, famously, the end of the Passenger pigeon. A recent exception to this was the extinction of one Hawaiian tree snail species (*Achatinella apexfulva*). The species declined for many years because of habitat loss and introduced predators. The last survivor, lonely "George" (named after the last surviving Pinta tortoise) died on January 1, 2019.[69] He was fourteen years old, in a captive breeding program—and, as is inevitably the case, he also finished a population extinction.

George was a player in one of the most complex stories of population differentiation and population extinction, that of Pacific Island land snails. Land snails have suffered more *recorded* species extinctions than any other group of animals—something we know because their beauty, ease of study, and fascinating evolutionary history has attracted the attention of many biologists.[70] The snails evolved into a vast array of species when they invaded the high oceanic islands of the Pacific because they had access to many different ecological niches there and faced few predators. Hawaii alone had some 750 species, and its forests were originally carpeted with the shells of dead snails.

But the snails proved to be a rich and abundant food source for exotic predators such as rats and a predatory snail from North America, and they suffered from habitat destruction as forests were converted into croplands and pastures for cattle grazing. In addition, exotic predacious snails were deliberately introduced as "biological control" agents on many islands in attempts to wipe out an invasive pest, the giant African snail (*Achatina fulica*). Rather than control the African snail, the predators tended to make short work of populations of the native land snails.[71]

Land snail populations outside of Hawaii have played prominent roles in solidifying human knowledge of evolution. Snails of the genus *Cepaea* have most famously provided examples of natural selection acting in nature. For example, English populations of the snail were preyed upon by thrushes, which ritually chose rocks, denoted "thrush anvils," on which to smash open the shells, thus allowing the birds to gobble down the soft parts.[72] This allowed comparison of the characteristics of the shells of the snails found living and those that had been eaten—showing selection favoring snails with color patterns that camouflaged them in local environments.

Sadly, *Cepaea* and their land snail and slug relatives are especially vulnerable to population extinctions caused by rapid climate change, subjecting them to different temperature and humidity regimes.[73] That degrades their important ecosystem functions as decomposers eating decaying vegetation, as prey of many different animals above them in food webs, supplying vital calcium to predators, especially birds needing it for their egg shells, and as predators controlling populations of soil organisms.

Land snails are not the only snails in deep trouble. Populations of some 550 extant species of predatory marine cone snails (*Conus*) live mostly on or near coral reefs and mud or sand flats associated with them. Their populations, which in many cases may actually consist of two or more "cryptic" species (ones with indistinguishable shells but unique biochemical make-ups including venoms)[74] are disappearing as their reef habitats are destroyed by climate disruption, dynamite and cyanide fishing, silting from onshore construction, sewage flows, and other hazards. Populations of the snails themselves yield millions of individuals for the ornamental collecting and trade for their gorgeous shells and, to a much lesser degree, for use in research.[75]

Research has also revealed the hunting techniques of cone snails. They are predators specializing on marine worms, or other snails, or fishes. Techniques differ with targeted prey, but generally, the cone snails crawl over reefs and mud flats and spear their prey with a long tongue tipped with a rapid-fire dart. The dart injects a

complex nerve-poison consisting primarily of a mix of polypep-
tides (proteins or protein fragments) evolutionarily selected for
efficacy in the usual prey. Single reefs may have populations of
thirty or more similar species, but they have evolved to not compete by each having special prey and venom evolved to be effective with each target.

The complexity and diversity of their venoms has attracted
much attention from neurobiologists trying to develop important
drugs for treating human problems. Supplying natural products of
value to medicine is an important ecosystem service. Those products, whether in snail venoms or plant defensive compounds, have,
through hundreds of millions of years of evolutionary development, attained astonishing variety and complexity. Searching for
them in cone snails or cone flowers are examples of what has been
coined "bioprospecting," and especially where indigenous groups
lay claim to populations of the source organisms, disputes arise
over ownership and "biopiracy"—unethical or criminal appropriation of biological materials.[76] *Conus* defense consists of trying to
sting perceived predators, and some human beings have been killed
by the beautiful snails. Amazingly, *Conus* have been collected in the
field in many places by Gary Vermeij, a brilliant evolutionary biologist, who is blind. Besides his bravery, Gary is famous for being able
to identify snails better by feel than most scientists can by sight.

One of the largest gastropods was the queen conch (*Lobatus gigas*), a marine snail of the Caribbean seas. It was an abundant and
successful herbivore feeding on sea grass, algae, and other marine
plant material. It had, however, one bit of very bad luck—these
snails taste and look good to an invasive predator, *Homo sapiens*. The
result: overharvesting populations for meat for the conch fritter
trade and for souvenir shells. In addition to overharvesting, their
populations were subject to general oceanic deterioration from
acidification and climate change. Ocean acidification refers to an increase in the degree of acidity of the ocean water, caused primarily

by uptake of carbon dioxide from the atmosphere. The queen conch is now much less abundant than sixty years ago, and some of their populations are now very rare. Fortunately, its trade is now regulated, which has allowed the species to begin slowly recovering.

Octopuses are apparently the smartest of all invertebrate animals.[77] Among the most famous stories about them is their amazing ability to escape from captivity—exemplified by one that escaped from its aquarium in a zoo, slithered to another aquarium, worked its way into it, devoured a fish, and then returned to get back into its home aquarium and look innocent. When we learned more about octopuses, all three of us, vegetarian or not, dropped octopus from our diets: too much like cannibalism. Despite this, a major threat to many octopus populations is predation by people—somewhere around 400,000 tonnes (around 440,000 tons) are captured and killed annually. They are also threatened, like many other marine organisms, by habitat destruction such as that caused by bottom trawling—commercial fishers dragging nets across the sea floor, essentially rototilling it.

There are now relatively successful efforts to "farm" octopuses for the food trade. This has led to a great debate over the ecological consequences of the farming, and on the ethics of raising masses of such intelligent animals in grim, captive conditions. Similar questions can and are raised about farming, killing, and eating another highly intelligent animal, pigs. We expect the many ethical issues around human diets and food distribution will become more intense as human populations continue to expand and wreak havoc on our life-support systems.

Other Invertebrates

A group of invertebrates especially important to humanity is earthworms, which play major roles as ecosystem engineers in keeping soils fertile.[78] Earthworms in floodplains are notoriously subject to natural population extinctions and reestablishment,[79] and they are similarly notoriously subject to anthropogenic extinction caused by agricultural operations.[80] Nearly two centuries ago, Charles Darwin recognized the great importance of earthworms to humanity.[81] But his insight has been largely forgotten—populations of earthworms, like most underground biodiversity, have not been appreciated and thus are barely monitored.[82] A start at reversing this trend has been made with new studies of the diversity and biogeography of earthworms[83] and soil biodiversity in general.[84] That includes nematode worms (roundworms)—the most abundant animals on Earth.[85]

These mostly minuscule organisms play many different roles in nature. Some are parasites, whereas many others are essential for the proper function of nature. In surface soils globally, it has been estimated the weight of these tiny creatures is some 0.3 gigatonnes—about the weight of a million Boeing 747 jet airliners. The aboveground nematodes are often tiny parasites of plants—so abundant that if all plants disappeared, their outlines would still be visible as nematodes. As the critical roles of soils in supplying ecosystem services to humanity gains appreciation, so will the need to understand population extinction and reestablishment in these obscure (to us) but ubiquitous animals.

It once seemed that, with very special exceptions, marine invertebrates were not in danger of extinction. One such exception, the eelgrass limpet (*Lottia alveus*), was confined to the leaves of eelgrass (*Zostera marina*) on the coast of North America from New York to Labrador. In the 1930s, there was a great die-off of eelgrass largely due to disease caused by a slime mold, and among other things, to increasing nitrate pollution. The only surviving populations of eelgrass were in estuarine refugia, but the limpet could not tolerate brackish water, and all its populations disappeared. It was the first historical oceanic invertebrate extinction.[86] Not unexpectedly, the disappearance of eelgrass populations on the east coast of

North America caused declines in populations of eelgrass-eating brant geese. The surviving brants changed their diets, increasingly grazing inland on grasses as did Canada geese, and according to brant hunters, have changed their flavor.

But at least one modern group of marine invertebrates now appears to be suffering from population declines and extinctions traceable to human overharvesting—abalones.[87] Although these quite sedentary and tasty mollusks release their gametes into the sea, there is evidence that those of the endangered white abalone (*Haliotis sorenseni*) do not travel far, and local collapses make it clear that there are separate abalone demographic units that are subject to extinction. Sadly, simply removing fishing pressure does not seem to allow populations to rebuild—outplanting (supplementing natural populations by introducing captive-bred individuals) seems necessary for permitting abalone populations to recover.

Overfishing, pollution, and habitat destruction can wipe out populations of marine invertebrates responsible for bioturbation, the critical mixing of near-shore sediments that, among other things, determines the oxygen level within the sediments and thus their community composition and the characteristics of the marine coastal ecosystems they underlie. The high productivity of those ecosystems, critical as "nursery" areas for harvested marine fishes, thus can be threatened by population extinctions among the invertebrates.[88] Declines in fisheries, in turn, can lead to food shortages in coastal communities which, especially in Africa, can have destructive effects on wildlife. Hungry people deprived of critical protein supplies from the sea naturally tend to increase bushmeat harvesting—to the detriment of terrestrial mammal, bird, and reptile populations.[89]

Considering the obviously huge scale and importance of the population extinction problem, we have far too little detailed information on the past (or even present) dynamics of populations. This is especially true of invertebrates that play crucial roles in delivering ecosystem services to humanity. One result of the paucity of information has been an underestimation of the rate of decay of biodiversity. We are largely dependent for understanding the scale of the ongoing sixth mass extinction event on records of species range shrinkage in big animals.[90] But there is one tiny invertebrate that turns this generality on its head because it is an ecosystem engineer—one that outdoes even the beaver in the role of creating and maintaining habitat for other animals: that little cylindrical relative of the jellyfish, the coral polyp.

Corals

Anybody that has been fortunate enough to snorkel or scuba dive on a coral reef in the tropics knows the incredible biological diversity and beauty of that ecosystem. Coral reefs are the most diverse ecosystems in the sea, equivalent to rainforests on land. But coral reefs are vanishing because the populations of reef-building corals are in deep trouble.[91] The coral polyps, mostly less than two and a half centimeters (one inch) in diameter, engineer giant stony reefs in cooperation with the photosynthesizing algae (zooxanthellae) that live within the coral polyp's body. Those reefs underlie only a minute portion of Earth's oceanic surface, about a tenth of a percent. But they are extremely important to *Homo sapiens*. Hundreds of millions of people live near them, and those creations of innumerable reef-building coral polyps may support as much as a quarter of marine biodiversity. Among other ecosystem services, the populations of fishes on the reefs are critical feeding bases for many human populations—supplying as much as 70% of global fisheries' catches.[92]

Populations of reef-building corals are under global assault from climate warming and acidification, threats that are everywhere in the oceans. Increasing sea surface temperatures cause some populations of coral polyps to expel the colorful zooxanthellae that lived within their tissues and supplied them with energy. In water above certain temperatures, the corals[93] can no longer supply the algae's needs, which in essence turns the algae into parasites.[94] The resulting expulsion leaves only the color of the polyp's light calcium carbonate skeletons—"coral bleaching," and bleached corals start to starve. Other threats range from sewage flows, silting, overexploitation of fishes and other resources, to tourist damage and explosions of starfish that devour corals.

If conditions were to change so the ocean cools a little, bleaching episodes could be reversed. The zooxanthellae can re-invade and save the coral, although there are complex relationships among environmental stressors, coral species, and clades of zooxanthellae.[95] But ocean conditions are trending to change in the wrong direction—thanks to ocean heating from greenhouse gases, compliments of the oil companies, human population growth, and expanding consumption, the oceans will warm further, become more acidic, suffer more overexploitation, and absorb more damage from human activities on the land.[96]

But scientists do not yet know enough to be able to predict when, if current trends continue, virtually *all* reef-building coral populations will go extinct, primarily from water heating and, likely, acidification. Corals can genetically evolve substantial high-temperature resistance (some already have) and also to a degree acclimate phenologically (changing according to environmental fluctuations) to rapid heating.[97] But projecting the effects of acidification is particularly difficult because of the complex interactions among the formation of skeletons, their dissolution, and water temperature.[98] In more acid water, marine organisms struggle to build their shells, thus becoming more vulnerable.

Patterns in the decline of coral populations under present levels of ocean warming and increasing acidity are sufficiently variable that a global disappearance of populations of reef-building corals seems rather unlikely in the very near future.[99] But a "canary in a coal mine" is already warning humanity of the demise of the corals and general collapse of oceanic ecosystems quite soon. "Sea butterflies" or pteropods—small sea snails that swim with wing-like feet—have long been noted for their sensitivity to acidification.[100] Now, scientists are finding pteropods with their shells dissolved in the arctic waters of the Gulf of Alaska and the Beaufort Sea.[101]

We believe the intrinsic interest of these fascinating (to us) creatures more than justifies humanity in trying to protect their populations described in this and other chapters. But of course, there are many practical reasons for putting in conservation efforts, especially in preserving what are important elements of the human food supply. One key reason is anchored in uncertainty. It is analogized in the so-called "rivet popper" hypothesis.[102] Suppose you are walking toward a commuter airliner to fly home when you notice a mechanic on a step-ladder working hard with a screwdriver on the wing:

"What are you doing?" you ask.

"Popping rivets out of the wing."

"What for?"

"MoneyAir sells them for $5.00 each—adds to profits."

"Are you crazy? You're weakening the wing!"

"Don't worry, wings are built with many more rivets than they need. There's lots of redundancy."

"But if you hit bad turbulence you might run out of redundancy. The wing could come off and ruin your whole day."

If you are smart, you'll of course skip the MoneyAir flight and re-book on another airline or drive. But human beings can't migrate to another planet or just depart on a spaceship. Only the ignorant believe that large numbers of people will leave the planet in the foreseeable future. There is redundancy in most ecosystems; often, populations of several different species can play much the same role. But it's equally clear if you push too many populations to extinction, civilization will disappear because one truth underlies our existence and survival: we are utterly dependent on the integrity of our too-little-understood life-support systems, the populations of our fellow inhabitants of Earth.

Chapter 7

Vanishing Green

Plants—Our Emerald Treasure

IN THIS CHAPTER, WE DESCRIBE why the green component of the terrestrial biological realm is not only a beautiful, distinctive feature of our planet, but also represents a critically endangered emerald treasure on which the survival of countless numbers of species, including our own, utterly depends.

Indeed, one can hardly overstate what a pivotal episode in the evolutionary saga of life on our planet is represented by the colonization of land by plants. When exactly this momentous change occurred has been a matter of exciting debate for a long time in the scientific community. Examination of the oldest fossils of terrestrial plants suggests that this watershed moment occurred some 420 million years ago, but because plants lack bones or hard shells that can be easily fossilized, suspicion of an earlier start has always remained.[1] It is now widely accepted that vascular plants (trees, shrubs, herbs, and ferns—the plants that currently dominate the planet and with which we are most familiar) entered the ecological theater scene sometime after the non-vascular plants (liverworts, hornworts, and mosses), although the precise sequence of entrances and exits among the latter on the terrestrial ecological theater is still a matter of debate.

Regardless, the early appearance of plants indicates that terrestrial plant life coincided with, and likely influenced, the explosive diversification of terrestrial *animal* life—the "Cambrian explosion" referred to in chapter 1. Such plant-animal relationships, over eons, undoubtedly made possible the ebullient, relentless biological diversification of the Phanerozoic period (the geological period spanning the last 550 million years). Beyond that, plant colonization of the land had other critical global implications, such as the cooling of the climate and the increase of the level of oxygen in the Earth's atmosphere—factors that, evolutionarily speaking, we owe our existence to and should ever be truly grateful for.

As previously stated, colonization of the land by vascular plants triggered a pulse of terrestrial plant diversification, in which the lineage of the Pteridophytes (ferns and allies) first became dominant, with a spotlight that lasted some 100 million years. At some point, they were joined by the Gymnosperm (conifers) lineage, which then became the preponderant group for some 150 million years.[2] The next act saw both Pteridophytes and Gymnosperms joined by the Angiosperms, flowering plants with seeds encapsulated within fruits, some 150 million years ago. Equipped with these evolutionary novelties, Angiosperms underwent an explosive diversification and became the Earth's dominant plant lineage, remaining so even into our present time.

Although Pteridophytes and Gymnosperms still appear on stage today, they are now upstaged, with their numbers much diminished from the times of their zenith. This emerald treasure, one of the most visible garments of the planet today, is a product largely of that hyper-successful group of plants, the flowering Angiosperms—whose ancestry, complex evolutionary trajectory, and explosive diversification Charles Darwin once referred to as the "abominable mystery" in a famous letter to his contemporary botanist and friend, Joseph Hooker, in July of 1879.[3] Let us now describe what the emerald treasure consists of, how it is distributed, what it does, and how it is endangered.

The Components of the Emerald Treasure

Plants, via photosynthesis, capture and mobilize the sun's energy and, with nutrients and water at hand, convert it into plant tissue (technically called biomass), which contains carbon—the central chemical currency of life. Plants therefore (together with photosynthetic microbial organisms) are life's primary producers. The rate at which biomass is stored by plants and made available for transfer to other levels of the food web in ecosystems and, collectively, the planet, is called net primary productivity or biological productivity. In other words, organisms that photosynthesize are the source of all food, fiber, and fuel upon which all other organisms, including humans, depend. As such, they are the first—and most crucial—step in the food chain.

Net primary productivity (NPP) is the process by which primary producers fundamentally determine the planet's habitability, i.e., Earth's life-supporting systems. NPP is measurable, estimated at an overall annual value of grams of carbon, representing the food available for the rest of the food chain. This net primary productivity is huge, equivalent to the weight of between 5 and 10 billion elephants (or, in technical terms, 56 petagrams, where 1 petagram is 1 billion tonnes, or over 1.1 billion tons).[4] In the 1980s, Paul, together with other Stanford colleagues, took up the formidable challenge of estimating the proportion of the Earth's net primary productivity that is appropriated by humanity. The team calculated that an astonishing 40% of net primary productivity is hijacked, much of it wasted, by a single species: humans!![5] Such a massive appropriation of the energy potentially available for all other living organisms has signified a massive conversion of land for crops, pastures, infrastructure, and so on (collectively called land use and cover change), which, together with direct harvesting, represents the main driver of populations and species extinctions of plants and other life companions.

In addition to plant biomass and NPP, another facet of the Earth's emerald treasure is the richness of plant species. The current number of plant species is suspected to be much larger than previous estimations of about 320,000 species, based on the number of valid, described species, plus the rate of description of new species at the beginning of this century. Subsequent estimates arrived at 450,000 in 2015,[6] although an early evaluation suggested that richness of vascular plants could even be as high as 500,000 species. It is now increasingly accepted that the number of terrestrial plant species can be estimated at around 400,000, 95% of which (i.e., 380,000) are vascular plants, with the vast majority of those (97.4%) being Angiosperms—the flowering plants that decorate nature and even most of our human-dominated landscapes.[7]

Another element that underscores the magnitude of our emerald treasure is that plant species are typically composed of a collection of populations, in some cases a plethoric collection of populations. The latter is particularly evident in the case of species that have a large geographical range. For example, Rodolfo has observed populations of one of his favorite tropical trees, *Brosimum alicastrum* (the remarkable, multi-use Maya "nut") ranging from the west coast of Mexico in seasonally dry tropical forest near the town of Chamela, to the east coast of the same country in tropical rainforest near the town of Catemaco. Across the latitudinal range, populations occur from the northerly state of San Luis Potosi, Mexico, near the town of Xilitla, down south to the Magdalena region of Colombia, down to the state of Acre, Brazil, and out to Jamaica and Cuba in the Caribbean.[8] Within Mexico, populations occur at elevations from close to sea level (e.g., Chamela) to 700 meters (e.g., Xilitla). Across this huge latitudinal, longitudinal, and elevation range, the populations of this tree exhibit variation in numerous

Jaguars

Jaguars (*Panthera onca*) are the most abundant species of large cat in the world, outnumbering lions, tigers, and leopards. The still-vast tropical forests of the Amazon basin and the Pantanal maintain the largest jaguar populations. In Mexico, their estimated population size increased from 4,000 to 4,800 between 2010 and 2018 due to increasing conservation measures and public awareness campaigns.

Cheetah

The cheetah (*Acinonyx jubatus*) was once widespread from Africa to India, but most of its populations disappeared because of habitat loss and hunting. A small population perhaps twelve individuals strong still exists in Iran. In Africa, there are some 7,000 cheetahs left. Habitat loss and the pet trade are important factors in their decline.

Giant river otter

The giant river otter (*Pteronura brasiliensis*) is the largest otter in the world. It is found in the Amazon basin and Pantanal (Brazil). Its decline was associated with the pelt trade in the 1960s and 1970s. Pollution and habitat loss also played a role in their decreasing populations. It is estimated that only a few thousand individuals still survive.

Sumatran orangutan

There are three species of orangutans: the Bornean (*Pongo pygmaeus*), the Sumatran (*P. abelii*), and the Tapanuli (*P. tapanulensis*). The Sumatran orangutan is only found in Indonesia's Sumatra Island. There are fewer than 7,000 left, making this orangutan one of the most endangered great apes. Restricted to tropical forests, their populations have declined due to the conversion of their habitat to palm oil plantations and other crops.

Cotton-top tamarin

Opposite: The Cotton-top tamarin (*Saguinus oediopus*) has one of the most restricted geographic distributions on the planet. It is confined to tropical forests in northeastern Colombia, between the Magdalena and Atrato Rivers. Like many other species of primates, habitat destruction is the most important threat to the species. The pet trade and domestic animal diseases pose addition threats.

Black rhino
A black rhino (*Diceros bicornis*) and its calf are a hopeful sight on a private reserve in South Africa. Hunted almost to extinction because of the illegal trade of its horn, the species survives due to heroic private and government conservation efforts. Like most endangered species, its fate relies almost exclusively on our actions.

Hyacinth macaw
Opposite: The extremely beautiful Hyacinth macaw (*Anodorhynchus hyacinthinus*), with its dark blue plumage and yellow facial marks, is exclusively found in South America. It is the largest flying parrot and one of the most majestic. It is threatened by the pet trade and habitat loss. Fortunately, there are still healthy populations in some regions in Pantanal, in southern Brazil.

Green honeycreeper
Tropical birds such as the Green honeycreeper are losing populations mainly because of loss of habitat and the pet trade. The massive losses experienced by tropical forests in the Amazon and the Congo basin in the Americas and Africa, respectively, as well as in Southeast Asia, has caused the decline of thousands of birds.

Adélie penguin
Opposite: The Adélie penguin (*Pygoscelis adeliae*) is found across the Antarctic continent. Fortunately, there are still some 7 million individuals, but climate change is starting to pose a serious threat to this as well as many other species of birds.

Galápagos giant tortoise

Giant tortoises were some of the first victims of the historic extinction crisis, disappearing from several islands such as Mauricio in the Indian Ocean. In the Galápagos archipelago, off the cost of Ecuador, some fifteen species (*Chelonoidis*) were alive when Charles Darwin visited the islands in the eighteenth century. However, three or four are currently extinct and most are considered endangered by habitat loss, tourism, and introduced rats.

Olive ridley sea turtle

Opposite: The populations of the seven species of sea turtles have declined drastically in the last decades. Some like the Loggerhead are critically endangered. The olive ridley sea turtle (*Lepidochelys olivacea*) is the most abundant of all sea turtles. Its populations have increased steadily in the last three decades, but only thanks to extensive conservation efforts.

Variable harlequin frog
The variable harlequin frog (*Atelopus varius*) is a critically endangered species that was once widely distributed in Costa Rica and Panama. The species vanished from most localities by the 1990s because of the deadly chytrid fungus disease. Once considered extinct, it was rediscovered in the early 2000s, and now is thought to survive in only two known populations.

Colorado pikeminnow

The Colorado pikeminnow is endemic to the warm waters of the large rivers of the Colorado River basin. Dams and water overexploitation of its habitat caused its decline; it disappeared from the lower Colorado River, and its populations became extirpated in Mexico. The Colorado and Green Rivers are home to the only remnant wild populations, although the species has been reintroduced to several rivers. The Colorado pikeminnow is one of hundreds of freshwater fishes that have become extinct or are endangered because of human activities, in what is now known as the "silent crisis."

Caribbean reef shark

It is estimated that only 2% of the number of sharks that were found the oceans in the 1970s are found today. Some 100 million sharks are caught every year. A few species, such as the Caribbean reef shark (*Carcharhinus perezi*) in Cuba's Jardines de la Reina National Park, are still very abundant.

Coral reefs

Coral reefs are among the most diverse ecosystems on the planet. Unfortunately, most coral reefs around the world are being severely affected by climate change. Warmer temperatures produce coral bleaching and diseases. When exposed to stress conditions such as changes in light or temperature, the coral expel their symbiotic algae and first turn white, then die.

Bay checkerspot butterfly

The Bay checkerspot (*Euphydryas editha bayensis*) population within Stanford University's Jasper Ridge Biological Preserve is an example of the complexity of population extinction. In the 1970s, there were fluctuating populations in the area, and they were gone by
1997. The loss of the populations was a gradual process in which changing climate played a
dominant role, but other factors were likely involved, including nitrogen deposition from
Bay-area automobile exhaust.

Poweshiek skipperling

The loss of insect populations can be very rapid. The Poweshiek skipperling (*Oarisma
poweshiek*) lost most of its populations in a decade (late 1990s and early 2000s), concurrent
with the development of neonicotinoid pesticides and the invasion of the Asian soybean
aphid, which the pesticide is sprayed against. This arc of points matches the northern
range of the corn-soy planting region and hence concentrated pesticide use of the US Midwest. Photo courtesy of David Pavlik.

This illustration by biologist and artist Mattias Lanas conveys the drama of defaunation, in this case in an African savanna. The pencil eraser exemplifies the human intervention leading to the demise of large vertebrates.   Illustration courtesy of Mattias Lanas.

botanical characteristics as well as equally varied uses by local human populations. This highlights the importance of local populations of plants, and the implications of the omnipresent reduction of their geographic range in this age of human impact—the Anthropocene.

Furthermore, in addition to the diversity of populations that characterize most plant species, there is another critical aspect of the diversity of the planet's emerald treasure, the diversity present within populations. Represented by the variation across individuals, genetic diversity within populations is a critical factor for the survival of a local population. The magnitude and significance of genetic diversity cannot be understated and is dramatically illustrated in the case of food crops, as we will discuss next.

Agrobiodiversity and Genetic Diversity

Surprising as it may seem, cabbage, cauliflowers, broccoli, kale, brussels sprouts, collard greens, kohlrabi, gailan, and calabrese are all *the same* plant species—selected and bred from variants of their wild ancestor, *Brassica oleracea*. Each of the aforementioned variants may, in turn, exhibit considerable variation in form, flavor, and other properties (e.g., red cabbage, cone cabbage, Jersey cabbage, marrow cabbage, and savoy cabbage). Similarly, peppers and chilies, including an astonishingly diverse array of variants, such as jalapeño, chipotle, paprika, cayenne, bell, chili mirasol, chiltepín, cascabel, chilaca, puya, pimiento morrón, cuaresmeño, serrano, poblano, chili de árbol, güero, and piquín, to name but a few,[9] are all domesticated variants of one single species—*Capsicum annum*. The genesis of such an explosive genetic diversity of these two crops is representative of a process called domestication, which results from the particular cultural idiosyncrasies and profound traditional ecological knowledge of ancestral peoples who, through countless experiments of trial and error and spanning multiple plant and human generations, transformed what initially were wild (in some cases unpalatable or even toxic) plant species into dietary staples. This is yet another expression of biological diversity, agrobiodiversity.

The impact of agrobiodiversity on humanity has been undeniably profound; currently there are some 2,500 species of plants that we use for food. Of these, the grass family (e.g., the cereals—maize, rice, wheat, sorghum, and oats) and the legume family (e.g., beans, peas, fabas/broad beans, soybeans, lentils, garbanzos, pigeon peas, peanuts, and so on) represent about 30% of all human-use crop species. Another ten plant families include anywhere from a few to several dozens of crops, and many other families have one or a few crop species. In addition, there are many semi-domesticated species, and many other undomesticated but edible species (fruits and spices, for example) that are harvested directly from the wild.

The implications of this aspect of plant diversity are multiple. On the one hand, industrial agribusinesses's emphasis on producing homogeneous crops of a few species in extensive monocultures to facilitate mechanized harvesting reduces the productivity of the land from a plethora of species to one or two per area, thus putting our global food supply at risk from increasingly extreme weather conditions and potential crop pests or pathogens. Mechanized harvesting of monocultures, coupled with the tunnel-vision focus on visual attractiveness of produce to the eyes of the consumer in supermarkets, work to impoverish the richness of our diet.

Compounding the effects of human-driven massive land-use change to accommodate agroindustry, we see, together with climate change, adverse effects on the remaining wild relatives—the ancestors—of our food crops, thus further driving the loss of their populations and with it, the obliteration of the precious gene pool that represents a safety net for humanity's food security in light of global change. The well-known, dramatic events of the Great Irish

Famine (1845-1852), which was triggered by a fungus that decimated the susceptible genotypes of the potato crop in that country, should serve as a stark reminder of the risks of relying so heavily on commercially driven monocultures.[10]

On a more optimistic note, and not surprisingly, the bastion of hope to meet the food-supply needs of a changing world might lie with the indigenous peoples and rural communities whom, in many parts of the world, are continuing their traditional agricultural practices, maintaining not only diversified, sustainable "agro-ecosystems" with multiple crops that safeguard the genetic diversity of the latter, but also caring for and safeguarding the wild relatives of some of our commercial crops. Additionally, these peoples, through their careful stewardship and legendary relationship with their environments, are continuously developing land races of several crops, adapted to a variety of ecological settings, such as drier and hot areas, thus creating a knowledge base for the rest of humanity to learn from and replicate given the increasing need to produce food in the new landscapes that are proliferating due to anthropogenic climate change. In effect, strategies to mitigate and combat the anthropogenic impacts we have inflicted upon our emerald treasure may be accessible to us, so long as we do not continue to neglect it or the lessons from the indigenous and rural communities that live in harmony with the land.

Neglected Losses

In contrast to the media and academic attention devoted to the severe threats to animal life that we have discussed so far in this book, plant life extinction is much more difficult to appreciate. There's an irony in such a myopic perception: over the last five decades or so, we have had access to increasingly sophisticated remote sensing technology (satellite imagery) that allows us to monitor the status of Earth's green carpet on a continuous and reliable basis. Such technology not only allows us to quantify the magnitude of lost vegetation cover (via a commonly used metric, deforestation rates) but helps us to describe the spatio-temporal patterns of plant life loss, model and project the trajectories of plant-life coverage, and provide vivid, compelling representation of such trends. Such pictorial depictions abound in the scientific literature, textbooks, magazines, and even newspapers, all of which can convey a powerful visual representation of the pace of shrinkage of plant communities and, by extension, loss of populations.

Although we've yet to globally quantify the loss of specific plant populations via deforestation rates, it is evident that vegetation cover has been steadily decreasing and therefore plant diversity must necessarily have diminished at the levels of ecosystems, communities of species, and plant populations worldwide. Recent research and large database compilations, including the International Union for the Conservation of Nature (IUCN)'s Red List, or the Botanic Gardens Conservation International (BGCI)'s ThreatSearch[11] show that the current and projected deforestation rates, i.e., plant diversity loss,[12] are heavily driven by land-use change (including the massive deforestation of the tropics to give way to oil palm and soy monocultures, grasslands for cattle ranching, urbanization, mining, and other types of human-made infrastructure).[13] Moreover, without urgent remedial action, we can expect that land-use change, in complex synergy with anthropogenic climate change, overexploitation (legal and illegal logging), pollution, and invasive species (including insect pests such as the catastrophic bark beetle, and fungal phytopathogens such as those causing Dutch elm disease and sudden oak death), will be increasingly hard to manage, thus giving rise to a subsequent mass extinction of plant populations.

Setting aside the synergies among the drivers of global change to focus on the effects of climatic change alone, a recent compilation of published reports on shifts in species ranges at the warm edge of their distributions found that out of a sample of 260 plant species, close to 40% had already contracted their range, implying a direct loss of many populations.[14] Considering these dramatic impacts on plant populations already occurring at the current observed levels of warming, which are below even optimistic projections for future warming, along with current and predicted increases in land-use change, and other anthropogenic drivers, one can augur, again, a massive pulse of plant population extinctions in the coming decades. Such extinction forecasts should be consistent with independent metrics of evaluating plant life loss, for example, the rate of species being classified as threatened. The IUCN's impressive efforts to increase the number of assessed species for threat classification to be included in the Red List, together with efforts of ThreatSearch and BGCI to match threat assessment with valid plant names, have led to revised estimates for threatened species of vascular plants of between 43.7% (in the case of IUCN)[15] and 37% (in the case of ThreatSearch).[16] Experts involved in these assessments conclude that two in five (40%) known plant species may be considered as threatened with extinction.[17] This represents a considerable increase over the oft-cited estimate of 14% for the Gymnosperms (the most thoroughly examined group of plants) in the early 2000s.

The accelerated rates of local decline of many species, leading to their range contraction and massive population loss, ultimately represent a prelude to global extinction. Global species extinction is now being documented, particularly in some areas of the world where more information is available. Such is the case of North America, where recent research indicates that sixty-five taxa (fifty-one species and fourteen subspecies) from thirty-one US states, the District of Columbia, and Ontario, Canada, have gone extinct since the beginning of European settlement.[18] These seemingly low numbers are likely an underestimate, considering that the number of still undescribed species is relatively high (a fact inferred by the number of new species annually described, for example, twelve in California alone).

As might be expected, a high proportion of the extinct plant species were endemic to a particular area. One might have thought that regions with unusually high levels of plant endemism would have been the target of stronger protection, and scientific research has indeed provided relevant insights to guide policy and develop such an important agenda. This is particularly well-illustrated by the concept of Global Biodiversity Hotspots (GBHs).[19] GBHs are areas where at least 1,500 plant species are endemic, present nowhere else on Earth. Contrary to what is known about the value of GBHs, 70% of those areas have been already impacted by land-use change. The significance of that fact is critical: protecting those remaining areas (about 2% of the global terrestrial area) would save approximately 40% of the Earth's plant species.

Sadly, though, recent research using sophisticated remote sensing technology reveals a disheartening picture: on average, each of the thirty-five GBHs have only retained 15% of their area that could be considered intact vegetation.[20] Given the uniqueness of the floras initially present there and their recent dramatic decline, we can expect a significant pulse of global *species* extinctions, in addition to the global massive loss of plant populations.

Beyond the dire situation of the GBHs, the cumulative number of estimated global plant species extinctions since Carl Linnaeus's times (ca. 1753), corrected for rediscovery and reclassification, is estimated to be around 600.[21] Estimates are that this massive loss is some 500 times higher than the expected pre-Anthropocene vascular plant extinctions that might have been part of nature's course.[22] In sum, although the number of globally extinct plant species is deceivingly low, the loss of local populations portents that the number of species rapidly approaching tipping points that could lead to extinction is appallingly large, underscoring the tremendous need for society to safeguard plants if we are to safeguard ourselves.

It could be argued that plant life may be less susceptible to population and species extinction given that plant seeds can remain viable in the soil for a long time, perhaps up to hundreds of years. Likewise, it is argued that plants can be conserved more easily than groups of animals by seed "banking" at low temperatures or tissue culture when seeds are not available or able to live for a long period (say, 100 years) under storage conditions. In support of this, recently, seeds of the extant species *Silene stenophylla* that were found buried for 32,000 years in the Russian permafrost were successfully germinated in the lab and produced beautiful plants with flowers, fruit, and viable seeds.[23] The authors of that research concluded that "At present, plants of *S. stenophylla* are the most ancient, viable, multicellular, living organisms."

Indeed, important efforts to maintain large seed banks of crops from around the world have been established in places such as Svalbard in Norway. This approach is now being used for the preservation of non-crop seeds, with a remarkable example in China, where out of the 30,910 native species to the country, 11,000 are maintained in the Germplasm Bank at the Kunming Institute of Botany. However, some plants do not need to depend on seed banks or great longevity for persistence. Some populations can potentially be regenerated from part of stumps or tubers that survive fires, floods, and other disturbances. In addition, some endangered plants are more readily preserved and propagated in botanical gardens and collections than animals are in zoos and conservation facilities.

Although these efforts are commendable and inspirational, their long-term success is difficult to ascertain because plants in nature do not live in an ecological vacuum—they are part of a complex biological network of neighboring plants (some of which can be competitors or facilitators), pollinators, seed dispersal agents and root-associated fungi critical for their capture of soil nutrients. Furthermore, plant populations are also immersed in a complex social context that synergizes with the biological context and makes conservation efforts even more challenging.

To take just one case, consider Tiehm's buckwheat (*Eriogonum tiemii*), which occupies a small area of Nevada with a soil rich in lithium and boron. This buckwheat is adapted to those elements, but they are also needed for electric car batteries and magnets that allow wind turbines to function. An Australian corporation wanted to mine those soils, but to do so would destroy much of the only known population of this buckwheat.[24] A petition to formally protect the population has been submitted, raising questions such as: Will the planned quarry destroy so much of the population that it will be driven to extinction? Is the species sufficiently distinct and important enough to block the mining of materials important to a transition to "green energy"?

Inevitably, over our scientific careers, we have seen dramatic declines in the abundance of plant species, several of them endemic, signaling their likelihood of extinction. Rodolfo has painfully noticed, for example, the local disappearance of magnolias, agaves, and cycads in Mexico. However, in the absence of thorough expedition searches conducted multiple times and covering a long time period, it is extremely hard to declare global extinctions, although in cases like these, one can palpably see the relentless loss of populations. In some instances, however, botanists have been able to document global plant extinctions. One plant extinction well-known to biologists is that of the St. Helena olive tree (*Nesiota elliptica*), a small tree sufficiently distinct that taxonomists placed it in a separate genus. Portuguese sailors discovered the isolated South Atlantic island of St. Helena in 1502. Little more than 300 years later, in 1815, the island gained notoriety as the place the British chose to exile Napoleon after his defeat at Waterloo until his death in 1821.

But the true source of the island's infamy was the early Portuguese introduction of goats to the island. As is their wont, the animals on the island, free of predators, quickly devoured much of the vegetation, which had not evolved defenses against large herbivores. Along with goats, human activity ended up destroying more than 99% of the island's original vegetation.[25] By around 1850, there were only a dozen or so surviving olive trees, too few to have much of an ecosystem function. The last specimens in cultivation died in December 2003 despite extensive efforts to save them.

Epilogue

Probably the single most important measure of "global change" that is of highest practical importance to humankind is the change in biological diversity. As we have discussed, plant populations (like invertebrates) are major casualties of the sixth mass extinction, but largely ignored by society's myopic focus largely on charismatic animals like pandas and Passenger pigeons. The most recent conservation efforts are centered, as usual, on "species," suggesting that plant populations are disappearing just like animal populations—although plant populations are more difficult to define than animal ones, and population extinctions, while inferable from range contractions, are hard to document. The fact that many species are extremely long-lived (trees, for example) complicates the assessments of plant species extinction at the global level.[26] Some individuals or small groups of a species may remain in fragments or even heavily degraded habitats. These plants, although physiologically functional, may represent what Dan Janzen[27] astutely labeled "living dead"—elements of genetically and demographically unviable populations. However, the fact that "plant extinctions take time"[28] may actually help us buy time and undertake action to rebuild viable populations, for example, through restoration of suitable habitats.

As we will discuss in the last chapter, the "cure" for the population extinction crisis lies in a complex set of societal changes and attitude shifts that would impact almost every facet of modern life. The alternative, however—massive plant and animal extinctions exacerbating the worst effects of global environmental change— will undoubtedly bring sweeping and devastating impacts to our way of life. It would seem better to take action now to preserve our biosphere and retain some control over the process than be forced to respond to a heavily impoverished planet—with a biodiversity crisis signaling a sixth mass extinction and a climate out of control. Moving plants away from the category of *neglected losses* and preventing the looming mass extinction of their populations is central for humanity to avoid the consequences of an unsustainable, ghastly future.

Chapter 8

Microbes

A Hidden World

MICROBIOLOGIST TIM CURTIS HAS REFERRED to the importance of microscopic life forms as follows: "I make no apologies for putting microorganisms on a pedestal above all other living things, for if the last blue whale choked to death on the last panda, it would be disastrous but not the end of the world. But if we accidentally poisoned the last two species of ammonia-oxidisers, that would be another matter. It could be happening now and we wouldn't even know."[1]

Thus, we now come to the least-known aspect of biodiversity—which may well be the most important. As suggested in the epigraph, civilization clearly can survive without populations of elephants or lions. But civilization is dependent on agriculture and fisheries, which in turn are dependent on tiny creatures not subject to much attention by conservation biologists. Bee populations, those crucial pollinators, themselves are dependent on a diversity of microbes that can, like their insect hosts, be driven to extinction.[2] Microorganisms are also subject to the vicissitudes of climate distortion and global change in general, but how and where are also very little understood.[3] And we as individuals are a joint enterprise among billions of *Homo sapiens* cells interacting with billions of bacterial cells, without which we could not exist.

The situation with agriculture is similarly complex. Conservation biologists often point out that agriculture is a major enemy of biodiversity.[4] But ironically, as Andrew Beattie has emphasized, "agriculture itself depends on biodiversity," mostly in the form of non-charismatic little organisms such as nitrogen-fixing bacteria.[5]

The situation with our bodies is also ironic. Since pathogenic microbes were discovered, the attitude of both scientists and the public has mostly focused on "germs." As a society, our approach to microorganisms has been to avoid or kill them—in other words, to *cause* population extinctions. But new tools for understanding our tiny companions in and on our bodies, in and on our farm fields, and out there in nature, are producing a much more nuanced view of microbes and an understanding of our total dependence upon them—in addition to the continuing strong need to pay close attention to those that can wreak havoc directly on us (e.g., malaria parasites, *Vibrio cholerae* bacteria, and corona viruses, among others), on our crops (e.g., the potato blight fungus),[6] or on our domestic animals (e.g., the rinderpest virus, which wiped out many populations of cattle and related animals until a vaccination program eradicated it early in the twentieth century).[7]

We need to recognize that perspectives on biodiversity depend on context. Most authors writing about the loss of biodiversity are really referring to a limited range of species, sometimes directly critical to humanity (e.g., certain bats, bees, birds, and fishes), often of importance in maintaining ecosystem function or of iconic interest or spiritual value (rhinos, antelopes, butterflies, or songbirds) that are the traditional focus of conservation biologists. That array could be called "conservation biodiversity," and it has been the principal topic of this book so far.

The biodiversity that supports global agricultural production, fisheries, and other industries, from pharmaceutical bioprospecting and engineering to medical biomimetics (trying to solve human problems mimicking nature solutions),[8] and biological pest control,[9] has been called by Beattie "production biodiversity." We here include in production biodiversity the bacteria that contribute to the productive functioning of human bodies. As Beattie has pointed out,[10] "production biodiversity is enormous and irreplaceable." It comprises at least 90% of all species—predominantly microbes and invertebrates (such as insect pollinators, soil nematodes, and earthworms),[11] but also the genetic diversity of crops (actual or potential) and crop relatives.

Production biodiversity harbors most of the planet's chemical, genetic, and metabolic diversity—and this may be an underestimate because there may be more than a trillion species of microbes.[12] The most recent "tree of life" that documents the variety of all species on Earth shows stunning arrays of microbes forming the main branches of the tree, whereas all the species embraced by conservation biodiversity occupy just three twigs.[13] As Beattie has pointed out, "global assessment of conservation biodiversity is crucial. But conservationists must acknowledge the greater part of biodiversity on which humanity depends, or the very word 'biodiversity' risks losing its scientific, economic, and social meaning."[14]

Microorganisms are clearly a key element of production biodiversity—the "microbiomes" of soils, plants, and people. A microbiome is the total collection of microorganisms in an environment and their aggregate genetic material. The characteristics of an area's microbial community determine the regulation of nature, "the biological, chemical, and physical processes that transform nutrients and energy within an ecosystem."[15]

A common early view of microbial biogeography was that everything was everywhere—that, for example, bacteria were so abundant, tough to kill, adaptable, and easily transported that they would show few, if any, biogeographic patterns. More recent work and advances in molecular genetics have demonstrated that this is not correct. Different habitat conditions and dispersal barriers produce complex patterns of microbial occurrence,[16] and scientists have discovered dramatic variation in the distribution of bacteria with different characteristics on a microgeographic scale. For instance, in one small meadow, eight different kinds of bacteria showed strikingly different abundances depending on the properties of the soil in various parts of the meadow.[17]

Similarly, within the Chihuahuan desert spread across parts of Mexico and the United States, the Cuatro Ciénegas system of springs, streams, and pools represents a "desert oasis" of microbial life that harbors highly diverse aquatic communities, including thirty-eight unique phylotypes from ten major prokaryotic lineages of bacteria and one from archaea.[18] A unique community-wide survey of root-associated fungi in a Mexican tropical rain-forest uncovered a striking, taxonomically and functionally diverse fungal community comprising 2,090 fungal Operational Taxonomic Units (OTUs, in some sense similar to the concept of species), with most OTUs being rare, encountered in just a few of the samples.[19]

Some microorganisms are known to differentiate locally in response to geographic barriers. For instance, archaeans (single-celled organisms that resemble bacteria in some ways but are unique or resemble plants and animals in some other biochemical characteristics) are restricted to hot springs, but have resistant spores and are found in widely dispersed sites.[20] The microbial communities in sediments from parts of a Maine river with different salinities differed in both composition and their impacts on ecosystem functioning.[21] Similarly, the composition of microbial communities helps determine the rates of ecosystem processes on land, such as litter decomposition.[22]

In broad geographic terms, plant and animal diversity correspond well (more in the tropics, less toward the poles). But fungi do not fit this pattern, except soil fungi that form mutually beneficial associations with plant roots—ectomycorrhizal root symbionts—which can be important to the fertilization of crops.[23] Indeed, fungal root symbionts can determine aboveground plant diversity and productivity.[24] Thus, it appears that fungal diversity is more responsive to soil and microclimatic factors than to the plant communities growing on the soil, although the limited data available suggest that both termites and large herbivores can have substantial impacts on mycorrhizal communities as well.[25] Such patterns and the tight connection of soil microorganisms to aboveground biodiversity and crops[26] raise the issue of the occurrence and significance of microorganism population extinctions.

Issues of population extinctions among microbes are complex, especially because we don't know much about how populations of microorganisms are best defined and delimited. This is especially the case among different kinds of bacteria that can rather freely exchange genes with one another (horizontal gene transfer). However, discrete groups of microbes and the functions they perform can be shown by genomics (the study of their complete sets of genes). For example, analyses comparing the soil biota of remnant tallgrass prairie with nearby agricultural fields showed that farming had strong negative impacts on the microbial community, including those elements that contribute to soil fertility.[27]

From an agricultural point of view, one thing is clear. The rampant use of artificial fertilizers and pesticides has greatly reduced the microbiomes and other elements of production biodiversity in soils in areas of industrial agriculture;[28] in other words, pesticides are directly destroying the biodiversity that helps maintain agriculture itself.[29] Scientists are now actively measuring microbial communities in soils with a view to ecological management and conservation.[30]

Parts of the Microbiome That We Wish Had Stayed Home

Humanity would have been much better off if some microorganisms had not traveled from their homelands. A classic example is the fungus that caused a lethal disease of potatoes, the "late blight" (*Phytophthora infestans*) mentioned in chapter 7. It migrated from the potato's Andean home, and, in conjunction with British callousness, caused the famous nineteenth-century Irish potato famine, in which about a million people died and some 2 million left Ireland. In some parts of the world, the blight is still a major problem,[31] causing approximately $3 billion in losses annually. Its negative impacts occur mostly far from where it evolved, and humanity is still seeking a way to extinguish the "late blight" populations.[32] Potatoes are hardly the only victims of populations of migrating fungi, however. Many scientists believe the current amphibian holocaust can be traced largely to mobile populations of a chytrid fungus, as discussed in chapter 5.

Fungal dispersal may play a big role in future human health. If, for example, a current population of *Coccidioides immitis* or *Coccidioides posadasii* fungi, found in California and the southwestern United States and which cause valley fever, was driven extinct and replaced by a more adaptable strain, that serious disease could become an even more widespread threat. After all, climate change is already starting to spread the specialized desiccated habitats favorable to the disease,[33] and human populations are still expanding in those habitats. Unfortunately, given the twin trends of population growth and climate change, this is likely to occur with increasing frequency for other fungi as well—with potentially devastating results.

In addition, one of the global changes in part caused by the population extinction crisis is exposing Earth's vast and super diverse fungal biota to temperature stress. As Emily Monosson details in her brilliant book *Blight*,[34] mammals evolved energetically expensive high core body temperatures likely as a "thermal barrier" against fungal pathogens, which generally thrive in cooler temperatures. But there are scary signs that fungal strains are evolving to be perfectly at home in the mammalian body temperature range. They are already killing many immune-compromised patients and likely will cause vast difficult-to-control fungal pandemics in the future.

Similarly, the population biology of pathogens involved in zoonoses (diseases transmissible to people by animals) such as SARS-CoV-2 ties population extinctions of various mammals directly to human health. COVID-19 (possibly) and many other infectious diseases such as Severe Acute Respiratory Syndrome (SARS), Middle East Respiratory Syndrome (MERS), Ebola, Lassa fever, and Hantavirus are of natural origin, but the destruction of natural environments, differentiated defaunation (chapter 9), and the illegal wildlife trade has brought humanity into greater contact with wildlife in much of the world.

Zoonoses are more abundant in human-dominated landscapes.[35] The COVID-19 pandemic was originally thought likely to have originated in the Wuhan, China, "wet" animal market.[36] Wet animal markets, common in Asia, sell live animals and meat in non-supermarket conditions, which are more prone to disease transmission between animals and from animals to humans. Genetic data point to an origin possibly in a species of horseshoe (*Rhinolophus*) bat commonly found in East and Southeast Asia. There is also, grimly, the possibility that the virus causing the COVID-19 deadly pandemic was an accidental escape from a Wuhan laboratory studying coronaviruses and doing controversial[37] and potentially dangerous gain-of-function research, which could increase the lethality of a virus.[38]

The outbreak of another coronavirus, SARS, in 2002–2003, which sickened 8,000 people and killed some 800, probably traveled from a bat to humans.[39] Saudi Arabia experienced a similar event, MERS, in 2012. Indeed, in the last four decades alone, roughly fifty emerging epidemic zoonotic diseases affecting humans have originated due to humans being exposed to the vectors because of habitat destruction, wildlife trade and selective defaunation, and poultry farming. Wildlife trade is absolutely gigantic and is of major concern because of the threats it poses to animal (and some plant) populations.[40] For example, between 2011 and 2020, there was a legal trade of approximately 80 million animals, and 90 million kilograms (almost 200 million pounds) of animal parts were legally traded. However, these vast amounts are likely dwarfed by the sheer scale of the illegal trade. This creates abundant contacts for a wide range of wild animals with each other and with human beings and thus vast new opportunities for some populations of pathogens to travel from one species to another.[41] From the viewpoint of humanity, the dynamics of population spread and population extinction of pathogens are mostly unknown but clearly can have potentially serious consequences. For instance, a large population (millions) of captive minks had to be wiped out when it started to exchange COVID-19 with people in Denmark, and the SARS-2 virus is now evolving and circulating in large deer populations in the northeastern United States.

Human Microbiome

Homo sapiens also cannot persist without the production biodiversity closest to us: the human microbiome. It is vast. Estimates differ, but there likely are slightly more bacterial cells than the 30 trillion human cells in our bodies (of the latter, red blood cells make up the majority).[42] For each human cell, there may be a dozen or so viruses in our bodies, and maybe a fungus cell for each ten of our own. Whatever the precise numbers (which would be constantly changing in any case), we contain literally a shitload of microbes—

and that term is accurate in the sense that many of the bacteria are in our guts, and a single defecation can carry sufficient bacterial individuals to alter their numerical dominance. The bottom line is that human beings carry a giant community of gut bacteria, among other kinds, and the structure of that community may be unique to each individual, as distinctive as (though somewhat more labile than) its genetic code.[43]

Microbiome investigation is one of the most recent areas of science to reveal surprises. Among them is that the collection of microorganisms that share our bodies plays many more roles in our health than previously thought. The human microbiome is not just involved in digestion, nutrition, and disease, but also influences the nervous system. For example, the microbiome in your gut, consisting of perhaps 1,000 species and 7,000 strains,[44] is mostly bacteria but also comprises viruses, archaeans, protozoa, and fungi. Startlingly, it has been found that an individual's gut biome can change the functioning of their brain. It does this by producing neurotransmitters that activate nerve pathways along the "brain-gut axis" that affect a person's chances of being stressed or even depressed.[45]

Microorganisms in your intestines can even activate the entire central nervous system. It is especially interesting, for instance, that stress is related to the composition of the gut microbiome[46] because stress is tied to a wide array of diseases.[47] It is not actually known, however, how changes in the composition of the gut microbiome influence mental health. For example, could the extinction of a population of a bacterial species bring on a bout of depression?

Microbial Population Extinctions with Potentially Lethal Consequences

One situation, increasingly serious on a global scale, is the extinction or depression of populations in the gut microbiome, which can have very serious consequences. That occurs when a person is prescribed intensive antibiotic treatments, particularly with broad-spectrum agents. Although these antibiotics are necessary to fight an infection elsewhere in the body, they do not discriminate in the microbes they kill. Our gut biomes are bystander casualties, and a course of antibiotics can kill enough gut microbes to dramatically alter the composition of that microbiome. Freed of competition from benign microorganism species in our guts, a virulent pathogenic bacterium, *Clostridium difficile*, may explode, producing toxic chemicals that attack the gut lining, producing inflamed patches that in turn cause frequent episodes of watery diarrhea and painful cramps, high fever, dehydration, kidney failure, and even death. Some strains of *C. difficile* are highly antibiotic resistant, presenting serious difficulties for treating them using traditional therapies.

However, successful and sometimes life-saving treatment of the condition has been achieved in an unusual way. Fecal transplantation, most recently using defrosted capsulized frozen fecal samples from unrelated donors,[48] has been found to work where antibiotics and other treatments could not. Fecal transplants seem to have the potential to help in the treatment of other digestive tract problems as well.[49] The fecal transplant procedure presumably restores the normal microbiome, replacing populations that the antibiotic regime had exterminated and rebuilding decimated populations of benign microbes. This once again demonstrates the importance of our microbial inhabitants—and the peril of their loss.

The parallels to the need to maintain and restore the microbiomes supporting production biodiversity of cropland are clear. It is also conceivable that some treatments for neurological and psychological problems will be developed that involve manipulating gut microbiomes. The mechanisms underpinning this crosstalk require further elaboration but may be related to the ability of the gut microbiota to control pathways related to the breakdown of a key amino acid, tryptophan, producing substances that can activate nerves, influence immune responses, and are sensitive to stress.

There are parallels in our own bodies to what goes on between a plant and its root microbiome. In the least-loved part of our gut— the colon—fiber-fermenting microbes produce chemical compounds with medicinal properties when we eat whole plant foods. Butyrate, for example, one such compound, helps to maintain the cells that form the colon wall. Butyrate can also help reduce low-level systemic inflammation that is involved in causing many chronic ailments. So, if we want both healthy crops and healthy people, the prescription is really pretty simple—cultivate microbial allies and be alert for mechanisms causing population extinctions. Microbiomes can help to keep both plants and us healthy. And conserving these important elements of production biodiversity means supplying our microbial allies with an appropriate diet— high levels of organic matter in soil as free as possible from synthetic organic chemicals,[50] and high levels of fiber-rich plant foods in our own diets.

The Way Forward

The incredible importance of microbes to all life, including ours, is matched only by our incredible ignorance of their diversity and population biology. Attempts to remedy this are underway, especially benefitting from the advances in molecular biology over the last seventy-five years. In that field, humanity has been helped by populations of a lowly gut microbe that plays diverse roles in the intestines of people and other mammals, and occasionally evolves into populations that kill us. It is so renowned for its part in developing human understanding of basic genetics that biology students learn its name, *Escherichia coli*, alongside that of *Drosophila*.

We know *Homo sapiens* cannot persist without the microorganisms of production biodiversity, but we have only a fragmentary view of the array of virus, bacterial, and fungal populations that are essential to maintain human populations.[51] There is still a very long way to go, but at least we can now be aware that there is an unseen, often intimate, and largely unknown element in the sixth mass extinction crisis to which it will be profitable to pay attention—the fate of unseen populations that surround us and upon which we depend, both inside and out.

Chapter 9

Defaunation

RODOLFO'S RESEARCH IN THE EARLY 1990s led to the adoption of the term "contemporary defaunation"[1] as scientists became concerned that the patterns and magnitude of loss of animal life—and its consequences for ecosystems—were being overlooked. Compared with the increasingly powerful technology of remote sensing that by the 1990s was vividly displaying the dramatic patterns of plant-life loss—aptly captured in images and measurements, as well as projections of deforestation—there has been no similar technology or sufficient fieldwork or databases to illustrate what was happening to Earth's contemporary animal life. In order to bring the extent of this often-invisible threat to mainstream consciousness, Rodolfo initially used the term "defaunation" to describe the absence of large mammals in Mexican tropical forests and the ecological consequences thereof for the plant communities.[2]

This attempt to bring visibility to the problem was splendidly advanced using the "empty forest syndrome" term that Kent Redford popularized,[3] along with discussions of its multi-faceted consequences.[4] The wealth of current research now allows us all to see that right now, right in front of us, there is an ongoing explosive sixth mass extinction episode signaled by vertebrate and invertebrate population losses and declines.[5]

In his book *Harvesting the Biosphere*, Vaclav Smil[6] provides a dramatic assessment of the changes of animal biomass at three points in time (10,000 years ago, 1900, and 2015). His data lead to a first rough yet vivid picture of Anthropocene defaunation from the perspective of wildlife. At the end of the ice ages, global animal biomass (i.e., the estimated weight of vertebrate animals) was close to 300 million tonnes (around 330.5 million tons), of which a tiny fraction corresponded to a human population of about 4 million persons with almost no domesticated animals. By the year 1900, when the human population had reached 1.6 billion, total animal biomass was still close to 300 million tonnes. Approximately 60% of that, however, corresponded to domesticated animals, about 23% was human biomass, and wildlife biomass had been reduced to less than 20%.

By 2015, total animal biomass had exploded to a whopping 1,850 million tonnes (around 2,040 million tons), but this was largely composed of domesticated animals, which comprised 76% of the total, followed by humans (7.3 billion humans by then), representing 23%—while wildlife was now reduced to a mere 1%. Even more sobering, this is actually an underestimate because the data did not include seals, sea lions, amphibians, and birds. But this only heightens the central, disturbing message. This biomass perspective of defaunation, however rough (in that it does not consider the kinds of species involved, the time-course and geographic distribution of the impact, and so on), speaks to a catastrophic global decline of wildlife and the domination of Earth's biomass by essentially a handful of species, in which two are particularly predominant: cows and humans!

Over the last few decades, scientists' field work in many parts of the world in both research stations and adjacent unprotected areas, and through local quantifications of poaching and hunting, have considerably improved our capacity to illustrate the defaunation crisis. More recently, with the deployment of long-term monitoring work, use of camera-trap technology, and more sophisticated analytical procedures and models, non-invasive genetic analyses, and increasingly more robust and multi-faceted databases, we are in a better position to delineate the general patterns of the current pulse of population extinctions.

A first important aspect lies in the framing of defaunation by considering three general interrelated facets of biological extinction processes: global species extinctions, geographical range contraction, and local decline in abundance of species populations. Among these three aspects, global species extinction—the complete disappearance of a kind of organism from the face of the Earth—is the one that has captured most of the attention, not only within the scientific community but also among the general public and even some policy-makers. Although this is an undeniably critical aspect of biological extinction, the strong emphasis placed on numbers of extinct species alone can lead to the misinterpretation that animal life is not immediately threatened but is just part of a slow episode of extinction.

For example, "only" 338 vertebrate species have been recorded as extinct over the last 520 years (i.e., since the year 1500). Such seemingly low numbers, however, become more meaningful if one considers that many of those extinctions (close to 60%) actually occurred over the last 120 years. Considering such an accelerated extinction rate and comparing it with the estimated rates of "background extinction" (i.e., those occurring at times when there is not a mass extinction event) of mammals, the recent rates turn out to be 100 to 1,000 times faster than background (depending on the vertebrate group). The mere counting of globally extinct species is important, but it must be understood that it only gives a limited view of the problem.

Species extinction numbers fail to expose the real nature of defaunation, for, as the chapters in this book document, they overlook *population* losses and declines in local abundance. Species are, in reality, a mosaic of populations distributed throughout their geographical range. Those populations normally represent the same species but with somewhat different genetic characteristics and, often, behaviors. From the point of view of the species' ecological roles within their natural communities and ecosystems, these somewhat differentiated local populations are what really matter.

Consider, for example, the case of the African savanna elephant (chapter 3), which has been exterminated in many areas of its original distribution range throughout the sub-Saharan area of that continent. This massacre means that many populations of this species have now been lost (a veritable major pulse of within-species biological extinction). It also means that the ecology of the savanna (in terms of the dynamics of fire, for example) is now disrupted in those localities; it may also represent an economic downturn for the local populations of humans who might have had, for example, an ecotourism business based on elephant sightings. These compounded population extinctions occur even though the entire species is not extinct.

This pattern of local population loss is consistent with that of other emblematic species in Africa (e.g., common hippo and black rhino), Latin America (e.g., jaguar, harpy eagle, and tapir), Asia (e.g., tiger, Sumatran rhinoceros, and orangutan), and North America (e.g., gray wolf, bison, and prairie dog).

The available information on the geographic shrinkage of many species such as elephants and rhinoceroses permits us to gauge the possible magnitude of population extinctions, and from those data, we have concluded that a process of "biological annihilation" of wildlife biodiversity is occurring globally at this very moment.[7] For example, using a sample of 177 mammal species for which sufficiently detailed information is available, we analyzed their range contraction over the last twenty-five years. The results revealed that almost 50% of the sampled species had experienced a range contraction of at least 80%, signaling a dramatic loss of local populations.

As would be expected from the differential defaunation observed in previous mass extinctions, many of the most affected species are those of large body size. For example, twenty-five of the largest herbivores currently occupy on average only about 19% of their historical ranges.[8]

When it comes to the decline in genetic diversity implied in contraction of the distribution range of species, a multi-species study (including plants, thirty-five vertebrates, and nineteen arthropods) suggested that, globally, the magnitude of extinction of genetic populations could range between 1.1 and 6.1 billion, depending on different deforestation rate scenarios.[9]

One form that range contraction takes has to do with the other facet of defaunation we referred to previously: the decline in the abundance of local species populations. Our own research on statistical estimates of local abundance declines using big data from a large sample of 14,000 populations of 3,500 species of vertebrates censused at different points in time yielded estimates of an overall decline of 25% in the period 1970–2010.[10] If we break this estimate down regionally, we find a 50% vertebrate decline in tropical regions over that same period. A more recent analysis with a larger sample (31,821 populations of 5,230 species), using the same approach, reports a dramatic, gloomier picture: an overall decline of 69% over the 1970–2018 period, with greatest declines in Latin America, Africa, and some other tropical regions of the world, compared with temperate regions.[11]

Elucidating the underlying reasons for this apparently slower rate of vertebrate decline in temperate regions might need further research in light of some high rates of decline in temperate regions reported in more recent publications. For example, a recent study of bird declines in North America analyzed trajectories for 529 species and found an astonishing net decline of 2.9 billion birds (29%)

since 1970.[12] This estimate does not seem to be driven, specifically, by the decline of rare (uncommon) species; in fact, the observed species in decline included once-common species. The decline in abundance of the local populations across the entire range of the Passenger pigeon described in chapter 4 dramatically illustrates the loss of even extremely abundant species.

The potential variation in decline rates depending on body size was not specifically tested in this big data–based study of North American birds and warrants further examination as well. However, our own observations in some local study sites reveal that, for example, the recent-decades declines have led to the complete local population loss of harpy eagles, scarlet macaws (currently being reintroduced), military macaws, and king vultures in the area of Los Tuxtlas (Veracruz, Mexico), even within a protected area in this locality.[13] These are representative examples of extreme decline leading to local population loss and geographic range contraction. In addition, many other birds (not necessarily the largest-bodied), particularly those of closed forest and wetlands, have declined considerably, whereas those of open habitats have increased. In Brazil's Atlantic Forest, major declines of large frugivorous birds have been documented, including toucans, toucanets, and large cotingas.[14]

Beyond birds, studies examining the situation of mammals at particular locations provide evidence of the catastrophic declines currently underway. In the tropics (where large hyper-diverse communities of mammals occur), for example, an initial estimate of annual hunting of mammals in the Brazilian Amazon reported 14 million individuals being removed annually.[15] An updated analysis of hunting at the same region revised the number of vertebrates lost to about 16 million per year.[16] Sadly, this is not an unusual picture, but a rather omnipresent pulse of defaunation across the areas of the world that used to have large wildlife until recent decades.[17]

In our Los Tuxtlas study site, for example, we have monitored the abundance of mammals over a couple of decades using intensive field sampling protocols of diurnal and nocturnal sightings.[18] We have documented a dramatic decline of numbers and even local population extinctions of several medium- and large-sized mammals such as jaguars, pumas, tapirs, white-tailed deer, and white-lipped peccaries—all historically present in the Biological Reserve of Mexico's National Autonomous University (UNAM), where our studies were carried out.

The local loss of both large carnivores and herbivores at Los Tuxtlas is consistent with the long history of anthropogenic impact in the region (overexploitation, habitat loss, and fragmentation). Shockingly, a recent study indicates that defaunation within protected areas is not unusual: in the period 1970–2005, the populations of large mammals in the protected areas of Africa declined by about 60%.[19] Furthermore, this is not an exclusive terrestrial catastrophe: the marine and freshwater realms also show high numbers of species with threatened populations.[20] There are even extreme cases where a few remaining individuals keep a species "alive" despite the fact that their populations may no longer be viable due to inbreeding or because they are incapable of finding partners for reproduction. A sad example of this is the vaquita as discussed in chapter 3, which epitomizes the case of species that is essentially "living dead."[21]

Our current understanding of defaunation now permits us to examine its consequences in terms of ecological processes, the functioning of ecosystems, and the implications thereof for critical ecosystem services and human well-being, including disease regulation. For example, our research shows that a consequence of differentiated defaunation is that, as large animals are declining, the contingent of small-bodied animals is thriving. The winners of the Anthropocene include many species, particularly within the hyper-species-rich group of rodents, which are hosts to numerous

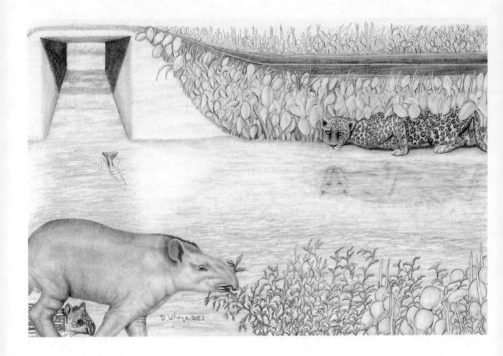

Tapir and jaguar

This drawing shows a tapir and a jaguar utilizing an aquatic corridor alongside and passing under a highway representing BR-262, which runs east–west, dividing the Brazilian *cerrado* (a tropical forest ecosystem). The tropical forest was once three times the size of Texas and was formerly considered to be unsuitable for farming. The annual whiplash of violent rains and parching droughts leached nitrogen, potassium, calcium, and other nutrients from the soil, leaving the surface acidic. But in the 1970s, after government researchers determined that adding lime and fertilizer to the soil would make it suitable for soybean production, farmers went crazy converting the area to agriculture. It would make Brazil the world's second largest soybean producer and lead the *cerrado* to disappear at a rate twice as fast as the Amazon.

In 1965, Brazil enacted its first Forest Code requiring landowners in the Amazon to keep a percentage of their land in its natural state. By 2012, however, the government loosened Forest Code restrictions and opened the way for legally clearing a California-size chunk of native *cerrado* (nearly 40.5 million hectares or 100 million acres) for cultivation. Although only 2% of critical jaguar habitat is effectively protected, studies showed that when corridors connect those areas, it takes just a few individuals to ease the big cat's extinction risk in small populations. Its prey, of course, must also survive. One such prey animal, the Brazilian tapir, is at a higher risk than jaguars. The tapir, the largest South American mammal, is a solitary vegetarian that forages mostly at night with the aid of a supple, trunk-like snout. It is classified as endangered in the *cerrado* and is at risk to habitat loss and hunting with only declining and isolated populations remaining. Jaguars preying on tapirs take mostly their young.

pathogens of zoonotic potential such as bartonellosis, leptospirosis, leishmaniasis, and bubonic pest, among many others.[22] This represents a red flag in light of the devastating historical pandemics such as the calamitous bubonic plague of Europe in the fourteenth century, and the devastating pandemic of SARS-COVID of the present century.

A better understanding of defaunation will hopefully spur us to become the stewards and survivors of our current ecological theatre rather than tragic Hellenic heroes thwarted by our own actions—and inactions. In other chapters of this book, we've examined the relationships between biodiversity loss (including defaunation) and another critical facet of extinction—the erosion or outright local extinction of natural ecological processes that we humans translate into services that represent multi-faceted existential values for *Homo sapiens*, be those of direct provisioning (e.g., food), regulation (e.g., pollination of plants and biological control of agricultural pests and some host- and vector-borne diseases), and last but not least, educational, emotional, and inspirational values. And the living tapestry of Earth is indeed inspiring. Together, our vast ecological support system not only keeps us alive—providing us with food and the air we breathe, as well regulating biological processes in ecosystems and within our own bodies. We owe it our survival, and paying the favor back is the least we can do—for the planet's beauty, its awe-inspiring inhabitants, and our own future.

Chapter 10

Drivers of Extinction

ONE DOES NOT NEED to be a scientist to understand why human population growth and the accompanying increase in humanity's total consumption of resources are root causes of the sixth mass extinction currently underway.[1] They are, indeed, root causes of virtually every current environmental problem. As Peter Gleick, leading expert on the water crisis, summarized it, "The size of the world's population, the nature of our consumption and economies, and the use of energy and water resources have combined to disrupt and threaten our very existence."[2] The scientific community, both national and world, has warned society about population-related issues for decades. For example, more than 15,000 scientists came together to sign on to a "World Scientists' Warning to Humanity: A Second Notice."[3] The human population's connection to biodiversity loss was recently emphasized again in a technical study led by Partha Dasgupta, one of the world's most distinguished economists, and sponsored by the British government.[4]

Despite the near unanimity among national and international scientific bodies focused on environmental issues regarding the key role of overpopulation in the human predicament, it is still largely ignored by policy-makers. But all they really need to take seriously is that every living being has evolved to have a set of specific habitat requirements. An organism cannot live where the temperature is too hot or too cold, or where there's too much or too little rain. If it lives in water, it requires not only an appropriate temperature range, but also water with appropriate salinity, acidity, oxygen concentration, and other chemical characteristics. If it is a butterfly, it must have access to plants suitable for its caterpillars to eat, and other plants that produce nectar that can fuel the adult butterfly's activities, including sperm or egg production. A lion requires plant-eaters to catch and devour. A tree needs a certain amount of sunlight and carbon dioxide and access to specific soil nutrients and water. A *falciparum* malaria parasite can't survive and reproduce without *Anopheles* mosquitos of an appropriate species in its habitat and a human bloodstream to feed on. These requirements are underlined for tropical fish enthusiasts every time an aquarium heater fails or for an orchid fancier whose watering schedule is interrupted. And they all have been widely altered by growth of the human population and efforts to sustain it.

The human population has grown so large that more than 40% of Earth's land is now farmed to feed people—and feed them none too well at that. Largely due to persistent problems with affordability and food distribution, almost 800 million people go to bed hungry, and between 1 and 2 billion suffer from micronutrient malnutrition (the latter exacerbated by the processed food industry). Even many of the rich suffer from an unhealthy diet. Because of its booming population size, *Homo sapiens* has taken much of the most fertile land to grow plants for its own consumption or for the diets of the animals people raise to later devour. But guess what? That cropland is now generally not rich in food plants suitable for the caterpillars of the 15,000 butterfly species with which people share the planet. Most butterflies cannot sustain themselves on the foliage of wheat, corn (maize), or rice plants on which human beings largely depend. From the viewpoint of most the Earth's wildlife populations, farming can be viewed as "habitat destruction."

This habitat destruction can also actually hurt the farming enterprise itself. In Costa Rica, for example scientists have found that clearing tropical forests for coffee cultivation destroys the habitat for populations of bees. The bees pollinate coffee plants, increasing the yield and quality of coffee beans—so that leaving areas in undisturbed forest, although reducing the area of coffee plants, leads to financial gains.[5] But tropical forests are still razed to graze the cattle to supply hamburgers for growing human populations[6] at a huge cost to biodiversity. In addition, widespread wetland drainage in response to the needs and desires of expanding humanity amounts to a defaunation of freshwater fishes and other aquatic organisms and a lethal attack on many bird populations.[7]

Unsurprisingly, few species of wildlife have evolved to live on highways, on vacant lots, or in strip malls, office buildings, kitchens, or sewers—unless you count Norway rats, house mice, bed bugs, European starlings, and German roaches as wildlife. Virtually everything expanding humanity constructs or destructs provides an example of habitat destruction from the viewpoint of all other organisms except a few "pests" or "weeds" adapted evolutionarily to the sorts of conditions people create.

Overpopulation is responsible for much more than habitat destruction. The more people there are, especially in rich countries and among the wealthy everywhere, the more products of nature they demand to meet their needs and wants: timber, grains, legumes, fruits, seafood, meat, gasoline, oil, rubber, metal ores, rare earths, elephant ivory, sand for concrete and glass, space into which to throw their wastes, and on and on. They also seek rare animals or plants to eat or to use for medicinal purposes. Increasing human demands cause both further habitat destruction and outright extermination of animal and plant populations by overexploitation. So, when you watch the expansion of the human enterprise, when you see buildings springing up, when you settle down to dinner at home or in a restaurant, when you become a proud frequent flyer, and, particularly, when you have more than one child, you are observing (and often participating in) the causes of the sixth mass extinction.

By far the one personal action that has had the strongest impact on an indicator of anthropogenic disturbance on nature (CO_2 emissions, which are closely linked to biodiversity loss), having one child less had, for citizens of rich countries, the strongest beneficial consequences.[8] In modern overpopulated and overconsuming societies, it is virtually impossible to avoid being personally involved in causing the sixth mass extinction. This is especially true given the grossly unfair distribution of the ability to consume without restraint both on national and international scales because it allows the few rich to employ ridiculously damaging technologies in their activities—space and deep-sea tourism, private jets, and super-yachts being obvious examples.

There has been a staggering increase in the human population in the last two centuries. When Paul was born in 1932, there were roughly 2 billion people, up from a few million before *Homo sapiens* adopted agriculture 10,000 years ago. Today, less than a century later, there are more than 8 billion people. Although the rate of increase has slowed down, the annual addition of people has hardly changed. The expanding human population has outright destroyed habitats, and where it did not, it altered them drastically to the detriment of wildlife and often people themselves. Over the last century, the more the population grew, the more greenhouse gases flowed into the atmosphere, and the greater the impacts on wildlife were, all of which required specific temperature ranges and other limited climatic conditions. And the more people there are, the more cities, roads, farm fields, fences, and other barriers there are preventing wildlife from living in or moving to areas of more favorable temperature or humidity in a rapidly changing climate.

The evidence meticulously gathered and analyzed by scientists from around the world (and reported in this book) clearly shows the major cause of the unusual and accelerating extermination of wildlife populations: expansion of the human enterprise. The same expansion is also driving accelerating climate disruption, which threatens both us *and* the rest of biodiversity. As we'll discuss in chapter 12, it will take a long time to humanely stop that growth and start the gradual shrinkage of the human population that is required if civilization is to persist and much of the rich life with which we share Earth is to survive. Determined efforts to limit per capita consumption among the rich while increasing needed consumption among the poor can have a more rapid effect than shrinking human population by limiting births—if there is a net reduction of consumption. Unhappily, like discussion of human population limits, at present neither consumption limits nor other steps to protect more populations of wild species seem politically feasible at the scale needed. For that matter, equitable distribution of resources among people, which is not a common feature of educational programs, also presently lacks the level of political support required.

Consistent with what we know in general about threats to biodiversity in the Anthropocene, two major kinds of drivers can be distinguished. One kind is the ultimate drivers, particularly the "population-consumption" nexus. Those ultimate drivers include not only the net numbers of humans, but the uneven distribution of resources among countries and sectors within countries, the outrageous overconsumption and waste of resources by the rich and middle classes, and the factors in our cultural-economic arrangements (such as corporations) that underlie society's self-destructive behavior.[9]

The ultimate drivers are exemplified by China's Belt and Road Initiative, which may be the single recent project that poses the greatest direct threat to biodiversity and civilization. It involves more than sixty countries, perhaps as much as two-thirds of the world's population, and a huge expansion of surface transportation—far,

far beyond that of the ancient Silk Road that serves as a symbolic model.[10] Many of the infrastructure projects, such as the Trans-Amazonian Highway, are planned for regions that are home to threatened and endangered species found nowhere else in the world. In addition, China's Belt and Road Initiative will provide greater access to populations of carnivores and other species already threatened by Chinese (often illegal) and Southeast Asian markets for them, especially for body parts with mythical medicinal powers.[11]

Although China's Belt and Road Initiative may eventually include measures that help biodiversity in some areas, on the global scale, these development plans are almost as certain to be a vast disaster for other animals and non-domesticated plants. The plan is the usual conglomeration of new roads, dams, factories, shopping malls, mineral extraction, and industrial agricultural expansion seen over so much of Earth already. This Western style of development will affect negatively local rural people in particular, Chinese people in general, and of course all the planet. It is being promoted by an authoritarian government that, despite some recent hopeful trends, represents a culture that already serves as a major source of biodiversity endangerment around the world (remember the rhinos, elephants, and pangolins!). Whether various kinds of international pressure might be applied to reduce the impacts is questionable, especially since most other nations, and especially the United States, are afflicted with a similar mania for growth.

We've already addressed the population and consumption issues that are the basic threats to civilization; now, we will focus on the proximate drivers. But before we do, we want to emphasize that these are where the real action is, where steps can be taken today to slow or reverse population extinctions without the massive (and slow) social change that ultimately will be required to reach sustainability.

The first proximate driver is land-use change (manifested as deforestation, fragmentation, urbanization, conversion to crop agriculture or grazing, and increasingly devastating and frequent forest fires). Second is overexploitation, such as hunting, fishing, poaching, and illegal trading for food, folk medicines, and aphrodisiacs. Third comes invasions by exotic organisms, often caused or aided by human activities (including incursions of disease-causing pathogens). Fourth is the global spread of toxic chemicals (natural and human-made) through the air, water, and soil. Last, but far from least, is climate disruption and its accompanying phenomena such as ocean acidification and coastal flooding.

The predominant approach to examine drivers, as it will be here, has been to consider them separately—not an ideal or particularly realistic approach. But given the complexity of the interactions of drivers and the limited research available, we consider our siloed approach as a first approximation, while highlighting that not only are there important variations in the relative importance of each driver, depending on the location and group of creatures examined, but that the drivers interact among themselves, so that, for examples, climate warming often helps mobilize toxic chemicals, and deforestation promotes climate disruption or facilitates over-exploitation of wildlife. The other drivers have considerably smaller representations in this database, with climatic change having the lowest representation (ranging from 5% in mammals to 13% in birds). Personally, we feel that climate change could overwhelm the other drivers soon if drastic steps are not taken by humanity to stop promoting it!

Land-Use Change

The prevalence of habitat loss and degradation as a force increasing population extinctions, including the many examples we've discussed in previous chapters, is not surprising when one considers that modern satellite imagery shows that by now, humanity has transformed at least 70% of ice-free land, much of it for the sake of agricultural production. Indeed, in its 2018 report, the Intergovernmental Platform on Biodiversity and Ecosystem Services[12] indicated that only a quarter of land on Earth is substantially free of the negative impacts of human activities. Such a colossal transformation of Earth's ecosystems implies that most of the animal communities that lived in the affected spaces either became homeless and migrated or died, or were able to survive in human-dominated landscapes if they had sufficient behavioral, physiological, and/or evolutionary flexibility to be able to do so. Some are now living dead, as explained in chapter 7. These doomed populations are part of the vortex of "extinction debt" that has become so omnipresent on our planet. That is to say, they are rapidly disappearing or already entrained but not yet completed.[13]

The destruction of habitats as a driver of extinction has become horrifyingly evident in these times of wildfire proliferation that devastates forests and shrublands in many parts of the world, a situation that is epitomized by the gigantic fires of places like Canada, California, Russia, Spain, Brazil, and Australia that have killed or damaged (physically and psychologically) so many people, destroyed their homes and, inevitably, exterminated vast numbers of animal and plant populations as well. In an account for Australia's *Evening Standard*, for example, Harriet Brewis noted that about 4 million hectares (almost 10 million acres) of wildlands and rural areas were burnt in the state of New South Wales alone in the 2019 to early 2020 fires.[14] The charred regions included nature reserves in the Blue Mountains World Heritage Area (one of the most beautiful wildlands we three have seen) and parts of the Gondwana rainforests. Millions of mammals, birds, and reptiles were killed, directly or indirectly, by the devastating fires after they began in September 2019. Amidst this tragedy, koalas, an emblematic animal for the country, suffered tremendous losses. Almost 8,000 koalas (which represent about a third of the population in its primary habitat) are believed to have burnt to death on the state's mid-north coast, while heartbreaking footage also shows kangaroos desperately attempting to hop away *en masse* from the advancing infernos.

Overexploitation

Overexploitation or overharvesting, as we have emphasized elsewhere, is harvesting a living resource at a rate that will eventually destroy it. It is a serious problem widely affecting terrestrial and marine plants and animals. The 2022 Living Planet Report (an initiative led by the World Wildlife Fund) has compiled data on 3,789 populations of vertebrates, based on the analysis of 5,059 threats. Their analysis found that for birds, mammals, and amphibians + reptiles, habitat loss/degradation is the main driver, represented by just shy of 50% of all the cases.[15] This is followed by overexploitation involved in a large portion of cases of threatened populations (range ~20% for birds to ~40% for mammals). This trend is different for fishes, for which the predominant negative force is overexploitation (~60%), followed by habitat loss and degradation (almost 30%).

For example, only about 2% of the populations of large marine fishes, such as sharks and tuna, now remain in the oceans compared with 1970. In 1945, the renowned writer and naturalist John Steinbeck made famous the then-flourishing sardine fisheries of California's Monterey coast in his book *Cannery Row*.[16] By the late 1940s, that resource had disappeared, making Cannery Row a ghost town. Sadly, more marine fisheries are on the same road. Close to 90% of the world's marine fisheries are fully exploited, overexploited, or depleted. Populations of many terrestrial animals have also been depleted by overexploitation. One of the most remarkable studies has shown that there were some 2,200 kilograms (4,850 pounds) of wild mammal biomass for every human being in the 1800s, and there are only two kilograms (almost four and a half pounds) today! This means that in just two centuries, humans have vaporized billions of individual wild mammals, almost analogous to the Chicxulub meteorite that vaporized the billions of dinosaurs.

A 2016 review reported that there were 301 species of mammals for which hunting was recognized as the main driver of population reduction, representing close to 26% of all terrestrial mammal species worldwide that the IUCN classified as "threatened."[17] Of species belonging to the main groups of mammals (taxonomically called "orders"), the highest percentages threatened by hunting were pangolins (order Pholidota) and elephants (Proboscidea) at 100%, followed by platypus and echidnas (60%), odd-toed ungulates like horses, zebras, and rhinos (Perissodactyla, 50%), apes and monkeys (primates, 31%), and even-toed ungulates such as deer and antelopes (Arctiodactyla, 30%).

Mammal species threatened by hunting consist predominantly of ungulates among large-sized mammals (bigger than ten kilograms or twenty pounds), primates among medium-sized mammals (one to ten kilograms, or around two to twenty pounds), and bats among small-sized mammals. More generally, close to 60% of the largest terrestrial mammals were classified as at risk of extinction from human exploitation for food or body parts.

Another study reported that at the end of the twentieth century, the number of primates consumed in the Brazilian Amazon alone amounted to several million individuals, producing some 14,500 tonnes (16,000 tons) of meat worth over $30 million annually. And these were not big primates like our African relatives. For example, "a single family of rubber tappers in a remote forest of western Brazilian Amazonia killed more than 200 woolly monkeys (*Lagothrix lagotricha*), 100 spider monkeys (*Ateles paniscus*), and 80 howlers (*Alouatta seniculus*) within 18 months." Yet another study "recorded the consumption of 203 brown capuchins (*Cebus apella*) and 99 bearded saki monkeys (*Chiropotes utahicki*) in a village of 133 Kayapó Indians over 324 days of study."[18]

From a geographical perspective, most of the overexploitation occurs in the tropics, particularly in economically challenged countries. From tropical rainforests, for example, bushmeat hunting annually removes over 6 million tonnes (approximately 6.6 million tons) of mammal meat. Of this, an estimated 75% of the trade occurs in Central Africa, followed by Asia and Latin America.[19] Information on illegal trading is scanty—not surprisingly, given that it is a clandestine activity. Some insights can be gained, however, using the (largely imperfect) data from police seizures, which suggest that close to 67% of the products seized originate from mammals (elephants, pangolins, rhinos, and tigers, in decreasing order of contribution), followed by reptiles with 20%, and birds with 7%, whereas mollusks, fish, amphibians, and insects make up the remaining 6%.[20]

The pangolin illegal trade offers some glimpses of the problem. Pangolins are medium-size mammals ranging in weight up to five kilograms (eleven pounds). Well-organized gangs, usually abetted by crooked politicians, murder unbelievable numbers of animals—to say nothing of forest trees. A single incident is illustrative. A ship traveling from Nigeria carrying slabs of frozen meat stopped in Hong Kong. There, hidden under the meat, inspectors discovered over eight tonnes (nine tons) of pangolin scales and 1,000 elephant tusks.

Sometimes, concealed among the traded animals, there are some most surprising findings. For example, among the pangolin scales confiscated between 2012 and 2019 in Hong Kong and Yunan Province in China, scientists found a new species of pangolin that they have named *Manis mysteria*—the mysterious pangolin.[21] Nobody

Saola

The face of the saola is visually striking. A dark line extends from each eye toward the mouth along the edges of two flaps. The saola flares those flaps to expose pores leading to a gland that emits a strong musk-like smell when scent marking. Multiple white facial bands and patches that also contain scent-emitting pores seem to break up the outline of the head when viewed in the dappled light of the creature's forest habitat.

Little is known about this very rare species, which was discovered by scientists in the Annamite Mountains of Laos and Vietnam in 1992. Sadly, however, it may already be extinct. If there are any surviving individuals, a captive breeding program and anti-poaching patrols are critical. But both efforts face significant challenges. Saola have yet to survive more than a few months in captivity, and snares like the one shown here are very common in their mountain habitat, although they are typically intended for other animals.

except the poachers, who were unaware of its scientific identity, have knowingly seen the species alive. To get a feel for the worldwide criminal network from desperately poor poachers risking their own lives to kill rhinos to filthy rich politicians making great profits, we highly recommend Rachel Nuwer's fine book, *Poached*.[22]

It is now clear that the effects of land-use change and overexploitation synergize to create a most powerful tsunami of defaunation. These drivers, however, interact in multiple and complex ways, including feedback and direct and indirect effects. For example, in several instances of defaunation, including in one of our research sites in Veracruz, Mexico, deforestation and fragmentation have reduced the suitable habitat needed to maintain viable populations of large animals, therefore leading to declines and eventual local loss. In addition, such deforestation and fragmentation also facilitate the access of poachers to sectors of the habitat that previously were inaccessible.

Invasive Alien Species

It is estimated that in about half of the cases of extinction or major decline of a species where the specific cause is known, an invasive alien species is involved. This is because most extinctions in previous centuries occurred on islands that were mostly free of predators and diseases. For example, 220 oceanic island bird populations became extinct because of exotic predatory mammal species established on those islands after European colonization.[23] Often-cited examples include the almost complete eradication of seabird species following the introduction of foxes to Arctic islands and the numerous cases of insular native species whose loss or decline can be attributed to rats carried to their islands by people.

However, continental systems have also suffered from the impacts of invasive species, with an increasingly visible prevalence of feral cats, currently probably the most important cause of bird mortality across multiple countries, including the United States. Beyond cases of total extinction, in recent years, the Global Invasive Species Database has indicated that an estimated 1,352 species of vertebrates are threatened by invasive species. Among all bird species classified as threatened, 26% of threats are attributed to invasive species, for amphibians the figure is 23%, 18% for reptiles, and 15% for mammals. The incidence of invasive species is greater on islands than on mainlands, and major hotspots of invasive species that affected native vertebrates can be found in the Americas and India, as well as island nations such as Indonesia, Australia, New Zealand, and Madagascar.

One of the most harmful groups of invasive species is that of pathogens, with dramatic examples such as the avian malaria that devastated the native avifauna of Hawaii, and rinderpest in East Africa, which led to declines of up to 90% in some ungulates. The story of avian malaria in Hawaii is fascinating. It is likely tied to the Manila Galleon trade route in the early nineteenth century between the Philippines and Mexico. Water casks on the sailing ships were refilled in Mexico, where there were plenty of mosquito larvae and pupae in the water. On the long voyage to Hawaii, adult Southern House Mosquitos (*Culex quinquefasciatus*) hatched and then colonized paradise. The Hawaiian forest birds had evolved in isolation, so they were vulnerable to introduced diseases and their vectors.

More recently a suite of fungi has been devastating populations of frogs (e.g., the now near-omnipresent chytridiomycosis) and bats (e.g., white nose syndrome) in the Americas. Globally, among species on the IUCN's Red List, the most threatening introduced disease is the *Batrachochytrium dendrobatidis* fungus, which affects thousands of species of amphibians (chapter 5) and is suspected of being the cause of some 200 amphibian species extinctions.[24] The fungus is closely followed by rats (which have a large cascading impact on birds and mammals) and cats as sources of populations extinctions.

One spectacular invasive species is the giant Wels catfish of eastern Europe, which is now invading countries westward.[25] The

Wels catfishes

In addition to taking waterbirds (especially chicks) from below, Wels catfishes (*Silurus glanis*) have been observed using killer whale—like beaching behavior to leap from the water and devour unwary land birds they drag back in. Here, one catfish grabs a pigeon on the shore of a river island in southern France. Catfish approach shorelines unobserved and take prey so quickly that nearby flock members may momentarily continue drinking unperturbed.

largest European freshwater fish, it is not only devouring threatened native fish populations but has the habit of leaping from the water and devouring unsuspecting pigeons and some other terrestrial birds as well as ducks, other waterbirds, and shorebirds.[26] Specialized angling programs are being considered as possible controls.[27] Perhaps the most famous story of an invasive fish causing population extinctions is that of the Nile perch (*Lates niloticus*), which was released into Lake Victoria in the 1950s and '60s to establish a fishery there. The lake held an amazing radiation of haplochromine cichlid fishes—more than 500 species adapted to many different niches such as surface, bottom, and midwater foragers (chapter 4). There was a little-understood delay in the introduction's effect on other species until about 1980, but then, the perch population exploded. Meanwhile the lake was becoming eutrophic (nutrient rich), and the combined effect was to push almost all the haplochromine populations to extinction.[28]

Plastic Pollution

The role of pollutants in reducing the size of populations has not received as much attention as other drivers in defaunation ecology studies. This is a dangerous oversight. The gigantic loads of plastics that civilization dumps into the oceans may be there permanently, being reduced to smaller and smaller particles by the grinding action of waves and currents and forming a surface layer that might alter basic properties of the ocean. Civilization's plastic trash has formed vast oceanic islands of plastic garbage, some bigger than Texas. The largest one, known as the Great Pacific Garbage Patch (GPGP),[29] is thought to contain between 72,500 and 118,000 tonnes (80,000 to 130,000 tons) of plastic floating in an area of 1.6 million square kilometers—around 620,000 square miles, or about the size of Alaska. It is estimated that soon, plastic trash will outweigh all the fishes in the sea, and it is already affecting populations of marine mammals, sea turtles, and sea birds directly.[30]

The tiny fragments of plastic also accumulate persistent organic pollutants (POPs—toxic chemicals resistant to environmental breakdown) on their surfaces and enter marine food chains by being ingested by tiny floating animals (zooplankton). They then carry the toxins back to us as we dine on plastic-laden seafood, sometimes on particles so fine they can cross the blood–brain barrier.[31] Those particles can then damage the human cells they are able to invade. These tiny particles also become carriers of microorganisms pathogenic to human beings.[32]

Toxification

A very dangerous but little-recognized environmental problem facing both people and wildlife is the increasing toxification of the entire planet with synthetic chemicals and rare earths. These can harm the nervous and reproductive systems, endocrinology, increase susceptibility to cancer, and cause many other problems, and their presence in the atmosphere is entirely attributable to human actions. Growing human populations want myriad more items of plastic that often leak toxic chemicals and end up as microscopic fragments (microplastics and nanoplastics) that are now common in animal bodies, including human brains. More and more cosmetics, cleansing compounds, insecticides, herbicides, preservatives, paint strippers, lubricants, medicines, drugs, and industrial chemicals are being produced and consumed every year. Many of these contain novel chemicals that mimic natural hormones, and even in tiny quantities, they are able to alter the development of young animals including human children, with potentially catastrophic consequences.[33]

Some information on this global toxification has hidden in specialized literature on animal toxicology, a field of which many students of biodiversity loss are largely ignorant. This results in part from the misplaced belief in traditional animal toxicology that high doses, rarely encountered in natural ecological settings, are required to have large adverse effects. High doses do occur, of course, sometimes with disastrous effects. Scientists, for example, have long been aware of the impact of toxicants since the initial reports of dichlorodiphenyltrichloroethane (DDT) turning birds' eggs into omelets and leading to avian declines worldwide.[34]

Pesticide exposure has also been an important perpetrator of pollinator declines, as well as many animal declines from aquatic environments, a literally sickening result of the accumulation of those chemicals as they are respired, drunk, and absorbed by a variety of aquatic animals. In 2023, a large study of birds in Europe found that "agricultural intensification, in particular pesticides and fertilizer use, is the main pressure for most bird population declines, especially for invertebrate feeders."[35]

In classical toxicology, dose makes the poison; the more of the poison you took, the worse off you were. For instance, if you eat enough common table salt (about one-thousandth of your body weight), it will kill you. Now, we know the opposite can also be true. For example, early in this century, brilliant research by our colleague Tyrone Hayes began to make evident the demasculinizing effect on frogs from exposure to the herbicide atrazine, even when exposed to low doses.[36] Hayes's research, and that of many others studying chemicals that interfere with hormone action, have made it clear that the levels of pollution in natural ecosystems around the world are already regularly occurring at concentrations capable of disrupting hormones.[37] No place on Earth can now be deemed pristine. And this is not a problem lying in wait in the future—it is already here.

Perhaps the most vivid recent example is a report demonstrating that concentrations of perfluorinated compounds (PFAS) are present in rainwater samples taken from around the world, including

Antarctica, at levels deemed hazardous by regulatory authorities.[38] First invented in the 1930s and known as "forever chemicals" because they do not degrade naturally, PFAS are accumulating in virtually all species tested. These synthetic endocrine-disrupting chemicals share characteristics of natural hormones and thus interfere with reproduction and development in vertebrates, cause cancers, and undermine immune system function, among multiple endpoints, each of which can undermine population health. The magnitude of the problem is truly alarming. A recent paper concluded that "a planetary boundary has been exceeded due to PFAS levels in environmental media being ubiquitously above guideline levels."[39]

In another example, high concentrations are regularly reported in wastewater being released from sewage treatment plants. Those chemicals are there because women don't metabolize all of their birth control medicine, releasing what's left when they pee. The lake treatment of the sewage feminized male minnows and reduced reproductive capacity in both males and females. The minnow population crashed by 99% over the seven-year treatment period. Populations of the minnow's main predators also shrank dramatically, not because of direct action on their hormones but because the population of their prey had shrunk so dramatically. Although experiments like this are rare in ecotoxicology, uncontrolled releases of toxicants capable of interfering with hormone action are ubiquitous. Toxification is likely to be contributing to biodiversity loss far more commonly than most conservation biologists realize.

The pervasive effect of toxicants on freshwater ecosystems has now become well-established, but the defaunation observed in these systems most likely has resulted from the synergies of several human-generated forces, including that between toxicants and habitat transformation (including the impacts of global warming).[40] Toxicant-related drivers of defaunation are considerable, not only on land and freshwaters, but in oceans too, particularly as related to a pervasive and increasingly worrisome plastic pollution, representing a long-lasting source of mortality and damage to marine fauna, as well as the seabirds that accidentally ingest it after mistaking some for prey. The magnitude of this defaunation force, although not thoroughly quantified yet, is reflected in the deplorable statistic that by 2050, the weight of oceanic plastic, a significant portion of it in the form of microplastics, may well surpass that of marine fish biomass.[41]

Climatic Disruption

We want to emphasize that climate change and a nuclear holocaust are currently the greatest existential threats to humanity—and unlike a nuclear holocaust, climate change is rapidly becoming beyond human ability to control. Anthropogenic climatic disruption is now potentially the most critical contributor to the population and species extinction tsunami.[42] There are literally thousands of books and articles about the ever-growing impacts of climate change on biodiversity, so we can only cover but a few examples here. We've already paid attention to it indirectly by looking at fires as a cause of both direct death and of land-use change, and in other ways where relevant at various points in the preceding chapters. And although the IUCN database placed climatic change as the least impactful extinction force, recent research is suggesting that the warming edges of various species' geographical ranges have been leading to local population extinctions there, and for a considerable number of organisms, this has resulted in range contraction. In addition, the impacts of recent increases in wildfire intensity and spread, resulting in part from climate disruption, have at this writing yet to be evaluated, and little is yet known directly about the impacts on marine populations of ocean warming and acidification.

Myriad populations of terrestrial, aquatic, and marine species are already suffering the negative effects of climate change. There is no aspect of an organism's environment that is not influenced by climate change. Populations are shifting in terms of geographic range, time of reproduction, habitat, prey selection, and so on, as responses to climate disruption. But the disruption can be so fast that many organisms cannot respond genetically, ecologically, or behaviorally to cope with the changes—with devastating consequences.

Climate change may be especially threatening to rare, highly specialized species, such as the Golden toad of Costa Rica and other amphibians (discussed in chapter 3). Another example is the Arabian sand cat. Once thought to be extinct, it was recently rediscovered in the Baynouna protected area of Abu Dhabi through the use of camera traps.[43] The cat is exquisitely adapted to living in the desert, with hairs arranged to keep sand out of its eyes and give its paws traction. It also has the ability to get much of the water it needs from the small animals on which it preys. The evidence suggests that the cats prefer to forage in the cooler parts of the night, so rising nighttime temperatures may well present a threat to the persistence of this population and to other sand cats that reside in the deserts of Asia and northern Africa.

A recent study including a large sample of 716 species of animals, both invertebrates (283, with 271 of them insects) and vertebrates (233 birds, 69 fishes, 40 mammals, 19 amphibians, and 12 reptiles), indicated that there have been more species showing local extinctions in tropical ecosystems (52%) than in temperate ecosystems (39%). Among all animal groups, the average amount of species with local extinctions was estimated at 42%, a considerable proportion.[44] Fish were by far the most affected, with 62% of the species losing entire populations, followed by insects with 57%, and marine invertebrates with 42%. The rest of the groups had lower values, with mammals being the lowest, with 36%. These results are astonishing because they suggest that climate change-related local population extinctions are already widespread even though the magnitude of temperature increase is comparatively low (~1.3°C, or around 2°F), relative to the expected increase in the next few years or decades.

Ultimate Causes Revisited

Returning to the ultimate causes of population extinction crisis brings us to the way human society is organized economically, which is dramatically symbolized by the role of corporations. Corporations have, in the last two decades, launched a highly funded campaign to convince the public and politicians they are socially responsible, will protect the environment, and should basically take over civilization's governance. As Joel Bakan's brilliant work has documented in detail, and many events have shown, this is dangerous nonsense both for democracy[45] and for biodiversity. Many of the administrative devices that go along with both corporate and government behavior, and which have been encouraged by corporate pressure and lobbying, are largely shams.

"Offsets" are a good example of a socio-political driver of extinction. Offsets describe corporations raping the environment in one area, but theoretically "balancing" the damage by spending money to protect the environment from rape elsewhere. But rarely do the two in fact balance—and even if they did in terms of, for example, carbon sinks versus carbon emissions, the harm is still done in the damaged area, including to the local human populations and other ecosystem services and benefits.

Another frequent sham is the certification of products or processes as "eco-friendly"—that is, manufactured or conducted in a way that reduces or avoids damage to the sustainability of ecological systems. These, and other industry ploys, including the abundant greenwashing propaganda that falsely claims products or processes are not environmentally harmful, all run up against one basic fact: for-profit corporations will generally only do good things that help increase or protect their bottom line and will generally prioritize economic growth—now the most lethal basic threat to biodiversity and other elements of human life-support systems.

Some non-profit corporations and corporate trusts must share part of the responsibility for this situation. Universities, for example, often are largely machines for churning out replaceable parts for the corporate growth machinery that is destroying the world. Many business schools represent the essence of this problem; business school professor Martin Parker's devastating critique of their general performance[46] applies pretty much to the entire university system: "too busy oiling the wheels to worry about where the engine is going." According to Parker, business schools "are places that teach people how to get money out of the pockets of ordinary people and keep it for themselves." To the degree their students are taught about the need to protect the environment, the public cost of corporate operations, and the failure to consider distributional equity and such, those topics tend to be afterthoughts. And even if these institutions did everything "right" in Parker's sense, we should remain vigilant because in the current corporate-dominated culture, some, maybe many of them, would undoubtedly still be promoting the myth of perpetual and ever-increasing growth, the fatal disease of modern civilization.

Corporations are, of course, not the only feature of humanity's financialized culture that contributes heavily to the annihilation of our only living companions. To get a feel for the more than century-long campaign of corporations to free themselves of any regulation by society at large, see the brilliant exposé *The Big Myth: How American Business Taught Us to Loathe the Government and Love the Free Market* by Harvard historians Naomi Oreskes and Erik Conway.[47]

Before we close this chapter, we wish to insist, again, that although it is important to understand the impacts of each of the major drivers of population extinctions and differentiated defaunation, it is even more important to realize the limited effectiveness of talking about these in isolation. It is true that each driver has its own negative effects, and creative innovation to reduce those impacts is desperately needed in all avenues in society.

The most important realization, however, must be that none of these drivers operates in isolation from the others. Most population extinction events are the result of the *combined* effects of *multiple* drivers. Failure to consider the full complement of forces at work can lead to dangerous underestimation of the extinction consequences of the situations we face. And that naturally feeds into a general tendency to underestimate the *costs* to us and our descendants of population extinctions. This is a dangerous and deadly mistake that we cannot afford to keep making.

Chapter 11

Nature's Decline

The Costs

BIODIVERSITY IS RAPIDLY DISAPPEARING—about that there is no question, as we have shown. But overall, what is this costing us? Sure, birdwatchers may need to travel further to see a rare species, or even a common one, and no amount of travel even today will get you a glimpse of a living Passenger pigeon. But so what? Birdwatchers, butterfly collectors, scuba divers, aquarists, and gardeners will miss some of their favorite sights, but most oil executives, gun manufacturers, cell phone moguls, space tourists, and drug barons won't. And if they do, they can afford ever-more realistic videos and simulations to replace the living things. With enough money they might even get a slick robo-*Tyrannosaurus* to eat their favorite lawyer.

There are many ethical, moral, cultural, religious, aesthetic, ecological, and evolutionary costs to losing so many populations and species. But there is also a much more practical "so what": we are all dependent for our lives on the natural services ("ecosystem services") that populations of other organisms supply to humanity.[1] What exactly are those services? The Millennium Ecosystem Assessment, a massive effort by thousands of scientists to raise environmental awareness,[2] divided the services that nature provides into four categories, supporting, provisioning, regulating, and cultural, each essential and interconnected.

Supporting services include the fundamental processes of biomass production (primary production, the basis of the food chain described in chapter 7), production of atmospheric oxygen, and formation and retention of soil, as well as the cycling of nutrients and water. *Supporting* also includes the production of organic chemical compounds that embed the energy that powers every single activity of both plants and animals. They support the *provisioning* services of supplying us with food, clean freshwater, fuel, fiber, chemical compounds of medicinal value (or models to produce them), wood, and other building materials.

The main *regulating* services are amelioration of climate extremes (wildfires, floods, droughts, and tornado damage), water purification, controlling of disease (such as reducing numbers of animals that spread pathogens, like mosquitoes), and pollination of wild and domesticated plants. *Cultural* services include esthetic and spiritual opportunities of the kinds we have already mentioned (enjoying the peace and beauty of a forest scene or listening to a bird song or attracting hummingbirds with sugar-water feeders), learning about ourselves and the natural world (the benefits of science and education), and skiing on clean powder, scuba diving on a coral reef, or hiking the Appalachian trail (recreational services).

Thus, a main cost to people of population extinctions is the damage they do to the delivery of all of these different categories of ecosystem services. The recent statement by the Intergovernmental Science-Policy Platform on Biodiversity and Ecosystem Services (IPBES) put the seriousness of the costs frankly: "Nature is declining globally at rates unprecedented in human history—and the rate of species [and population] extinctions is accelerating, with grave impacts on people around the world now likely."[3]

Biodiversity and the ecosystem services it supplies have been a major focus of ecological research, and that has led to the general conclusion that decimation and extinction of populations has a negative impact on ecosystem service delivery.[4] Scientists are just beginning to look carefully at a more detailed question, namely, what is the form of the relationship between the reduction in abundance in many natural populations and their extinctions, and the delivery of ecosystem services to humanity? A drug peddler or Wall Street economist may not care about robins or maple trees personally, but they're more likely to be mentally healthy if they live on a tree-lined street with abundant birds.[5] A first crack at this issue showed that in general there was a non-linear positive relationship between abundance of birds and service delivery.[6] In plain English, that means the more individual birds, the more services, but doubling population size does not necessarily double delivery of a service.

In one circumstance, a reduction in population size may, counter-intuitively, temporarily *increase* services as population size declines.[7] That occurs when cultural services are enhanced by rarity—something relatively frequent in birds, it is said. Paul has put in great effort just to see rare birds such as Kirtland's warbler and has been thrilled when successful. Paul and Gerardo jointly sought the ground pangolin in southern Africa. The pangolin was stunningly hard to see because of the staggeringly high number murdered for their scales, which are used in phony Far East medicines. The sighting was only possible because one individual had been radio-tagged for study by a dedicated young biologist. Rodolfo had the privilege of conducting research surrounded by herds of elephants in plain sight on the savannas of north-central Kenya, a region where these animals are now rare due to their dramatic ongoing decline. Another time, he walked over canopy structures to watch the gorgeous but rare resplendent quetzal (*Pharomachrus mocinno*) in the cloud forests of Monte Verde, Costa Rica.

There is a simple satisfaction gained by seeing a rare example of beauty and intricacy. That is an attraction that cannot be provided by superabundant organisms such as house sparrows or cabbage butterflies, or a widespread northern coniferous forest ecosystem. When Paul first saw prairie dogs in the southwest of the United States in 1947, they were considered pests, and he paid little attention to them. Now that most US populations are gone, he was thrilled to visit a giant colony in northern Mexico with Gerardo and had the bonus of seeing the rare Worthen's sparrow—which finds the habitat of dog "towns" especially suitable. In short, being a biologist seems to enhance a human curiosity about diversity and rarity—both supplied to us as a valuable aesthetic ecosystem service. That service is on the one hand expanding as more and more organisms are driven to rarity, and on the other shrinking as the rarities are driven to extinction.

The contradictory forces at work can also be seen in the case of rare butterflies, where scientists struggled to maintain populations of rare species such as Florida's Schaus' swallowtail and England's large blue for the esthetic service they provided, while criminals strove to illegally capture and sell specimens to collectors entranced by both beauty and rarity.[8] In both the United States and the United Kingdom, the increasing rarity of most butterflies represents a cost in a lost ecosystem service—diminished pleasure for those *Homo sapiens* who once enjoyed populations of large numbers of those beautiful insects in flower-filled fields and loved collecting and displaying them as a hobby.

Supporting Services

Let's look in more detail at how populations deliver ecosystem services. First, let us consider "supporting services," which are basically the functions of diverse populations that allow ecosystems to maintain themselves—and us. As mentioned earlier, they include

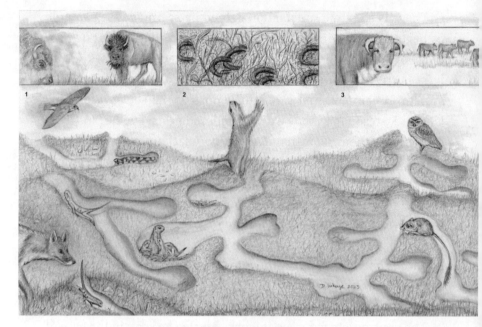

Prairie dog and bison

Prairie dogs are considered a keystone species and ecosystem engineer. (An ecosystem engineer is any species that significantly modifies, maintains, or destroys a habitat.) Prairie dogs live in grasslands where bison used to thrive. Both species played important roles in ecosystem services by maintaining the grasslands' function and biodiversity. Encouragingly, efforts to restore both species across their geographic range have been quite successful.

recycling nutrients, without which life would grind to a halt, maintaining and generating the soils required by agriculture and forestry, and sustaining local populations of other animals as well as populations of trees and other plants.

Dispersing the seeds of trees, especially in the tropics, is one of the most important supporting services that birds supply. In some tropical forests, birds disperse the seeds of most woody plant species. At one count,[9] these included 85 species of trees of value as timber, 135 of medicinal value, 182 used for food, 146 ornamentals, and 84 genera with other economic values for humanity. African mahoganies, for example, depend on, in some cases almost exclusively, populations of very specialized avian dispersers. Some tropical trees, such as African populations of the mulberry relative *Antiaris toxicaria* have multiple uses and big seeds, for which large fruit-eating birds are major dispersers. Thus, the disappearance of populations of big birds such as curassows and hornbills can have serious economic consequences.

Seed dispersal can be an important supporting service in temperate ecosystems as well. For example, the whitebark pine is a critical element in montane ecosystems in western North America. Among the ecosystem services it provides to humanity is the preservation of critical summer water flows by shading high-altitude snow fields. The pine's seeds are dispersed by Clark's nutcrackers; the service provided by that specialist bird is estimated to be worth about $10 billion across the pine's range.[10]

It is not widely recognized that populations of living organisms are often crucial to sustaining freshwater flows that are essential to organisms that live outside of the oceans. According to one estimate, transpiration (the pumping of water from the ground into the atmosphere accomplished by trees and other plants) moves more water in the global hydrological cycle "than all the world's rivers combined."[11] The extermination of tree populations in a forest can thus greatly diminish rainfall downwind and quite distant from the forest. Deforestation in the Amazon could pass a tipping point where the absence of repeated recycling of water through the forest's transpiration would lead to the death of that incredible reservoir of tropic biodiversity, replaced by a relatively sterile savanna. The ongoing clearing of millions of hectares (1 million hectares is around 2.5 million acres) of the Amazon forest for crops may lead to loss of the remainder and, ironically, the humidity that allows the crops to be grown.

Populations of organisms are not just central to many rainfall and freshwater systems; they are also critical to preserving the productivity of marine ecosystems (including fisheries yields), water quality, and stability of the oceans.[12] Populations of corals are an outstanding example of organisms supplying supporting services, forming reefs that underpin highly productive ecosystems. Coral reefs, for instance, can help to protect coastline facilities from storm waves or tsunamis. One study showed that coral reefs substantially reduced the worldwide flooding risks for an estimated 100 million people.[13] Populations of mangrove trees and other salt marsh vegetation play similar roles.[14] As is so often the case, marine ecosystems can provide more than a single class of service. For example, coastal wetlands can serve as nurseries for commercially harvested marine fishes, thus providing a provisioning service[15] as well as a regulating service.

Coral reefs also supply a key provisioning service to humanity, providing critical dietary protein to perhaps a billion people. Supporting services include sustaining organic diversity, which amounts to a vast genetic library that allows plants, animals, microorganisms, and the ecosystems they comprise to evolve in response to environmental changes. They then can underpin the lives of human beings who themselves are elements of ecosystems (chapter 6).

Closer to home, populations of microorganisms as noted in chapter 8 provide supporting services in human bodies, most noticeably in the human gut—as prevention of invasions of *Clostridium difficile* so dramatically demonstrates. Supporting services, be they in the body or worldwide, are rarely obvious, but they are nonetheless absolutely essential to the persistence of civilization. But they are currently so taken for granted that their influence only makes headlines when they are disrupted, allowing regulating services such as flood control to collapse.

Regulating Services

"Regulating services"—those that provide human benefits by controlling how ecosystems function—can make their absence publicly and dramatically known. For example, wildfires removing populations of plants from hillsides in southern California destroy the flood-prevention service provided by those populations, resulting in floods and mudslides that endanger lives and cause large economic losses.

Such loss of regulating services can be much more devastating in poor countries. In 1998, Hurricane Mitch dealt a horrifying blow to the people of densely populated Honduras. The country was already facing great economic inequity, and poor people were often forced to live on precipitous slopes, frequently largely denuded of tree populations by overharvesting. The loss of the forests' flood control service there, in neighboring Nicaragua, and in much of the rest of Central America resulted in Mitch causing about 20,000 deaths and many sad economic losses.[16]

In Australia, extinction of many tree populations by that nation's vigorous program of destroying its forests has, along with climate disruption to which Australia contributes mightily, helped produce both massive fires and massive water shortages. People have died because of such deforestation, from burning to death during fast-moving bushfires or suffering lung problems from inhaling the smoke that blankets vast areas. Property is destroyed, and many wildlife populations go extinct.

In 2020, people and wildlife in California and much of the rest of the United States were confronted with climate-related disasters like those in Australia, ranging from wildfires to floods to excessive heat and drought. Like Australians, Americans are heavy contributors to climate disruption, and from 2017 to 2021, they had an administration actively encouraging making Earth uninhabitable for most *Homo sapiens*.

If tree populations (and other vegetation) could have saved thousands of lives in Central America during Mitch, they could save millions upon millions of human lives worldwide in the future. Earth's populations of trees and the soils they help build actively sequester huge amounts of the carbon from humanity's ongoing CO_2 emissions.[17] The numbers of lives saved, and the financial costs avoided by the carbon sequestering function of tree populations in the future, will depend on a wide variety of factors ranging from forestry practices to the treatment of women (and that treatment's impact on fertility rates). But in any case, tree populations are invaluable allies in the fight to preserve civilization.

Urban tree and shrub populations do not just help control global climate, they also reduce air pollution in cities.[18] Tree populations lower temperatures in hot urban locations and seasons by shading buildings (thus reducing CO_2 emissions associated with electricity generation for air conditioning), and by direct cooling through transpiration. They also remove particulate pollution by dry deposition on leaf surfaces—acting as filters. This may be only temporary because wind can return the particles to the atmosphere, but they may be washed off by rain and removed from access to human lungs. Gaseous pollutants may be carried to the interior of plants through their stomata—leaf pores that evolved for intake of CO_2. It has been estimated that in 2006, more than 700,000 tonnes (771,618 tons) of pollutants were removed from urban air in the United States by the regulating services that urban tree and shrub populations provided, representing an estimated economic savings of almost $4 billion.

Many pollutants are kept from reaching human beings by the regulating ecosystem service of water purification. Water running through plant populations in wetlands, for example, does more than just remove silt. It also filters out most metal contaminants and most of the entering nitrogen that can cause excessive growth of algae and disrupt many processes important to people. Similar water-cleaning services are performed by populations of trees and other vegetation growing along waterways. Throughout those and other water systems, populations of microorganisms are hard at work decomposing toxins and excess nutrients.

Disrupting the water purification function by exterminating populations of organisms that play key roles in providing that function has high costs besides those obvious ones connected with human health. It adds energy-expensive purification needs where pure water is necessary for industrial processes, and it damages water-based recreation from swimming and fishing to boating.

As we've already seen, populations of honey bees play a critical role in human nutrition by regulating the productivity of important crops. In the United States, crops that require pollination services include almonds, alfalfa, melons, plums, and squash, among many others. Almost all fruit trees require pollination to produce their crop. Kill off pollinator populations and there is less food, lower quality food, or both.

Another important agriculture-related regulating service is biological control, the suppression of crop enemies or other pests by natural (or introduced) predators. That's a service, for example, classically delivered by bats. Guesswork estimates suggest that bats save global agriculture many billions of dollars per year, although the exact magnitude of the service needs more extensive investigation.[19] The same goes for regulating services supplied by insects. One estimate[20] is that their value in the United States is about $8 billion ($0.38 billion for dung burial, $3.07 billion for pollination, and $4.49 billion for pest control of native herbivores). These numbers are clearly rough estimates in view of ongoing debates over the value of pollination services and the role of wild versus domesticated (honey bee) pollinators.[21] But there is no doubt that the economic values are very large.

The key regulating service of pest control is most often observed by birders, who frequently watch beautiful birds scarfing down caterpillars, mealy bugs, beetles, grasshoppers, and the like. Bird populations protect crops,[22] and often do yeoman work in helping suppress outbreaks of forest pests.[23] To cite an anecdotal example, the 1958 extermination campaign in China against the rice-eating

Eurasian Tree Sparrow (*Passer montanus*) ultimately expanded outbreaks of insect pest populations rather than increasing rice yields, contributing to a famine that killed millions of people, considered one of the greatest human-caused disasters in history. This clearly shows the pest control ecosystem service supplied by these sparrows.[24]

Provisioning Services

Ecosystem functions that provide humanity with required/desired materials or energy are called "provisioning services." Perhaps the best-known provisioning ecosystem services are those providing food to *Homo sapiens*. Agroecosystems—farms—are the most obvious examples of ecosystems that nourish humanity and its domestic animals. The costs of damage to these services, such as when drought largely removes a dominant plant population from an agroecosystem, can be immediately obvious—often as a financial cost that occurs when, say, the price of wheat skyrockets, sometimes as a catastrophic famine if global food distribution systems are inadequate or not brought to bear. Lumber (roundwood) and precursors of synthetic industrial products, including medicines, are other major provisioning services supplied by plant populations.

The importance of provisioning services to humanity is reflected in the increasing efforts to map those services.[25] And that underlines the importance of population extinctions. Humanity doesn't just depend on the existence of species like *Oryza sativa* (rice), *Zea mays* (maize/corn), *Triticum* species (several kinds of wheat), *Quercus robur* (English or European oak), and so on, but on having many, many robust populations of them.

The need for many healthy populations of nature's diversity is obvious when it comes to the crops, domestic animals, bushmeat, marine fishes, and other seafood that provide us with nourishment.

In the past, it was also obvious when it came to many non-food materials. For instance, necessary "materials to build ships in the Age of Sail were vast quantities of wood, iron, canvas, hemp, pitch and tar," and those key resources "had a direct impact on Britain's national strategy and policy decisions."[26]

Human needs for provisioning services to support non-food enterprises are not as obvious now compared with, say, the needs of the British navy several centuries ago for tall, straight oaks to make masts for the ships of the line, but they are still substantial. Humanity needs a vast variety of organic chemical compounds of plant origin including natural rubber and a vast assortment of medicines, oils, spices, fragrances, gums, tannins, waxes, dyes, and so on. In addition, most plant populations have never been surveyed chemically—humanity has only begun to explore the products plants have evolved for their own purposes (often to repel or poison plant-eaters), and many of those may underlie products of the future.

One cost people pay whenever they exterminate a plant population is thus the opportunity cost of exploring it for chemicals that could be of human benefit. In fact, about a quarter of all the medicines prescribed around the world originate in plant populations.[27] Not only that, but many of the leads that scientists trying to develop new medicines follow come from chemical compounds found in plants. If we add fungi to plants to assess provisioning services, the costs of population extinctions grow even more.[28] Just consider that penicillin is a product originally discovered in a fungus (bread mold) population, and its original production in quantity depended on the discovery of another population of a more productive mold strain.

Cultural Services

Cultural services, as previously noted, are non-material benefits people obtain from ecosystems, including recreation, aesthetic pleasure, satisfaction of curiosity, scientific understanding, education, spiritual fulfillment, life fulfillment, and so on.[29] The demand for nature tourism is one indication of the value of cultural ecosystem services, and one in which the critical importance of biodiversity is clear—as clearly exemplified by the negative impact on African nature tourism of the continuing extermination of populations of elephants, rhinos, leopards, lions, cheetahs, wild dogs, and so on. The decline in cultural values would be substantial, and to a degree could be expressed as economic loss, certainly for those in the tourist industry.

But one does not need to go to Africa to see the economic value of being able to observe animals. In the United States in the late twentieth century, for example, birding was estimated to generate more than $20 billion of economic activity.[30] More recently, at the turn of the twenty-first century, Americans were reported to spent $40 billion watching wildlife.[31]

Cultural services are generally considered important by the public, even though they are often difficult to define. The beauty of a shimmering lake in a montane setting can provide many of us with a most pleasing feeling, for example, but is it proper to define that as a service of an ecosystem?[32] Like virtually any scheme for classifying natural objects, the fourfold system for ecosystem service, namely, supporting, regulating, provisioning, and cultural services, is not perfect, and services often overlap.

For example, bees naturally pollinating a coffee crop could be reasonably seen as supplying a supporting service; populations of bees in one area of New Zealand provide Manuka honey, thought to have health-supplying properties. Bees provided a great cultural service to many scientists and laypersons, such as when Karl von Frisch in the 1960s demonstrated that honey bees talked to each other in a "dance language."[33] How that language evolved and how it might change in the altered world of the twenty-first century fascinates scientists today—a continuing cultural service dependent on the persistence of many different bee populations.

Financial estimates such as those given above are merely indicative of the magnitudes of costs that accompany the extermination of populations. Many of the costs are difficult to estimate—such as the decline of nutrient value of the food of poorer people if tree nuts rise in price when pollination services become more expensive or falter.[34] What costs should be assigned to loss of populations being harvested for bushmeat? Should they be valued by the costs of replacement by commercial meat, or perhaps by some measure of nutritional deficits in consumers who cannot afford to purchase the meat? We should remember that many services do not fall into neat categories, and above all that humanity depends on the vast interacting sweep of them produced largely by myriad hardworking, thriving populations.

Careful evaluation of likely trajectories for ecosystem services suggests that the gigantic costs of their disruption will not be equitably shared across humanity. Further, it looks as if the people directly in need of certain services are facing the most dramatic declines.[35] For instance, as many as 5 billion poor people will have to drink more polluted water and will suffer inadequate nutrition from lack of pollination services by 2050, especially in Africa and South Asia. Hundreds of millions of people in the Americas, Eurasia, Oceania, and Africa will be more threatened by coastal disasters, and many of the less affluent will lack the means to take remedial action or move successfully to higher ground.

End of the Game

One of the neglected costs of population and species extinction is the loss of highly interested branches of the tree of life. We have called it the Mutilation of the Tree of Life.

This mutilation of Darwin's Tree of Life[36] is part and parcel of what we previously called biological annihilation or the annihilation of nature. Such mutilation is causing serious damage by the rapid removal of branches from the tree of life by human activities, and it is significantly changing the systems in which human beings and all other living organisms have evolved.[37] The Passenger pigeon and the gastric-brooding frog represent extremes on the widespread and very restricted species that provided important ecosystem services to humanity. The case of penicillin shows the way tiny organisms can be absolutely critical to humanity. A worldwide search for different populations of the mold *Penicillium notatum* turned up a strain that helped greatly increase production, which was ironically found on a cantaloupe in a market in Peoria, Illinois, near the investigating laboratory.[38] If there had been a mass extinction of molds and only the strain originally found by Fleming on his petri dish had survived, many human lives would have been lost.

Gerardo and Paul have emphasized that "Earth has already lost and is now missing significant twigs and branches of the tree of life, losing evolutionary morphologies, ecological roles, and ecosystem functions depending on them, among many other things. This mass extinction is transforming the whole biosphere, possibly into a state in which it may be impossible for our current civilization to persist. The mutilation is eroding the genetic library, with consequences on ecosystem functioning and services, including primary productivity, the biogeochemical cycles, and species interactions, among many others."[39] Humanity alone is in the position to stop this unprecedented damage we are unleashing, but in order to save not just untold fellow inhabitants of Earth, but ourselves as well, we must act now.

Chapter 12

The Cure

A Bittersweet Pill

DEFINING HUMANITY'S DILEMMA regarding the conservation of nature is relatively simple. Human activities are actively mutilating the tree of life, causing the extinction of millions of populations and eroding the conditions that make life on Earth possible. However, solving the dilemma is extremely complex because it involves all the economic, political, and social aspects of the human enterprise, as we have seen in the book's preceding chapters. In this chapter, we present a series of solutions, as an example of the many available, to reduce the impact of some of the proximate factors and thus offer hope in fighting extinction. We will not deal with the *ultimate* causes of overpopulation, overconsumption, or inequity because they have been covered elsewhere and have little prospect of progress in the immediate future. But we can emphasize that the root cause is "growth mania": the dramatic increase in human population size and its continuing growth, which is intertwined with gender, economic inequity, and use of environmentally malign technologies.[1]

So, it is easy to understand that the most basic answer to "How can we solve the erosion of wild population diversity?" is to reduce the scale of the human enterprise. The *proximate causes* of population and species extinctions include habitat loss and fragmentation, overexploitation, legal and illegal trade, pollution, toxification, and invasive species (chapters 10 and 11). Solutions for these problems include a wide variety of cures, which we have divided into policy and education solutions and conservation actions. Because of the scale and magnitude of the solutions and actions occurring at all scales, from local to global, we here provide a sample of some of the most instructive or interesting ones. After all, there are literally hundreds of books on these subjects, and our aim here is to shine a spotlight on the ones offering the best chance for success.

One thing to remember about our current ongoing population extinction crisis is that many organizations, scientists, and members of the public are deeply concerned about the crisis and are seeking and implementing solutions. There are many different approaches to that task, and as one might expect, many different degrees of success. The negative caveat is that despite all the well-meaning efforts and local successes, the overall crisis is growing worse. That makes it all the more imperative that the suggestions presented in the rest of this chapter are implemented without delay.

Policy and Education Solutions

Protecting populations and species from extinction requires enacting global conservation policies to curb the direct destruction of populations and their habitats, and limit the indirect impact of other human activities such as the burning of fossil fuels and spreading of toxic chemicals. A major hindering factor is that, contrary to what many people believe, there is no single global governance institution or non-governmental organization that is organizing and implementing a conservation agenda adequate to the scale of the problem.

Additionally, lack of funding makes responding to the extinction crises even more difficult.[2] It has been estimated that around $1 billion is required for direct wildlife conservation each year,[3] which is a tiny fraction of current US annual military spending. However, to deal with the fundamental indirect cause of the crisis and other existential threats—reducing the gigantic scale of the human enterprise—would require many trillions of dollars.

International Union for Conservation of Nature

Globally, one of the most important legal steps that has been taken to identify species at risk was the establishment of the International Union for Conservation of Nature (IUCN) in 1948. In its own, quite accurate words, the "IUCN has become the global authority on the status of the natural world and the measures needed to safeguard it. The knowledge and the tools IUCN provide are critical for ensuring that human progress, economic development and nature conservation take place together."[4] The Union is made up of representatives of governments and more than 1,400 civil society member organizations of 170 nations.

Perhaps the IUCN's greatest contribution to conservation is its Red List of Endangered Species. Created in 1964, the Red List is regularly updated to serve as the world's most comprehensive source on the global extinction risk status of wild species[5] and, by default, the general picture of population extinctions.[6] It has so far accomplished an impressive assessment of more than 134,000 species of animals, plants, and fungi, and in so doing, allowed scientists and policy-makers to have an incomplete but very solid basis to act. It is incomplete because we only have data on about 2% of all the wild living described species on Earth,[7] and the status of the known species is dynamic and thus needs to be reassessed regularly (chapter 2).[8]

To help preserve biodiversity, the IUCN's operations could be greatly expanded with additional funding. In one direction, more scientists could be engaged, for example, to assess a new category, "Endangered (Vulnerable) Ecosystems." In another direction, more support is needed to increase IUCN's public relations (PR) capabilities and engender more interest among members of the news media. The relatively trivial fluctuations of financial markets are usually included in TV news and always in the newspapers. Yet statistics on biodiversity, a vital indicator of civilization's future, are virtually never transmitted to the public. Doing so effectively could not only raise awareness of the urgency of the tasks facing us but also hopefully create political and social will for the changes to our way of life that must be made.

Convention on International Trade in Endangered Species

In the international arena, the flagship agreement that came into force in 1975 is the Convention on International Trade in Endangered Species of Wild Fauna and Flora (CITES). Its goal is to monitor and regulate the international trade in animals and plants to ensure it does not threaten any endangered species. Countries (and now organizations like the European Union) that voluntarily join CITES must tailor their domestic legislation to support CITES' goals. Although it may seem the Convention has few enforcement powers, conclusions by the signatories as a group can be effective, and enforcement does happen.[9]

CITES has been a very important convention in dealing with one of the most significant threats to wildlife. There are criticisms of it that we agree with. For instance, its bureaucratic procedures make listing species a very lengthy process, and it has failed, for example, to list giraffes, whose numbers are rapidly declining and who are now becoming heavily involved in wildlife trade,[10] both for giraffe parts as trinkets and live animals. There are now fewer giraffes in Africa than elephants, but CITES and so few others are noting it that the decimation of their populations has been called a "silent extinction."

Other CITES problems include corruption that often allows illegal trade of endangered species, lack of enforcement to verify, for instance, trade of rare species or species parts such as snake skins, and so on.[11] Some experts have suggested "that the Convention's entire approach is fundamentally misconceived, and that it is in urgent need of substantial reconfiguration in order to meet the conservation requirements of the 21st century."[12] Clearly, CITES needs to be "modernized" by streamlining its bureaucracy, adopting a more realistic approach that understands that identifying many threatened species or their parts is basically impossible under current conditions, implementing stringent regulations and substantial fines where possible, and relying on effective governance where it exists.

Legal National Protection

Many countries have national legislation to provide legal protection to endangered species. Gerardo published a paper on endangered Mexican mammals[13] that prompted him to work with colleagues to create the Mexican Norm on Protected Species. In 1994, the Mexican government published the new legislation that now protects more than 3,000 endangered species of plants and animals.[14] Although this Norm has shortcomings and problems in implementation, it has become the stronger policy action to protect endangered species in Mexico.

Many other countries have adopted similar endangered species legislations. The federal Endangered Species Act (ESA of 1973 as amended) of the United States is one of the better known globally, and we discuss it here in some detail. The Act offers little to suggest that its good conservation intentions might reach imperiled populations or that it might facilitate conservation of the habitats that imperiled populations absolutely require for their survival. However, the statute's congressional authors were surprisingly mindful of the essential pathways to conserve biological diversity.

The Act's statement of purpose is "to provide a means whereby the ecosystems upon which endangered species and threatened species depend may be conserved, to provide a program for the conservation of such endangered species and threatened species." Thus, not just species but habitats are afforded the attentions of the US Fish and Wildlife Service and National Marine Fisheries Service, the federal agencies empowered to protect biodiversity. The Act goes on to define an imperiled species worthy of formal legal protection, or listing, as an *endangered species*, "any species which is in danger of extinction throughout all or a significant portion of its range," or *threatened species*, "any species which is likely to become an endangered species within the foreseeable future throughout all or a significant portion of its range."

Some of the most imperiled species at the time of authorization of the ESA fifty years ago—the whooping crane and California condor, for example—were already down in distribution and numbers to just a small remnant population or two at immediate risk of extinction. For those species, the conservation challenge, by default, is a focused effort to save a population.

On their face, the terse purpose statement and definitions in the Act seem to miss the salient fact that extinction operates at the level of *populations* of imperiled species; usually well before a species goes extinct, its constituent populations will disappear in sequence and warrant protection well before a final few remain. Although the authors of the statute did not explicitly recognize that salient ecological fact, they made notice that essential ecological variation and evolutionary potential exists in demographic units within species and can be worthy of targeted conservation attention.

Accordingly, they expanded the standard definition of "species" to include "any subspecies of fish or wildlife or plants, and any distinct population segment of any species of vertebrate fish or wildlife which interbreeds when mature." A distinct population segment, or DPS, "is a vertebrate population or group of populations that is discrete from other populations of the species and significant in relation to the entire species."

Operationally that language allows the federal wildlife agencies to confer protection to individual populations or meta-populations of imperiled vertebrates recognized as a DPS.[15] Invertebrates do not enjoy protection under the Act at the population level, but they can be listed as endangered or threatened as subspecies. And, not surprisingly, by the time an invertebrate subspecies is conferred protection, invariably just one or a very few populations remain. About 1,700 vertebrate and invertebrate species are listed and ostensibly protected under the Act—noting that some species that once were listed have been "recovered" and delisted, and eleven species once listed have disappointingly been lost to extinction.

Of the currently listed species, 170 are subspecies and another 141 are vertebrate DPSs. Many of those at-risk subspecies and DPSs persist with multiple populations, some across very large landscape areas, but some are very narrowly distributed, including eighteen salmon and steelhead runs that can be assumed to survive as single populations. A few vertebrate subspecies apparently survive as just remnants of just a single population. The Pacific pocket mouse survives on one to two hectares (three to five acres) directly adjacent to development on a coastal bluff in southern California and is immediately threatened with further loss of habitat on the most expensive private land in the state.

Although invertebrates cannot receive protection as populations, listings in certain taxonomic groups include numerous subspecies, and many of those subspecies exist as just one or several populations or meta-populations and thus are at immediate risk of extinction. Of ninety-eight insects listed under the ESA, thirty-seven are butterflies. Thirty of those listed butterflies are subspecies, and at least a third of those subspecies survive as a single or a very few populations or meta-populations.[16] Year-to-year variation in environmental conditions, including ever more frequent drought and warming, put every one of the listed butterflies at ever-intensifying risk of population extinction even without changes in land uses or proposals for land conversion or development.

Thus, the Endangered Species Act can and does target populations with its prohibitive policies. But it must be appreciated that "protection" under the ESA is not in full and unequivocal. Although "recovery" and delisting are primary objectives for listed species, the ESA allows for "incidental take," including the killing or displacement of listed species and the degradation or destruction of habitat incidental to otherwise lawful activities, under permits that allow for "consultation" between federal agencies and habitat conservation plans by non-federal entities. In those circumstances, mitigation, minimization, and targeted management actions can be required to compensate for losses of listed species and the extent and quality of their habitats. Given that so many listed species, subspecies, and DPSs are just a remnant few populations at immediate risk of extinction, this means that incidental take findings and management-planning decisions by the wildlife agencies need to be especially well-informed. Unfortunately, all too often, the wildlife agencies make such determinations using surmise and assertion, that is, educated guesswork.

The immediate risk to listed species with few populations makes it increasingly imperative that the ESA's decisions that allow for losses of individuals from endangered or threatened populations be informed by "the best available scientific and commercial data," also known as the best available science directive.[17] Defensible decisions by the agencies are informed by a structured effects analysis, which is an analysis that predicts the effects or impacts of an action on the listed population, for example from a timber harvest plan or a housing development. Effects analysis must use data on species–habitat relationships, including the effects of environmental stressors on the at-risk population(s).[18]

Analyses employ demographic modeling, following professional standards of practice in population viability analysis, to predict population persistence times under alternative environmental scenarios.[19] Some commitment to land management or restoration nearly always needs to accompany permitted actions, which should be implemented in an adaptive resource management plan that allows for "learning by doing." Well-designed population and habitat monitoring in a sampling design (experimental frame) adequate to pick up a signal of conservation planning success or lesser outcomes is requisite.[20] Regrettably, however, the rigorous monitoring necessary to track the status and trends of imperiled populations and to understand environmental, demographic, or genetic causes of population declines is seldom implemented.

Science and Education

The conservation of our planet's biodiversity is now more a political and education problem than a scientific one. There is more than enough scientific knowledge to guide us on how to act now. Better scientific knowledge will obviously help us to take better decisions in the future, but the deficiency of effective, global action is more a reflection of the lack of understanding by the public and governments of the magnitude and impact of the current extinction crises on ecosystem function and human well-being, which we have described in previous chapters. While the catastrophic loss of biodiversity has been given increasing attention recently by the biological community,[21] the general public and decision-makers are in near-total ignorance of its existence and its potential consequences. This could be a side effect of the media-generated view that environmental problems are not all that important—with the possible growing exception of climate disruption. This was dramatically demonstrated in the 2016 and 2020 US presidential elections, in which there was almost no debate on the existential threat that those problems pose to civilization.

Therefore, conservation biologists need to focus on demonstrating the scale and importance of population extinctions. Unhappily, time is very short or already run out to end population extinctions, although long-term studies are important to understanding the causes and rates of those extinctions. More such studies should be started, for often, they produce useful results early on—for example, the highly restricted mobility of some butterflies was clear in the first year of a study of checkerspot populations.[22] That proved to be important in understanding problems of their conservation, for instance that recolonization of sites of extinct populations would be rare and slow.

Whatever the original questions that stimulated the initiation of long-term studies, the payoffs have almost always been impressive for analyzing population responses to anthropogenic threats. For instance, decades of studies on yellow-bellied marmot populations[23] have shown that those rodents can survive in a very wide variety of habitats through an altitudinal gradient of some 2,000 meters (around 6,500 feet) and appear to be resilient to climate disruption. But similar data for other marmot species would be most helpful in forecasting the fate of montane rodent populations under climate disruption, as has been done for desert rodent communities.[24] Such knowledge can also inform steps to preserve those populations.

New long-term studies are not always essential, however, when detailed historical data sets are available for comparison. A classic example of the latter is a study of plant-pollinator populations in a Midwestern US deciduous forest understory. The system was first studied at the end of the nineteenth century. Laura Burkle and her colleagues[25] redid the study recently and discovered that in ~120 years, the plant-pollinator network had been disrupted by many factors, including extinction of half of the bee populations originally involved. At a much less detailed level, Paul was able to extract evidence of extensive population extinctions since 1940 from "square-bashing" censuses of British butterflies.[26] In Lepidoptera, the order of insects most attended to by naturalists, records showing their decimation in Europe goes much further back.[27] But considering the obviously huge scale of the problem, detailed information on the past (or even present) of population extinctions is rare indeed. We are largely dependent for understanding the scale of the problem on records of species range shrinkage.[28]

But that does not include another problem in tabulating extinctions, especially population extinctions: that of "zombies"— populations or species that are counted as extant but have so few individuals left or so little habitat remaining that they can no longer play an important role in an ecosystem and are doomed to extinction. They are the living dead that we described in chapter 7. Ilkka Hanski and colleagues were instrumental in bringing the meta-population version of this problem to attention.[29] This means that many estimates of extinction rates are conservatively low. This is especially so in plants.[30] Think of the "specimen" rainforest trees, left growing isolated in tropical pastures to give shade for livestock. Such long-lived individuals may survive for decades, but without hope of reproducing because there are no conspecific trees for reproduction or there is no habitat for their offspring.

Plants in general are more likely to become zombies than animals. Besides having a long life expectancy, they can also leave seed banks in soils that last for years, producing relaxation times (the period from when extinction of a population is guaranteed until the last individual dies) of even centuries.[31] That generates a huge "extinction debt"[32] before the final mature individuals pass from the scene.

Citizen Science

Until the last century, much of science was done by amateurs— think Newton and Darwin, neither of whom had tenure at a university. Even more recently, citizen science has made important contributions,[33] and has great potential. Paul and Peter Raven's work on coevolution would not have been possible without the massive amount of information assembled on butterfly food plants, much of it by citizen scientists.[34] Butterflies also show great potential for important citizen science contributions by butterfly enthusiasts with the discovery by Darryl Wheye that many of their wing patterns are representations of aposematic caterpillars.[35]

The potential for investigating distributions, population dynamics, food plant choices, and coevolutionary interactions seem great, yet require little more than careful observation and a camera. Aquarium enthusiasts, who may incidentally threaten some tropical freshwater fishes with population extinctions and threaten ecosystems by release of invasive species, equally could and do, as citizen scientists, help with population protection.[36] Many rare or little-known species are kept and studied by specialist segments of the gigantic (~1 billion people) hobby.

Aquaria can also introduce the public to living organisms in habitats—something that many people almost never encounter in the increasingly urban global human population. Aquaria can provide dramatic examples of the importance of environmental conditions in maintaining populations—as anyone whose heater fails in their "community aquarium" of tropical fishes can testify.

In general, citizen scientists can make important contributions to the mitigating of the population extinction crisis by helping monitor its course through established mechanisms such as Audubon bird counts or British "square-bashing" exercises, or by encouraging new mechanisms. And simply by becoming informed and communicating with friends, hobbyist groups, and politicians, they can help make the public more aware of its peril.

But the search for cures is very difficult in some of the worst situations we have described, such as war-ravaged Vietnam. One estimate has 10% of Vietnam's total animal species diversity facing extinction.[37] Sadder yet, Vietnam has become a center of criminal wildlife trade. Poaching is so intense in some areas, with tens of thousands of wire snares, that even some national parks are considered too dangerous as sites for release of captive-bred endangered species. Add in illegal mist nets capturing birds for the pet trade and as bushmeat for restaurants, and you can begin to appreciate the meaning of "defaunation." But there is some hope. Education for Nature–Vietnam, a non-governmental organization (NGO), is having some success against this tide of defaunation. It has managed to greatly reduce the proportion of restaurants illegally serving avian bushmeat, for example. Other campaigns are attempting to change attitudes on the use of rhino horn, again with at least some limited success. But the future is not bright because human demand for nature and natural products continues to grow.

Conservation Actions

There are endless conservation solutions for the problems that are causing population and species losses. Because of the scale and magnitude of the solutions, there are actions that can be carried out at all levels, from individuals, families, neighborhoods, and cities, to countries and the whole planet. Here, we provide a sample of some of the most instructive or interesting ones, with the understanding that there are literally hundreds of books on these subjects that provide more details and calls to action.

Protecting Natural Ecosystems

One of the most obvious ways to slow the erosion of biodiversity and preserve ecosystem services is to put aside tracts of land or sea where certain human activities, such as hunting, commercial fishing, timbering, mining, camping, use of motor vehicles, and so on, are excluded.[38] Different kinds of protected areas, both governmental and private, include national parks, biosphere reserves, indigenous conservation areas, national forests, state parks, and many more.[39] There is a huge variety of nature reserves around the world, ranging from great national parks like Yellowstone in the United States, Great Barrier Reef National Park in Australia, and Cape of Good Hope Nature Reserve in South Africa, and the huge indigenous lands in the Amazon forests, to small patches of private land protected to one degree or another by private owners.

Some of these protected areas have no people living inside, but others, such as biosphere reserves and indigenous conservation areas, have been designed specifically to maintain local communities in their lands. A general principle for the management of protected areas is that they have core and buffer zones. In the core zones, most human activities, with the exception of scientific research and management, are limited to ensure that habitat and species are exposed to minimum human impacts. Buffer zones, on the contrary, offer the possibility to compaginate human activities and nature conservation.

There is a well-established ecological principle that all else being equal, the larger the reserve, the more biodiversity will be accommodated.[40] That's why conservation biologist E. O. Wilson, recognizing the absolute need of humanity for the services provided by other organisms, recommended putting aside half of Earth "for nature," that is for our living companions.[41] The number, size, and configuration of reserves has been a central topic in the scientific literature of conservation.[42]

There have also been many scientific efforts to classify and prioritize the conservation of limited resources via protected areas at different scales.[43] For example, the great conservation scientist Norman Myers, together with a group of collaborators, developed an intriguing notion based on plants, which is relevant in the context of endemism and plant conservation, "Global Biodiversity Hotspots" that we described in chapter 7.[44] Briefly, these regions are defined as (1) areas in which at least 1,500 plant species are endemic, and (2) where only 30% of their original coverage remains. A total of twenty-four regions around the globe meet these criteria, and areas that stand out include islands (mostly tropical island nations such as New Caledonia and Madagascar), Mediterranean regions, south-central China, and tropical forests.

Regionally, the Americas stand out with seven hotspots, mostly in the tropics, including Brazil's Atlantic forests, the Tropical Andes, Mesoamerica, California, and the Mediterranean region of Chile. Amazingly, Myers and collaborators calculated that these hotspots were home to 44% of the global plant species in an area of just 2% of the planet's land. These areas include a great number of endemic species, which are organisms whose distribution is restricted to a particular region, country, or location. Not only are these areas geographically unique, but the species that are endemic to them would be at risk of disappearance from the face of the Earth should those places be impacted or destroyed.

A prime example of plant endemism in oceanic islands is found in the archipelago of New Caledonia, an island chain of about 18,000 square kilometers (almost 7,000 square miles) which is home to a flora of 3,371 species, 74% of which are endemic, along with 107 endemic genera and five endemic families.[45]

As an example of a non-oceanic territory with high endemism, Mexico stands out, with 51% of its flora constituted by endemic species. Mexican plant endemism varies by ecosystem type, with a remarkable 70% endemic species in the case of temperate forests, and 60% in the case of arid and semi-arid ecosystems.[46] A remarkable, extreme case of endemism for a plant species is that of *Lacandonia schismatica*,[47] a unique achlorophyllus (meaning it does not produce chlorophyll) plant. It is restricted to a single location in southeast Mexico, present in only one or two populations, and exclusively associated, as a parasite, with the roots of certain trees. Should anything happen to the localities where these populations exist, or to the trees that *Lacandonia* depends on, this unique species (and indeed a unique evolutionary plant lineage) would be tragically lost forever.

Because of the present ubiquity of human activities, there is extreme diversity not just in sizes and shapes of reserves, but also in the kind and degree of threats they face. For the same reason, *any* expansion of reserves can be politically difficult and frequently economically expensive. As a simple example, it took about a decade and much political effort by biology faculty to make reasonably secure a mere 483-hectare (~1,193-acre) reserve, the Jasper Ridge Biological Preserve, on the campus of Stanford University.[48]

Furthermore, many human groups (sometimes Indigenous) oppose or misuse reserves out of ignorance or self-interest,[49] a circumstance where social scientists could be doing much more to find ways to ease the protection of endangered populations[50] while improving the well-being of human populations.

The situation of Indigenous people in wildlife areas has been especially fraught with ethical issues and difficulties. Do (did) kings, colonial powers, or modern legislatures have the right to take land away from people and "give" it to wildlife? Can people be barred from hunting traditional game on traditional hunting grounds, especially if that hunting is part of the essence of an Indigenous culture, part of the group's religion? In practice, much of global reserve areas were taken from the control of their original occupiers, the people themselves sometimes ejected, subjected to outside control, and/or deprived of access to resources.[51] Biodiversity is as important, maybe more important, to Indigenous peoples as it is to invaders.

One might thus argue that these sad outcomes should be viewed as a contribution to the general good, preserving resources the natives need. But the record of appropriate use of reserves that were once controlled by local Indigenous peoples is mixed, to say the least. Among the negative results have often been loss of their role as defenders of their land and ecosystems from "environmentally destructive settlement, extractive industries, and large-scale infrastructure development."[52] Around the world, reserves from the Arctic National Wildlife Refuge in Alaska to the Yasuni Biosphere Reserve in Ecuador are currently often under assault to develop them for gas and oil, forestry, and other resources.[53]

The Yasuni case is instructive. A large oil deposit—almost a billion barrels—was discovered under Ecuador's Yasuni National Park in the Amazon. That discovery seemed to foretell yet another disaster for Earth's disappearing populations of plants and animals. The Ishpingo-Tambococha-Tiputinia (ITT) area of Yasuni is extraordinarily rich in biodiversity, a world Biosphere Reserve supporting more tree species in each hectare (~2.5 acres) than all of North America, as well as an extraordinary diversity of other plants and animals. Many globally threatened mammal species shelter in the forest, which has already suffered substantial damage in some areas from oil exploration.

But there was some hope. At a UN General Assembly meeting in 2007, Ecuadorian President Rafael Correa proposed to ban oil exploration in the Yasuni-ITT in return for a payment by the international community of $3.6 billion over thirteen years, one-half of the estimated value of the oil that would be left in the ground. It looked like a win-win solution achieved in a process of mind-boggling complexity.[54] Ecuador would gain half the cash value of about a fifth of its known oil reserves—and it would avoid damage to the ecotourism value of the forest, to the lives of several Indigenous groups that live there, and to the potential income stream from the area's sustainable timber production and harvest of other forest products. The dollar value of the latter is difficult to estimate, but certainly it was more than $1 billion.

And, of course, Ecuador would be paid not just for preserving a natural heritage of all humanity, but also for keeping over 350 million tonnes (around 400 million tons) of the greenhouse gas carbon dioxide out of the atmosphere. That benefit, especially if matched by comparable greenhouse gas emissions around the world, could help prevent the Yasuni from being destroyed by climate disruption as well as saving Ecuador an estimated $3.1 billion *annually* in damages from flooding of the port of Guayaquil alone, to say nothing of the costs of other possible climate-related disasters.

All of these benefits, however, were foregone.[55] In August of 2013, President Correa announced abandonment of the plan because donations to the Yasuni-ITT trust fund had fallen far short of expectations. Actual donations amounted to only $13 million of the planned $3.6 billion, less than one-half of 1% of the goal. This failure represents utterly unethical behavior by the international community, but also business as usual, and apparently some nasty politics involving Correa.[56] Rich countries have contributed most of the greenhouse gases added to the atmosphere since the industrial revolution and have a strong moral responsibility to take and finance remedial steps. The rich also have a large share of responsibility for launching the sixth great extinction crisis, which also threatens the ecosystem services upon which rich people, as well as the rest of humanity, are utterly dependent.

Thus, the failure of governments of the United States and other rich nations to make substantial donations to Ecuador is, in our view, not only highly unethical, but also deeply stupid and self-destructive. After all, a tax of one cent per liter (three cents per gallon) on gasoline in the United States alone would have raised the money requested by Ecuador in one year (and help reduce the US's carbon footprint).

This is especially short-sighted because the United States is already experiencing substantial losses related to climate disruption and seems on track to suffer much more.[57] One can hardly blame President Correa for taking the step he did, even though Ecuador is suffering considerably since that deal was scrapped and destruction of the Yasuni is proceeding.[58] But the reversal was a symbolic disaster, as well as a real one, for the world by apparently ending one of the most promising experiments in international cooperation to help avoid a collapse of civilization. Fortunately, as we were finishing writing this book, the people of Ecuador passed a national referendum to halt oil exploitation in Yasuni!

In the United States, the oil and gas industry frequently opposes efforts to preserve or protect biodiversity. A main battleground is the large Arctic National Wildlife Refuge (ANWR), where the fossil fuel industry has placed continuous pressure on both national and Alaskan governments to allow drilling, thereby destroying the refuge. In typical short-term Western analysis, whether to drill or not is framed in purely dollar terms.[59] In 2021, the Republicans looked near victory in their long campaign of destroying biodiversity for the sake of petroleum profits, as ANWR was opened for oil exploration, but that decision was temporarily reversed in June of that year when President Joe Biden suspended drilling. Successful extraction of ANWR's petroleum would not just exterminate many local populations—it would add to the potentially lethal effects of climate disruption globally.

Despite these near-ubiquitous problems with reserves, there are many more extant populations of species today than there would have been in their absence. But reserves are clearly not enough. Much more effort should be put into countryside biogeography, i.e., efforts to understand and conserve nature in our human-dominated landscapes[60] and general efforts to improve the quality of habitats that are shared with human activities.[61] Landscape connectivity can be increased to encourage the persistence of meta-populations (networks of small population units), shifts to organic farming encouraged to reduce the spreading of toxic substances without lessening productivity,[62] and more efforts made to identify and legally protect high-quality habitats.

A major question, of course, is whether we will be able to mobilize the political and social action needed to save populations of widely recognized and even beloved creatures. In the face of the decimation of many North American bird populations, a series of

physically and technologically possible steps could be taken to reduce the forces moving more and more of them toward extinction.[63] Whether many of the steps will be possible from cultural or
political standpoints is another matter. Can we greatly reduce the
populations of feral cats that amount to killing machines mowing
down birds and other elements of biodiversity? Will states and cities legally limit the amount of glass in tall buildings and the height
of vegetation planted near them, both factors that can make them
deadly to birds? Can we live without leaving lights on in skyscrapers and other tall structures at night, murdering millions of songbirds at migration time?

Similar questions can be asked about the decimation of both
feathered and non-feathered elements of biodiversity. Can people
be taught to consider occasional cosmetic damage to fruit (or even
the occasional worm in an apple) as signs of long-term safety for
themselves and biodiversity? That could allow agricultural operations to evolve toward more environmentally sane cropping
systems with less pesticide use. So could allowing wilder habitats
to remain between farm fields and on road verges—in many
cases leading to long-term benefits to agriculture.

Better forest management from taiga to tropics is essential, including science-based wildfire policies—what happens outside of
the United States in the tropics not only influences the nation's climate, but also the conservation of many migratory birds. To the
north, the fate of the Canadian tundra has similar impacts, with
the addition of a huge increase in greenhouse gas emissions as
melting permafrost releases vast amounts of methane (which long
ago may have contributed to the massive population extinctions at
the end of the Permian). Stopping the frequent overgrazing of tundra grasslands would help sparrows and some gamebirds, as well
as precious soils.

Many of these actions could be helped by tightening regulations
on pesticides, forest exploitation, and agriculture in general, among

other practices. At the moment, there are strong anti-regulatory factions in the world, especially in the United States where Republican Party adherents have been attempting to rescind, circumvent, or forbid what sensible regulations are in place.[64] But much will depend on the development of popular political organization and enthusiasm for regulation.[65] The alternative, among other disasters, will be increasing costs associated with lost ecosystem services and infrastructure repairs.

Protecting Biodiversity Outside Reserves

The European Union has regulations designed to reduce population extinction on private lands by mandating "offsets" for the landowner's development activities. In the face of the expected development impacts of the expanding human enterprise, various other nations have begun to adopt programs of "offsets." Offsets are designed to enhance biodiversity. In the United States, hundreds of millions of taxpayer dollars are put into conservation efforts on private lands, and one of the results is some 15 million hectares (around 40 million acres) of conservation easements—land that owners have agreed to put aside for conservation purposes.

Often, this is land set aside by developers under a Habitat Conservation Plan (HCP)[66] for protecting populations of endangered species as compensation when another project threatens the same species. HCPs clearly benefit endangered species, but there is little transparency in the results reported, so it is hard to know how effectively the funds are spent. Often, the Fish and Wildlife Service cannot even locate the easements. For example, an HCP to protect the endangered Mission blue butterfly (*Plebejus icarioides missionenesis*) near San Francisco was negotiated, but the easement had not even been recorded in the municipal land records.[67] A recent survey of peer-reviewed literature on conservation in private lands shows a need for greater geographic spread (four-fifths of investigations were in the United States, Canada,

South Africa, and Australia), examination of the effectiveness of conservation tools beyond easements, and attention on how best to engage various stakeholders.[68]

Biological Corridors

There is ample evidence that inserting exotic organisms into established ecosystems can dramatically disrupt them and lead to highly negative results—large-scale population and species extinctions. One only need consider those abundant extinctions caused by the movement of rats onto Pacific islands. But careful studies have shown substantially more subtle and not widely recognized impacts of invasives in systems such as the deciduous forests of eastern North America.[69] Nonetheless, there are circumstances in which movement can enhance the chances of population survival. Because habitat fragmentation is recognized as a major cause of population extinctions, it is hardly surprising that a substantial body of conservation literature has arisen on the issue of reconnecting fragments with corridors, protected passages that allow animals to travel safely from one area to another.

Studies made of animals from juvenile cougars[70] to adult butterflies[71] have shown that, in general, populations could be enlarged, increasing breeding and population survival, and extinct populations re-established through the appropriate use of corridors.[72] Even corridors of less-than-ideal habitat can function reasonably well.[73] Corridors can have drawbacks, though, and perhaps the one most important to guard against is that they can serve as conduits of epidemic diseases.[74] Others include spreading of fire, weeds, and undesirable native organisms, such as the Australian aggressive honeyeater, the noisy miner.[75] It may prove, however, that the greatest need for corridors in the future is to allow organisms that are relatively fragile to migrate in response to global change.[76] Short uphill corridors in the tropics and longer south-north corridors in temperate regions will need to be considered by all land-use managers.

Saving the Most Endangered Species

One of the obvious requisites of impeding the pace of population extinction is determining which organisms are most likely to be vulnerable, to be extinction prone. This is a complex issue.[77] Although most scientists think that risk should be highest for species with small populations, small geographic ranges, and poor dispersal ability,[78] the way in which life history characteristics affect the risk of extinction is still only partially understood. But if the kinds of measures we discuss next are to be effective, sound decisions must be made about where to target limited funds and effort, which means understanding both the kinds of species most likely to lose populations and the types of threats and their distributions among vulnerable populations. In this, it is important that more of the public and decision-makers be informed on these issues. In addition, documenting where and why populations are declining, as well as helping to restore them, are major areas in which citizen science can help and citizen participation is essential.

Manipulating Biota

There are naturally many instances where conservation efforts will involve either encouraging some populations that may help other populations under threat or trying to suppress populations of invasive species in order to conserve native populations of endangered species. An important group of encouraging populations that can serve to help threatened organisms here are top predators—lions, wolves, big eagles, sharks, and the like. These are creatures that have few if any other animals to fear and may, for example, be able to suppress smaller predatory animals that in turn eat endangered species. Natural experiments, such as the transformation of hills into islands when lakes form behind new dams, have shown how ecological "meltdown" can occur on those islands where top predators are absent.[79] Populations lower on food chains, freed from predation, can "explode" and wreak havoc on those on the chain below them (chapters 9 and 10).

How to deal with predators, some of whom occasionally eat members of that premier invasive species, *Homo sapiens*, often presents both ecological and political quandaries. The dingo is a case in point. Australia, as we have seen, is a mammalian disaster zone. Some twenty small mammals have gone extinct there in recent centuries, largely because of predation by feral cats and introduced foxes; many more species are now threatened by that pair. Australia's native animals have had no evolutionary experience with feline predators because felines never naturally reached that island continent. Some prey animals may not even recognize a cat as a threat. It is estimated that millions of feral cats roam Australia now and annually kill some hundreds of millions of each of mammals, birds, and reptiles. Many studies have documented that this assault wiped out twenty-seven species since European settlement and are moving scores more toward extinction.[80]

Dingoes, which are the size of a shepherd dog, eat feral cats and foxes. Therefore, many Australian conservation biologists would like to build dingo populations, especially reintroducing them to degraded lands where they are now absent. Graziers, whose animals may fall prey to dingoes, are naturally less enthusiastic. Other scientists, however, question this "encourage dingoes" tactic. The relationship between dingoes and cats can be complex. For instance, although dingoes kill and eat cats, in some areas, cats may benefit from dingo predation on the foxes that compete with cats. But it is worth looking closely at cats because they can be a major conservation factor where no dingo has ever been present.

It's ironic that globally, the most destructive invasive species outside of *Homo sapiens* today may be the one of its "comfort animals"—a critter that provides emotional support but, as we've seen, brings sudden death to many other animals—*Felis catus*. In Australia, depending on the climate in a given year, between some 1 and 6 million feral pussy cats are hunting. They are now competing with red foxes to be the deadliest enemies of the continent's unique small mammal fauna, to say nothing of its birds.[81] Australia's land mammal fauna had substantial prehistoric losses, at least in part due to the invasion of people about 60,000 years ago, and the arrival of the dingo just a few thousand years ago.

As mentioned above, since the first settlers, that fauna has been more than decimated by cats, suffering the loss of over 10% of the 273 endemic terrestrial mammal species during the last ~200 years. This contrasts with only one native terrestrial mammal from continental North America becoming extinct since Columbus landed in 1492. An additional fifth of Australia's native land mammal species are now threatened, indicating that the rate of population loss leading to one or two extinctions per decade is doomed to continue in the face of a frequently growth-manic government. The question is why?

Australia, with its vast deserts, seems comparatively free of the factors driving extinctions on other continents—habitat destruction and alteration, deliberate persecution (e.g., "predator control"), and overexploitation. Australia is losing its mammal fauna more like a large island than a small continent—suffering from the depredations of two exotic predators, cats and, often more importantly, red foxes, and the complex impacts of human-changed fire regimes, especially in response to plant agriculture and grazing and deforestation.[82] Bushfires in the early summer of 2019–2020 were estimated to have killed nearly 500 million mammals, birds, and reptiles and almost a third of the koalas of New South Wales.[83] If projections of climate disruption prove correct, much of Australia's habitat for *Homo sapiens* will likely disappear over the next few decades.

Native Aussie animals suffer greatly from the attacks of cats. Estimating the actual number killed is difficult, but a large number of scientific studies indicate that conservatively something approaching 400 million individual birds are killed in Australia each year.[84] Where they exist, estimates around the world are generally even higher than in Australia. In the United States, the estimated numbers of birds slaughtered by feral cats ranged in the low billions, and mammal losses are thought to be over 20 billion.[85]

In a broader view, cats and foxes are joined by other invasive mammalian predators, namely, dogs, pigs, and rodents. Their joint impact on biodiversity globally has been impressive. Conservatively they were involved in the extinction of eighty-seven bird, forty-five mammal, and ten reptile species. That means they account for over half the contemporary species extinctions of these groups worldwide, along with innumerable population extinctions. Invasive mammalian predators endanger a further 596 species at risk of extinction, with cats, rats, dogs, and pigs threatening the most species overall.[86] The major role of these predators in causing extinctions is attested to by the success of reintroductions of species when the predators are fenced out of the recipient areas.

The cures for the depredation of feral and domestic cats are pretty obvious. Laws must be passed, and more importantly, enforced, to keep pet cats neutered and from roaming wild. Laws should require bells on all pet cats so, if they should escape, potential prey would be alarmed by their approach. But even a simple measure like that can arouse resistance—in Britain, for example, some cat owners are opposed to belling their cats.[87] Besides economic issues, the largest barrier to making such changes would be a general ethical issue, at least in Western societies. There tends to be only a partial overlap between people deeply concerned with animal rights (especially of individual comfort animals) and those concerned with biodiversity maintenance, ecosystem functioning, and the welfare and the rights of wild animals and all human beings.

In 2015, the Australian government announced a continent-wide program to exterminate feral cats:

When the policy was announced, it was met in some quarters with apoplexy. More than 160,000 signatures appeared on half a dozen online petitions entreating Australia to spare the cats. Brigitte Bardot wrote a letter—in English, but with an unmistakably French cadence—beseeching the environment minister to stop what she called animal genocide. The singer Morrissey, formerly of the Smiths, lamented that "idiots rule the earth" and said the plan was akin to killing two million miniature Cecil the Lions. Despite anger from some animal rights groups and worries about the potential effects on pet cats, Australia went ahead with its plan, and the threatened-species commissioner replied by mail to both Bardot and Morrissey, politely describing the cat's victims as "delightful creatures already lost to the world."[88]

Hundreds of thousands of feral cats have since been shot with guns or bows and arrows, poisoned, and otherwise done away with. Animal welfare groups have been less vocal in opposition there than they would be in the United States or Europe. Few people are enthusiastic about slaughtering large numbers of animals (except by default for those of us who eat meat), but Australians seem to have a greater love of their unique native fauna than citizens of most countries. In the absence of a native feline on their island continent, the array of small marsupials is impressive, and they tend to be "cute." Australians are proud of their native animals—kangaroos are in many ways considered emblematic of the nation, and one paired with an emu decorates the country's coat of arms. Thus, the campaign to save its native animals easily garnered wide support.

There is even more support in Australia for controlling the introduced red fox, which may not only be a greater destroyer of native animals than the cat, but which also preys on poultry and is another bane of graziers, killing almost a third of lambs in some areas and creating millions of dollars in losses.[89] Introduced late in the nineteenth century, foxes now live on about three-quarters of Australia's land, having been recently and possibly deliberately introduced to Tasmania (previously suffering relatively less from introduced predators than the mainland).[90] Paul was impressed by a recent baiting of foxes in Sydney, which resulted in a great increase in populations of brush turkeys—a fascinating large bird whose males scrape together huge mounds of dry vegetation in which the heat of decomposition incubates the eggs. These birds, in a group designated "Megapodes," are "superprecocial"— the young hatch virtually ready to fly.

Captive Breeding and Reintroductions

When a species is reduced to a single population or a few small populations and is considered highly likely to go extinct, as we have seen, efforts are sometimes made to remove individuals from the wild and breed them in captivity. The goal is to preserve its unique genetic heritage and, often, to provide a source of animals to eventually reintroduce into the wild. Ordinarily reintroductions will be tried if dangers have been reduced, "empty" habitat is available, and the captive population had become large enough to give a reasonable life expectancy for the reintroduced group.

When the Arabian oryx, a beautiful antelope, was being hunted to extinction, a captive breeding population had already begun at the Los Angeles Zoo. It was so successful that in 1982, a decade after its extinction in the wild, it was possible to start reintroducing the oryx back into the desert. Unhappily, though, a standard risk for reintroductions struck home. A factor that contributed to the original extinction, in this case poaching and the opening of desert lands to prospecting, has again started to reduce the population.[91]

A classic and long-running example of a reintroduction is that of the California condor, a remnant of a population that thrived during the Pleistocene over California and much of the rest of

North America.[92] A combination of captive breeding and releases has reestablished some small populations of this huge vulture, and the persistence of the condor in the wild looks hopeful,[93] in part because of management practices that have reduced the chances of lead poisoning from hunter's bullets and shot, a major former source of condor mortality. Lead ammunition has been banned in the condor's range, and now in some other parts of the world where it is involved in poisoning wildlife. The extent of DDT contamination causing lethal egg-shell thinning seems also likely to decline[94] because of bans on its use.

So far, however, the impact of the levels of lead contamination in the projected condor range (largely from ammunition) is quite uncertain,[95] and the ultimate fate of introduced populations will likely depend on trends in anthropogenic threats such as junk ingestion by nestlings, which can also cause mortality.[96] The scale of the literature and scientific effort going into the California condor situation has highlighted the great cost of reintroducing one (admittedly charismatic) population and indicates the impracticality of using such a tool to counter the vast wave of population extinctions now entrained virtually everywhere and in most taxonomic groups.

Another example is the efforts to save the kakapo, New Zealand's amazing flightless ground parrot. This plump bird once was common in New Zealand, hunted only by a large eagle, from which it was camouflaged. Then Polynesians arrived, bringing with them rats, dogs, appetites, and an attraction to pretty feathers. Later, a mess of other mammalian predators, including cats, weasels, and possums, were introduced. The kakapo's natural defense of "freezing" (so its only original predator wouldn't spot it from the sky) was a disadvantage in dealing with these enemies, which hunted by smell. By 1995, only fifty-one kakapos survived. But then conservation biologists intervened with technology. Four small islands were cleaned of predators, the survivors were radio-tracked, artificial insemination was introduced to try to deal with an infertility problem (with sperm samples moved around on drones), sick birds were captured and medicated, and diets were scientifically supplemented.[97]

The population has rebounded to over 200, and two more areas are being cleansed of predators, one a mainland peninsula, to allow for further expansion. Indeed, the government has announced a massive program to make New Zealand "predator free" for the conservation of the local birds by 2050. That's not going to happen, but a determined campaign could reduce predator densities to a level where the kakapo and other of the some four-fifths of New Zealand's bird species now declining might have a chance.

Be it *Euphydryas gillettii*, the Arabian oryx, the California condor, or any other of the many populations and species conservation biologists would like to raise in captivity and move around to protect them, the challenges are daunting. They range from local resistance (ranchers not wanting wolves for neighbors), inbreeding and other problems among captives, and government stupidity (failure to protect recipient areas and ban poaching and collateral loss), to logistic problems (cost of helicopters) and climate disruption.

The variety of difficulties can be seen clearly in the problems the Chinese government has had in trying to use captive breeding to protect populations of the two highly endangered species of giant softshell turtles.[98] The Chinese experiences with the turtles show that climate disruption is just one aspect of the way *Homo sapiens* is rapidly transforming environments (hydroelectric damming is another), although perhaps climate disruption is the most problematic for most organisms. It makes it difficult to predict how secure from lethal change a potential recipient area will be. For instance, if an attempt had been made to translocate Bramble Cay rats to the Fly delta, how elevated should that area be to give a chance of at least a century of security? As we turn to the issue of de-extinction, the issue of security of the recipient area comes into sharp focus.

Killing for Conservation

It is sometimes claimed to be necessary to kill individuals to save populations. This has generated extreme controversy around the culling (systematic killing) of African elephants where population densities have been deemed by some too high.[99] A large number of issues are in dispute. What constitutes "too high?" Are the elephants overeating the vegetation? A herd can kill virtually all the trees in an area very rapidly; how does that affect other elements of biodiversity? Are they raiding villagers' gardens? A subsistence farmer's livelihood can be wiped out in minutes by a few elephants. To either save valued vegetation and other animals or to prevent locals from slaughtering entire herds, action to reduce the size of populations and thus preserve them is sometimes initiated. That action may be recommended to reduce population size if the elephants are far above the local carrying capacity and threatened with starvation.

Mixed in with these concerns are the ethical and national and international political/economic issues of the ivory trade.[100] Despite CITES regulations, that trade remains a major cause of elephant mortality.[101] Some people think that establishing and tightly regulating a legal ivory trade can save the African elephant in the wild[102]; others believe that will never work in the corrupt world that now exists.[103] However, a little progress is being made at one source of the problem. A new Chinese law banning ivory trading and processing, which went into force at the beginning of 2018, has produced some results. Early in 2019, there was an exhibit in Chengdu featuring "green collecting." It showcased intricate carvings by former ivory artists, now working in new media—olive and peach pits and walnuts! The hope is to sustain both a great artistic tradition *and* elephants.

Many people, including some scientists, are partially or fully opposed to elephant culling. Elephants are incredibly intelligent and long-lived animals, thought to share many human emotions[104] and shown to be deeply disturbed by human activities such as culling operations and moving individuals to unfamiliar locations.[105] Fortunately, molecular biology may be coming to the rescue. An immunocontraceptive (one that uses the body's defenses to prevent pregnancy) that is safe and effective has been developed for pigs (pZP) and is also effective on elephants.[106] It can be delivered in darts from helicopters with minimal herd disturbance.[107] Contraception instead of culling is now being tried in many wildlife species in small reserves,[108] but progress has been relatively slow because of needs for testing of long-term agents, and a mess of socio-political, economic, and ethical factors.[109] We suspect the main constraint on its use will be cost: helicopters are expensive.

Nonetheless, reports from long-term use make us hopeful[110] that contraception is better for elephants than translocation or export overseas for often unknown uses.[111] It is obviously better for individuals than culling, and should help save many wildlife populations from disaster. Ironically, the same is likely true for the human population, where contraception is one answer to reducing the likely enormous human death rate being entrained by civilization's failure to deal with the rapidly building environmental crisis.

Translocation

With climate bands shifting ever more rapidly and sea level rising, some reserves are simply going to become uninhabitable by the organisms that the reserves were supposed to save. A classic example was the Bramble Cay rat (chapter 2); there is no way it was going to become permanently aquatic so making the cay a reserve would not have helped. It might have been easy and at least temporarily successful to move enough individuals to one of the islands in the Fly delta and establish a population there. We'll never know because the species was gone before an attempt could be made. Future candidates for translocation in the region could include the desert pupfish populations whose home pools may be drying up. In these cases, climate disruption is the likely villain, and if it continues, the Fly delta will be submerged and all the pupfish ponds may be gone.

Such translocation experiments, like all conservation actions, require a careful cost-benefit analysis to avoid putting resources into obviously very short-term solutions. Calculating costs and benefits can be difficult even under nearly ideal circumstances of translocation. For example, in 1977, an attempt was made to translocate a checkerspot butterfly (*Euphydryas gillettii*) from Wyoming into the Colorado Rocky Mountains. The recipient area had extensive habitat that seemed appropriate, possessing both larval food plants and flowers that would serve as nectar sources for the adult butterflies, just as were available in the species' native range in the northern Rockies. It was thought the species had moved into North America from Siberia and previously only lacked access to Colorado because of a desert gap in its habitat.[112] The results were very complex, including population expansions and contractions.[113] Forty years later, it was still not clear whether the transplant was permanently successful, and there is no way to predict what the future responses of *E. gillettii* populations will be to climate disruption and other anthropogenic assaults.

De-extinction

Wouldn't it be great to have vast herds of mammoths roaming the Canadian tundra, or a thrill to see flocks of hundreds of millions of Passenger pigeons settling in Michigan forests once again to gobble down vast amounts of beech mast and supply succulent squabs to Chicago restaurants? Or maybe enjoy watching flights of Carolina parakeets over southern farms, or at least observe a living pair of saber-toothed cats in a cage in a zoo? Of course, being able to rent a pair of velociraptors to add spice to the "reality" TV show you're directing would be nice too. It is an appealing picture to many at first glance: *Jurassic Park* in reality, bringing vanished animals back to life, made possible by spectacular progress in molecular biology.

After all, isn't *Homo sapiens* destined to use its fine brains to engineer the entire planet (or universe)? But let us restrict our daydreaming to recreating organisms that *Homo sapiens* has itself exterminated. Surely, if that's an achievable goal, and we want to do it, humanity should go full speed ahead, resurrect the creatures we have wiped from Earth, and reintroduce them to the wild.[114] Or should we? But first, *could* we? Would it be possible?

It seems likely that in some cases, a simulacrum—perhaps a quite reasonable simulacrum—of an extinct organism can be produced. And one would be foolish to predict that even making a fully successful reconstruction of an extinct species is impossible. Science has come a long way in genetics, genomics, and development in a very short time; much that can be done today seemed impossible in 1960. So, even though we suspect the resurrectionists generally underrate the genetic, epigenetic, and environmental dimensions of the problem they recommend tackling, for the purposes of this discussion, let's assume that reconstructing extinct species eventually will be practical at some level, behavioral traits and all. Still, we suspect that the resurrectionists have been fooled by a cultural misrepresentation of nature and science—as in *Jurassic Park*, *Avatar*, and *X-Men*—traceable perhaps to Mary Shelley's *Frankenstein*.

So, what are the objections to trying to make amends for anthropogenic extinctions by restoring populations of them to life? The soundest scientific objection is misallocation of effort. It is much more sensible to put the limited resources for science and conservation into *preventing* extinctions by tackling the causes of demise. Spending millions of dollars trying to recreate populations of a few species will not compensate for the billions of populations and thousands of species that have been lost due to human activities, to say nothing of restoring the natural functions of their former habitats.

Resurrecting a population and then reinserting it into habitats where it could supply the ecosystem services of its predecessor is a monumentally bigger project than recreating a couple of pseudo-mammoths to wander around in a zoo. The Passenger pigeon is often mentioned as a target for de-extinction. Passenger pigeons once supplied people with abundant meat and likely helped to suppress Lyme disease. To create even a single viable population might well require fabricating a million birds or so because the species apparently survived by a strategy of predator saturation. And if the swarm were synthesized, where could it be introduced?

Recipient area size, appropriateness, and security would be big factors. The vast forests the pigeons required are partly gone and badly fragmented at best, and one of the birds' food sources, the American chestnut, is functionally extinct. The Passenger pigeon's previous habitat is utterly transformed, and if humanity does not very quickly and substantially curb greenhouse gas releases, the pigeon's old homeland will likely be completely unrecognizable in less than a century. In practical terms, in the near future in which action is required, extinction of the Passenger pigeon is certainly "forever."

Reintroductions of surviving endangered species (which are vastly more important than attempted de-extinctions) illustrate the complexity and scale of the task. Culturing and reinserting animals into nature is already known often to require intense and expensive effort (consider the California condor), and even human-instigated "invasions" of exotic species (such as the first two introductions of starlings to North America) often fail to "take." Zoos are already overwhelmed trying to breed endangered species for reintroduction and thus facing triage conundrums about which species to save and which to let go. Allocating more effort there is far more efficient than research into restoring a few prominent elements of Earth's past biodiversity with laboratory-created resurrections.

De-extinction thus seems far-fetched, financially problematic, and extremely unlikely to succeed on a planet being continually—and often detrimentally—transformed by human action. There are also risks beyond failure. Previously benign organisms could become pests in new environments post-resurrection; they might prove ideal reservoirs or vectors of nasty plagues, or might even harbor dangerous retroviruses in their genomes. But such problems will probably prove minor compared to another problem, which is "moral hazard."

Moral hazard is a term invented by economists for a situation where one becomes more willing to take a risk when the potential costs will be partly—perhaps largely—borne by others. For example, if a person can get government flood insurance, that person is more likely to build a beachfront home, worrying less about the risks of sea level rise. The problem is that if people begin to take a *Jurassic Park* future seriously, they will do even less to stem the ongoing sixth great mass extinction event. We are already seeing species extinctions occurring at a rate at least two orders of magnitude above prehistoric "background" rates (those outside of the past five mass extinction events), and that gives weight to the extreme seriousness of the current population extinction aspect of the crisis. And while the critical problem of climate disruption tends to occupy the attention of environmentally concerned people, the erosion of biodiversity is potentially equally crucial. The weather disasters to be caused by climate disruption likely could be resolved in a few hundred thousand years; recovery from a sixth mass extinction could easily take 5 or 10 *million* years—and yet, history tells us that the recovered biodiversity will likely be very different from the one existing pre-mass extinction. Remember that Earth was dominated by mega-reptiles before the meteorite-driven mass extinction;

what came after that is a very different biodiversity—one dominated by mammals and a branch of the dinos (our current birds).

Scientists interested in trying to resurrect extinct species should surely be free to pursue their interests if they can get the needed support. Perhaps there will be some significant scientific positive fallout, and maybe we'll be pleased to have some interesting results in a century (if civilization persists). But if de-extinction advocates are really concerned about the state of biodiversity, they should not be holding meetings or debates about de-extinction, or publicly dreaming about turning wood pigeons into replacement Passenger pigeons. They should be putting much of their time into such efforts as keeping plastics and persistent organic pollutants out of the environment and reducing or eliminating the production of both, stopping mineral exploration in places like Yasuni National Park in Ecuador and Murchison Falls National Park in Uganda, trying to suppress the ivory trade, pushing reductions in meat eating, and educating decision-makers about the roles biodiversity plays in human lives.

Above all, de-extinction scientists should be struggling to get a rapid transition to renewable energy, promoting a stop-at-one goal for family planning, pressing hard for equity for all, and generally seeking ways to reduce the scale of the human enterprise. Failing in those areas will make all discussions of de-extinction moot, even in the long term.

Controlling Ourselves

Saving biodiversity is no longer a scientific issue; it is now a political issue. We have the knowledge, but what we lack is collective will. This has been dramatically and definitively demonstrated by the presidency of Donald Trump, during which he succeeded in rolling back many elements of the conservation progress that had been made in the United States in order to further enrich himself and his wealthy donors.

Unhappily, although Trump is a horrific example of extreme greed mixed with bottomless ignorance and extraordinary lack of ethics, he does not stand alone in the world of politics. For example, Jair Bolsonaro, the right-wing ex-president of Brazil, seems to have vied to overtake Trump in the annals of conservation evil. After his election, his government successfully promoted the destruction of the Amazon rainforest by scaling back efforts to control ranching, logging, and mining. Meanwhile Brazil's largest city within the Amazon, Manaus, is setting an example of heedless destruction as it marches steadily into the Amazon forest.[115] This, of course, does not just threaten the livelihoods and lives of future generations of Brazilians, but of all human beings because of that forest's major role in regulating Earth's climate. In addition, the failing Amazon represents a deterioration of humanity's precious genetic library, as embodied in biodiversity.

Interestingly, Bolsonaro's excuse was that protected lands were an obstacle to economic growth—underlining the responsibility of old-time economists who are helping speed the collapse of civilization by promoting growth rather than sustainability, equity, or quality of life as a prime goal of society. Fortunately, the election of President Lula da Silva has dramatically reversed the horrible policies of Bolsonaro on the Amazon forests and Native tribes.

Efforts by Colombia and Peru to protect populations in their parts of the Amazon also tend to falter under the pressure of criminal gangs, criminal politicians, and the efforts of the Chinese, especially in view of competition from the United States,[116] to open up pristine forests to the production of cattle and soybeans they want to feed their gigantic population. Of course, China's "invasion" of Africa[117] through its Belt and Road Initiative (BRI)[118] seems destined to promote more unsustainable growth and exterminate more populations of non-human organisms than any other nation.

There are many aspects of human behavior that could help preserve nature's populations, including ones that are relatively simple to change by persuading individuals or passing laws. For example, convincing people to get rid of lawns and plant native species instead will help boost the diversity of populations of animals and plants. Passing and enforcing laws limiting ship speeds in certain waters can help save the North Atlantic right whale population. Voting against growth-manic politicians can be very important; instead, voting for those politicians that have a strong sense of climate change and other environmental issues could be very beneficial to the planet's long-term well-being.

Simply restricting human presence in as much native habitat as possible would preserve some populations. Human populations, even if not directly attacking predators such as wolves, can destroy their populations simply by creating a "landscape of fear."[119] Individuals can contribute to preserving populations by limiting their own consumption, especially of meat, plastics, or fossil fuels. Perhaps the most helpful and possible thing that rich and middle-class people can do is to have a maximum of one child.[120]

There is no single "magic bullet" cure, and no offloading the fix to specialists or "others." All of us must change our behavior in both small and large ways, acting both collectively and individually. Numerous small individual actions *can* help, although we admit that the three of us, like others, will find many of them tough to do. But every little bit will help undo the great harm that humanity has unleased upon Earth, and these changes are not just the least we can do, they are the least we *must* do.

End of the Story

In our view, all those dedicated to the preservation of biodiversity should be at the leading edge of efforts to inform decision-makers and the public about the basic drivers of the annihilation of nature[121] and the speed with which those drivers are getting the job done. Second, they should be pressing civil society to demand change on an emergency basis—despite the risks, especially to the profits of industry, including agribusiness. It is all very well, for example, to call for additional scientific research to detail the degree and causes of insect decline,[122] but there is already more than enough information to start rapidly taking action. One does not need to count and classify the grains of sand to know if a beach is eroding.

Decision-makers and the public must understand and act upon the basic causes of biodiversity erosion: human overpopulation and continuing population growth, overconsumption by the rich and middle classes, use of malign technologies in agriculture and elsewhere, and perpetuation of gross economic and social inequities. The standard approach of citing projected human population figures in scientific papers as if they are exogenous variables (ones controlled from outside the system) handed down from heaven and not susceptible to change through human action must end. It is a fiction that must be exposed—before surges in death rates clarify it for even the most obtuse. It is far too late to leave to politicians the monumental task of avoiding a collapse of civilization.[123] It will require concerted and democratically agreed-upon action by global civil society.

In addition, no one certainly should take the position that a collapse of human civilization would be a good thing, taking the pressure off the non-human parts of the biosphere. That might be true if the collapse were caused by the crumbling of the debt pyramid. On the other hand, if the collapse involves a large-scale nuclear war, as is all too possible,[124] or global climate change, as is currently underway, the results could be as catastrophic for the rest of biodiversity as for humanity.

Conservation biologists should be as involved as possible in public discourse over policy issues, including in education, where their expertise is pertinent.[125] Being a scientist does not debar people from acting as concerned citizens. Conservation biologists should be pioneers in seeking ways to reduce economic and other inequities and reduce consumption among the already middle-class and rich. To do otherwise is for them to give up on conservation and face the fact that most of our other battles—to outlaw transport of endangered species, to protect individual endangered organisms, to set up reserves, and to harvest sustainably—are at best only delaying actions. As conservation biologists, we must always remind the public and decision-makers that continuous growth is the "creed of the cancer cell." If we and other conservation biologists are not frank, forceful, and extreme in our defense of biodiversity, including humanity, how can we expect civil society to take the needed cures, "bitter" as they may be?

The massive loss of populations and extinctions, as Gerardo once mentioned, reflects our lack of empathy with all the wild species that have been our companions since our origins. It is a prelude to the disappearance of many more species and the decline of natural systems that make civilization possible. In addition to threatening the existence of civilization, allowing them to disappear in thrall to our own greed and short-sightedness is just plain morally wrong.

Notes

Foreword

1. Diamond, 1982.
2. Martin and Klein, eds., 1984.
3. Worthy and Holdaway, 2002.
4. Thomson, 1991.
5. Diamond, 1985.
6. Fuller, 1999.
7. Terborgh, 1989.
8. Fuller, 1987.
9. Martin and Klein, eds., 1984; Steadman, 2006.
10. Fanning et al., 1982.
11. Kroeber, 1961.
12. Jenniskens, 2019.

Chapter 1. From Origins to Extinction

1. Britannica, 2015.
2. Steller and Frost, 1988.
3. Estes et al., 2015.
4. Almond et al., 2022.
5. Ceballos et al., 2015a.
6. Ceballos et al., 2015a; Dirzo and Raven, 2003; Hughes et al., 1997.
7. Ceballos et al., 2015b.
8. Pimm et al., 2014.
9. Finn et al., 2023.
10. Bretts et al., 2018; Dodd et al., 2017; Schopf et al., 2017.
11. Bosak et al., 2013.
12. Chen et al., 2002.
13. Knoll and Nowak, 2017.

14. Morris, 1989.
15. Zhang et al., 2014.
16. Ivantsov et al., 2019.
17. Wood et al., 2019.
18. Linnaeus, 1758.
19. Darwin, 1859.
20. Koch et al., 2014.
21. Ehrlich, 1961.
22. Hughes et al., 1997.
23. Ceballos and Ehrlich, 2009.
24. Erwin, 1982.
25. Ødegaard, 2000; Hamilton et al., 2010; Stork et al., 2018.
26. Mora et al., 2011.
27. Larsen et al., 2017.
28. Locey and Lennon, 2016.

Chapter 2. The Sixth Mass Extinction

1. Steadman, 2006.
2. Steadman, 1989.
3. Fuller, 2000.
4. Nicholls, 2006.
5. Roberts and Solow, 2003.
6. Fulton, 2017.
7. Woinarski et al., 2015.
8. Tashiro et al., 2017.
9. Barnosky et al., 2011; Pimm et al., 2014; Peters and Foote, 2002.
10. Álvarez et al., 1984; Barnosky et al., 2011; Bond and Grasby, 2017; Erwin, 1998; Jablonski, 1986, 2005; Raup, 1986; Raup and Sepkoski, 1982.
11. Black, 2012.
12. Darwin, 1859.
13. Schulte et al., 2010.
14. Barnosky, 2014.
15. Molina et al., 1986.
16. Brannen, 2017.
17. Schoene et al., 2015.
18. Sallan and Galimberti, 2015.
19. Lowery and Fraass, 2019.
20. Hallock et al., 2003.
21. Lyson et al., 2019.
22. Hammarlund, 2012.
23. Jones et al., 2017.
24. Algeo et al., 2008; Murphy et al., 2000; Hammarlund, 2012.
25. Shen, 2011.
26. Pietsch et al., 2016.

27. Blackburn, 2013.
28. DePalma, 2019.
29. Pyron, 2017.
30. Barnosky et al., 2011; Pimm and Jenkins, 2010.
31. Ceballos et al., 2015b.
32. Barnosky et al., 2011.
33. Beebe, 1906.
34. Bradshaw et al., 2021.

Chapter 3. Lost and Vanishing Mammals

1. Beard, 1965.
2. Beard, 1965.
3. Finn et al., 2023; Almond et al., 2022.
4. IUCN (International Union for Conservation of Nature), 2022.
5. Prowse et al., 2014.
6. Fillios et al., 2012.
7. Ceballos et al., 2015a; IUCN, 2022.
8. Alcover et al., 1998; Johnson et al., 1989; Woolley et al., 2019.
9. Ceballos et al., 2015a; Dickman, 1996; Rosser and Mainka, 2002.
10. Gill, 2013.
11. Cahill, 2018; Edwards, 2011; Hailer et al., 2012.
12. Pagano et al., 2018.
13. Hunter et al., 2010; Kelly et al., 2010; Regehr et al., 2016.
14. Pongracz et al., 2017.
15. Fortin et al., 2005.
16. Crooks, 1999.
17. Swaisgood et al., 2023.
18. Liu et al., 2016.
19. Dallaire, 2007.
20. Xue et al., 2015.
21. Fossey, 1983.
22. Hanes, 2006.
23. IUCN, 2022.
24. IUCN, 2022.
25. Voigt et al., 2018.
26. Davis, 2013.
27. Davis, 2013.
28. King and Wilson, 1975.
29. Gruber and Norscia, 2016.
30. Coghlan, 2018.
31. Kouakou et al., 2009.
32. IUCN, 2021.
33. Clark, 2018.
34. Waal and Lanting, 1997.

35. Furuichi, 2011.
36. Coghlan, 2018.
37. Herrmann, 2010.
38. Beaune, 2015.
39. Coghlan, 2018.
40. Hickey, 2013.
41. McLennan et al., 2016.
42. Clark, 2018.
43. Narat, 2015.
44. Daily et al., 2003; Ceballos et al., 2019.
45. Ceballos et al., 2019.
46. Ceballos et al., 2019.
47. Estrada et al., 2017.
48. Estrada, 2017.
49. Pellegrini et al., 2016.
50. Chase et al., 2016.
51. Eggert, 2002.
52. Miao et al., 2022.
53. Nuwer, 2018.
54. Maisels et al., 2013.
55. Safina, 2015.
56. Menon and Tiwari, 2019.
57. Sukumar, 2006; IUCN, 2022.
58. Anthony and Spence, 2012.
59. *The Telegraph*, 2012.
60. Moodley et al., 2017.
61. Ceballos, 2011.
62. Zschokke et al., 2011.
63. Nuwer, 2020.
64. Challenger et al., 2019.
65. Ceballos et al., 2015a; Di Silverio, 2023.
66. Hornaday, 1889.
67. Ceballos et al., 2015a.
68. Di Silverio, 2023; Hornaday, 1889.
69. Mearns, 1907.
70. Ceballos and Pacheco, 2021; Martínez-Estévez et al., 2013.
71. Estes, 2006; Nicol, 2010.
72. Laidre et al., 2018.
73. Noel et al., 2018.
74. Breed, 2017.
75. Lah and Moya, 2010.
76. Goldfarb, 2016.
77. Servick, 2017.
78. Jaramillo-Legorreta et al., 2023.

79. Boyles et al., 2011.

80. Balmer, 2015.

81. Hayes, 2013.

82. Arnett et al., 2010.

83. Baker, 2003.

Chapter 4. Bird Songs Long Gone and Declining

1. Bodsworth, 1952.

2. Ceballos et al., 2015a.

3. Fuller, 2000; IUCN (International Union for Conservation of Nature), 2022.

4. Pimm et al., 2006; Ceballos and Ehrlich, 2023.

5. Ceballos et al., 2015b.

6. Halliday, 1978; Fuller, 2001.

7. Abbott, 1933.

8. Vice et al., 2005.

9. Snyder et al., 2009.

10. Fuller, 2001, p. 8; LaBastille, 1983.

11. Gelabert et al., 2020.

12. Walter, 2017.

13. Daily et al., 1993.

14. Kepler et al., 1996.

15. Rockwell et al., 2012.

16. Rosenberg et al., 2019.

17. Klem, Jr., 2009.

18. King and Finch, 2013.

19. Birdlife International, 2023.

20. Brooke, 2019.

21. McCauley et al., 2012.

22. Animal Welfare Institute, 2023.

23. Mulawka, 2014.

24. Gill, 2023.

25. Root et al., 2006.

26. Gill, 2023.

Chapter 5. The Silent Crisis

1. Cox et al., 2022; IUCN (International Union for Conservation of Nature), 2022; Lötters et al., 2023.

2. Weins, 2016.

3. Wake and Vredenburg, 2008.

4. Martel et al., 2013.

5. Martel et al., 2014.

6. IUCN, 2022; Diem, 1973; Tyler et al., 1983.

7. Fanning et al., 1982.

8. Crump, 1992.

9. Savage, 1966.
10. Crump, 2000.
11. Wake and Vredenburg, 2008; IUCN, 2022.
12. Crump, 2020.
13. Barrionuevo et al., 2008.
14. Clulow, 2019.
15. IUCN, 2022.
16. Stoddart et al., 1979.
17. Cox et al., 2022.
18. Böhm et al., 2013; IUCN, 2022.
19. Cox et al., 2022.
20. Marshall et al., 2020.
21. IUCN, 2022.
22. IUCN, 2022.
23. Esterman, 2023.
24. Sinervo et al., 2010.
25. Sinervo et al., 2017.
26. Cox et al., 2022.
27. Burkhead, 2012.
28. IUCN, 2022.
29. Ceballos et al., 2016.
30. González et al., 2020.
31. Maceda-Veiga et al., 2014.
32. Ceballos et al., 2018.
33. Salzburger et al., 2014.
34. Futuyma, 1986.
35. Ogutu-Ohwayo et al., 1997.
36. Hutchings and Reynolds, 2004.

Chapter 6. Ignored Victims

1. Goulson, 2021.
2. Forister, 2019.
3. FAO (Food and Agriculture Organization of the United Nations), 2019.
4. Salih et al., 2020.
5. Gaston and Fuller, 2008.
6. Lockwood et al., 2000.
7. Forister et al., 2021.
8. Forister et al., 2016.
9. Conrad et al., 2006; Fox, 2012.
10. McCarthy, 2016.
11. Ehrlich and Hanski, 2004.
12. Ehrlich, 2023.
13. Weiss, 1999.
14. Ehrlich and Raven, 1965.

15. Breedlove and Ehrlich, 1968.
16. Ehrlich et al., 1972.
17. Ehrlich, 1958.
18. Diffendorfer et al., 2020.
19. Boyle et al., 2019.
20. Diffendorfer et al., 2020.
21. Sáenz-Romero et al., 2012.
22. Kesler, 2019.
23. Zipkin et al., 2012.
24. Lemoine, 2015.
25. Sieff, 2020.
26. Kellogg et al., 2003.
27. Wagner, 2012.
28. Oberhauser et al., 2017.
29. Altizer et al., 2000, 2015; Lindsey and Altizer, 2008.
30. Gowler et al., 2015.
31. Brower et al., 2002; Pleasants and Oberhauser, 2013; Satterfield et al., 2015.
32. Pecenka and Lundgren, 2015.
33. Brower et al., 2011; Inamine et al., 2016; Vidal and Rendón-Salinas, 2014; Vidal et al., 2013.
34. Zhan et al., 2014.
35. Leach, 2013.
36. Wepprich et al., 2019.
37. iNaturalist.
38. Ehrlich, 1995.
39. Habel et al., 2016; Thomas, 2016.
40. Habel et al., 2019b; Warren, 2019; Haddad, 2019.
41. Burkle et al., 2013.
42. Eilers et al., 2011; Klein et al., 2006; Steffan-Dewenter et al., 2005; Potts et al., 2016.
43. Chaplin-Kramer et al., 2014.
44. Kleijn et al., 2015.
45. vanEngelsdorp et al., 2017; Watson and Stallins, 2016.
46. Lu et al., 2014; Crall et al., 2018.
47. Camero, 2019.
48. McMenamin and Genersch, 2015.
49. Dennis and Kemp, 2016.
50. Moritz and Erler, 2016.
51. Bauer and Wing, 2010; Garibaldi et al., 2011.
52. Ricketts et al., 2004.
53. Sánchez-Bayo and Wyckhuys, 2019.
54. Orford et al., 2015.
55. Hughes et al., 2000.

56. Hutchinson, 1959.
57. Klein, 1989.
58. Carrington, 2017; Milman, 2022.
59. Habel et al., 2019a.
60. Hallmann et al., 2017.
61. Leather, 2017.
62. Sánchez-Bayo and Wyckhuys, 2019.
63. Leroy et al., 2013.
64. Kawai et al., 2015.
65. Simon, 2011.
66. Kawai et al., 2015.
67. US Fish and Wildlife Service, 2011, obtained III-5-19.
68. Beck et al., 2011.
69. Wilcox, 2019.
70. Chiba and Cowie, 2016.
71. Civeyrel and Simberloff, 1996.
72. Cain and Sheppard, 1954; Sheppard, 1952.
73. Nicolai and Ansart, 2017.
74. Duda et al., 2009.
75. Chivian et al., 2003; Dutertre and Lewis, 2012.
76. Efferth et al., 2016.
77. Montgomery, 2016.
78. Jones et al., 1994.
79. Klok et al., 2006.
80. e.g., Curry et al., 2002; Spurgeon and Hopkin, 1996.
81. Darwin, 1892.
82. Ghazoul, 2015.
83. Phillips et al., 2019.
84. Fierer, 2019.
85. Van Den Hoogen et al., 2019.
86. Carlton et al., 1991.
87. Stierhoff et al., 2012.
88. Solan et al., 2004.
89. Brashares et al., 2004.
90. Ceballos et al., 2015b.
91. Cinner et al., 2012; Hoegh-Guldberg et al., 2007; Hughes et al., 2010; Jouffray et al., 2019; Williams et al., 2019.
92. Kolding et al., 2014.
93. Riegl et al., 2011.
94. Baker et al., 2018.
95. Baums et al., 2010; LaJeunesse et al., 2009.
96. Veron et al., 2009.
97. Palumbi et al., 2014.
98. Rodolfo-Metalpa et al., 2011.

99. Pandolfi et al., 2011.
100. Bednaršek et al., 2016.
101. McKittrick, 2019.
102. Ehrlich and Ehrlich, 1981.

Chapter 7. Vanishing Green

1. Morris et al., 2018.
2. Dirzo and Raven, 2003.
3. Friedman, 2009.
4. Field et al., 1998; Running et al., 2000.
5. Vitousek et al., 1986.
6. Pimm and Joppa, 2015.
7. Cheek et al., 2020.
8. Berg, 1972.
9. Aguirre-Hernández and Muñoz-Ocotero, 2015.
10. Toibin and Ferriter, 2002.
11. BGCI (Botanic Gardens Conservation International), 2022.
12. FAO (Food and Agriculture Organization of the United Nations), 2016; Hansen et al., 2013.
13. Eimera et al., 2020.
14. Wiens, 2016.
15. Bachman et al., 2018.
16. Lughadha et al., 2020.
17. Lughadha et al., 2020.
18. Knapp et al., 2020.
19. Myers et al., 2000.
20. Sloan et al., 2014.
21. Humphreys et al., 2019.
22. DeVos et al., 2015.
23. Yashina et al., 2012.
24. Morse, 2020.
25. Cronk, 1993.
26. Cronk, 2016.
27. Janzen, 2001.
28. Cronk, 2016.

Chapter 8. Microbes

1. Curtis, 2006.
2. Cerqueira et al., 2021.
3. Allison and Martiny, 2008.
4. Maxwell et al., 2016.
5. Beattie, 2017.
6. Bhattacharya, 2017.
7. Mariner et al., 2012.

8. Green and Elisseeff, 2016.
9. Beattie et al., 2005.
10. Beattie, 2016.
11. Jäger, 2017.
12. Locey and Lennon, 2016.
13. Hug et al., 2016.
14. Beattie, 2016.
15. Martiny et al., 2015.
16. Foissner, 2006.
17. Philippot et al., 2009.
18. Souza et al., 2006.
19. Schroeder et al., 2019.
20. Whitaker et al., 2003.
21. Reed and Martiny, 2012.
22. Matulich and Martiny, 2015.
23. Miransari, 2013.
24. Van der Heijden et al., 1998.
25. Petipas and Brody, 2014.
26. Wardle et al., 2004.
27. Fierer et al., 2013.
28. Mäder et al., 2002.
29. Brussaard et al., 2007.
30. Zimmerman et al., 2014.
31. Haverkort et al., 2008; Cooke, 2011.
32. CABI (Centre for Agriculture and Bioscience International), 2021.
33. Baptista-Rosas et al., 2007.
34. Monosson, 2023.
35. Suzán et al., 2008.
36. Gou et al., 2020; Zhang et al., 2020.
37. Sharples et al., 2015.
38. Ridley and Chan, 2023.
39. Ridley and Chan, 2023.
40. Nuwer, 2018.
41. Karesh et al., 2012.
42. Hesman Saey, 2016.
43. Nair, 2012.
44. Grenham et al., 2011.
45. Evrensel and Ceylan, 2015.
46. Foster and Neufeld, 2013.
47. Sapolsky, 2004.
48. Youngster et al., 2014a.
49. Malnick et al., 2021.
50. Tamburini et al., 2020.
51. Singh et al., 2010.

Chapter 9: Defaunation

1. Dirzo and Miranda, 1990.
2. Dirzo and Miranda, 1991.
3. Redford, 1992.
4. Dirzo et al., 2014.
5. Ceballos et al., 2017.
6. Smil, 2013.
7. Ceballos et al., 2017.
8. Ripple et al., 2015.
9. Hughes et al., 1997.
10. Dirzo et al., 2014.
11. Almond et al., 2022.
12. Rosenberg et al., 2019.
13. Vogt et al., 1997.
14. Galetti et al., 2013.
15. Redford, 1992.
16. Alberts, 2020.
17. Dirzo et al., 2014.
18. Dirzo and Miranda, 1991.
19. Craigie et al., 2010.
20. Young et al., 2016.
21. Janzen, 2001.
22. Young et al., 2014; Glidden et al., 2021.

Chapter 10. Drivers of Extinction

1. Paez, 2022.
2. Gleick, 2023.
3. Ripple et al., 2017; Crist et al., 2022; Union of Concerned Scientists, 1993; National Academy of Sciences USA, 1993; Wynes and Nicholas, 2017.
4. Dasgupta, 2021.
5. Ricketts et al., 2004.
6. Austin, 2010.
7. Gleick, 2023; Haig et al., 2019; Hughes, 2021; Kirby et al., 2008.
8. Wynes and Nicholas, 2017.
9. Ehrlich and Ehrlich, 2022; Oreskes and Conway, 2023.
10. Ascensão et al., 2018.
11. Farhadinia et al., 2019.
12. IPBES (Intergovernmental Platform on Biodiversity and Ecosystem Services), 2018.
13. Tilman et al., 1994.
14. Brewis, 2019.
15. Almond et al., 2022.
16. Steinbeck, 1994.
17. Ripple et al., 2016.

18. Chapman, 2001.
19. Fa et al., 2002.
20. Young et al., 2016.
21. Gu et al., 2023.
22. Nuwer, 2018.
23. Blackburn et al., 2004.
24. IUCN (International Union for Conservation of Nature), 2022.
25. Krishnaveni et al., 2021.
26. Milardi et al., 2022.
27. Vejřík et al., 2019.
28. Taabu-Munyaho et al., 2016; van Zwieten et al., 2016.
29. Lebreton et al., 2018.
30. Lebreton et al., 2018; Wilcox et al., 2015.
31. Mattsson et al., 2017.
32. Vethaak and Legler, 2021.
33. Colborn et al., 1996; Swan and Colino, 2022.
34. Lundholm, 1997.
35. Rigal et al., 2023.
36. Hayes et al., 2002.
37. Cousins et al., 2022; Landrigan et al., 2020; Gore et al., 2015; Kidd et al., 2014.
38. Cousins et al., 2022.
39. Collins et al., 2022.
40. Dudgeon et al., 2006.
41. WEF (World Economic Forum), 2016.
42. Bellard et al., 2012; Ceballos et al., 2015a; Habibullah et al., 2021; Stuart-Smith et al., 2022; Wiens, 2016.
43. Ahmed et al., 2016.
44. Weins, 2016.
45. Bakan, 2020.
46. Parker, 2018.
47. Oreskes and Conway, 2023.

Chapter 11. Nature's Decline

1. Daily, 1997.
2. Millennium Ecosystem Assessment, 2005.
3. Díaz et al., 2019.
4. Hooper et al., 2012.
5. Bratman et al., 2015.
6. Gaston et al., 2018.
7. Booth et al., 2011.
8. Haddad, 2019.
9. Sekercioglu, 2017.
10. Sekercioglu, 2017.

11. Pearce, 2019.
12. Worm et al., 2006.
13. Ehrlich and Harte, 2018.
14. Dahl, 2009.
15. Washington and Kopnina, 2018.
16. Morris et al., 2002.
17. Pan et al., 2011.
18. Nowak et al., 2006.
19. Wiens, 2016.
20. Losey and Vaughan, 2006.
21. Corbet et al., 1991; Gill, 1990; Ollerton et al., 2012; Potts et al., 2016.
22. Karp et al., 2013.
23. Sekercioglu, 2017.
24. Wenny et al., 2011.
25. Burkhard et al., 2013.
26. Oster, 2015.
27. Rates, 2001.
28. Harvey, 2008.
29. Daily, 1999; Milcu et al., 2013.
30. Hvenegaard et al., 1989.
31. Losey and Vaughan, 2006.
32. Daniel et al., 2012; Kirchoff, 2012.
33. Frisch, 1993.
34. Nicole, 2015.
35. Chaplin-Kramer et al., 2019.
36. Ceballos and Ehrlich, 2023.
37. Ceballos and Ehrlich, 2023.
38. Ligon, 2004.
39. Ceballos and Ehrlich, 2023.

Chapter 12. The Cure

1. Bradshaw et al., 2021; Dasgupta, 2021; Diamond, 2001; Dirzo et al., 2021; Holdren and Ehrlich, 1974; Ehrlich and Ehrlich, 2013.
2. White et al., 2022.
3. Deutz et al., 2021.
4. IUCN, 2021a.
5. IUCN, 2021b.
6. IUCN, 2021b.
7. Zedan et al., 2005.
8. Ceballos et al., 2015a; Ceballos and Ehrlich, 2023; Bridgewater, 2016; Butchart, 2007; Rodrigues, 2006; Willoughby et al., 2015.
9. Reeve, 2014; Sims and Frost, 2019.
10. López-Bao, 2019.
11. Bowman, 2013; Mak and Song, 2018; Martin, 2000; Shepherd et al., 2020.

12. Bowman, 2013.
13. Ceballos and Navarro, 1991.
14. Secretariat of Environment and Natural Resources (SEMARNAT), 1994.
15. US Fish and Wildlife Service, 1993.
16. US Fish and Wildlife Service, 1993.
17. Murphy and Weiland, 2016.
18. Murphy and Weiland, 2016.
19. Beissinger and McCullough, 2002.
20. Reynolds et al., 2016.
21. Barnosky et al., 2011; Ceballos et al., 2015b; Cronk, 2016; Maxwell et al., 2016.
22. Ehrlich, 1961.
23. Armitage, 2014.
24. Thibault et al., 2010.
25. Burkle et al., 2013.
26. Ehrlich, 1995.
27. Habel et al., 2016; Thomas, 2016.
28. Ceballos et al., 2017.
29. Hanski and Ovaskainen, 2002.
30. Cronk, 2016.
31. Diamond, 1972.
32. Tilman et al., 1994.
33. Hannibal, 2016.
34. Ehrlich and Raven, 1964.
35. Wheye and Ehrlich, 2015.
36. Maceda-Veiga et al., 2016.
37. Nash, 2019.
38. Franklin, 1993.
39. IUCN, 2022.
40. Yamaura et al., 2008.
41. Wilson, 2016.
42. Allan et al., 2022; Brennan, 2022; McGuire and Shipley, 2022.
43. Carwardine et al., 2008.
44. Myers et al., 2000.
45. Morat et al., 2012.
46. Rzedowski, 1991.
47. Martinez and Ramos, 1989.
48. Ehrlich, 2023.
49. Stevens, 2014.
50. Bennett et al., 2017.
51. Stevens, 2014.
52. Stevens, 2014.
53. Eilperin, 2019; Watts, 2018.
54. Espinosa, 2013.

55. Facchinelli et al., 2019.
56. Martinez-Alier et al., 2013.
57. Wang et al., 2018.
58. Brown, 2019; Watts, 2018.
59. Kotchen and Burger, 2007.
60. Daily, 1997; Daily et al., 2001; Ricketts et al., 2001.
61. Habel et al., 2019a.
62. Kremen and Miles, 2012.
63. Beehler, 2019.
64. Eilperin, 2019; Popovich et al., 2021.
65. Bauchner et al., 2017.
66. Langpap and Kerkvliet, 2012.
67. Conniff, 2019.
68. Cortés-Capano et al., 2019.
69. Mahon et al., 2019; Mahon and Crist, 2019.
70. Beier, 1995.
71. Haddad and Tewksbury, 2005.
72. Beier and Noss, 1998; Haddad and Baum, 1999.
73. Haddad and Tewksbury, 2005.
74. Hess, 1996; McCallum and Dobson, 2002; Simberloff and Cox, 1987.
75. Clarke et al., 2007; Grey et al., 1998; Hastings and Beattie, 2006; Low, 2013.
76. Colwell et al., 2008; Dyer, 1994; Low, 2013; McKelvey et al., 2011; Rose and Burton, 2009.
77. Hernández-Yáñez et al., 2022; Gonzalez-Voyer et al., 2016; Isaac and Cowlishaw, 2004; González-Suárez et al., 2013; Sanders, 2007.
78. Beissinger, 2000.
79. Diamond, 2001; Terborgh et al., 2001.
80. Losos, 2019.
81. Legge et al., 2017.
82. Kutt and Woinarski, 2007; Paltridge et al., 1997; Saunders et al., 2010; Woinarski et al., 2015.
83. Brewis, 2019.
84. Woinarski et al., 2017.
85. Loss et al., 2013a, 2013b.
86. Doherty et al., 2016.
87. Thomas et al., 2012.
88. Aguirre, 2019.
89. Saunders et al., 2010.
90. Sarre et al., 2012.
91. Morell, 2008.
92. Ceballos et al., 2015a.
93. Bakker et al., 2016.
94. Burnett et al., 2013.
95. West et al., 2017.

96. Mee et al., 2007.
97. Lazarus, 2019.
98. Wang et al., 2021.
99. Dickson and Adams, 2009; van Aarde et al., 1999.
100. Gillson and Lindsay, 2003.
101. Wittemyer et al., 2014.
102. Stiles, 2015.
103. Bennett, 2014.
104. Safina, 2015.
105. Shannon et al., 2013.
106. Perdok, De Boer, and Stout, 2007.
107. Delsink et al., 2007; Kerley and Shrader, 2007.
108. Kirkpatrick et al., 2011; Druce, 2011.
109. Rutberg, 2013.
110. Delsink et al., 2013.
111. Zietsman, 2019.
112. Holdren and Ehrlich, 1981.
113. Boggs et al., 2006.
114. Jørgensen, 2013.
115. Cowie, 2020.
116. Watts, 2018.
117. French, 2014.
118. Chatzky and McBride, 2020; Turschwel et al., 2020.
119. Suraci et al., 2019.
120. Wynes and Nicholas, 2017.
121. Ceballos et al., 2020.
122. Habel et al., 2019a.
123. Ehrlich and Ehrlich, 2013.
124. Kristensen, 2015; Perry, 2015; Toon et al., 2014.
125. Hanski, 2002.

References

Abbott, C. G. (1933). Closing history of the Guadalupe Caracara. *Ornithological Applications Prev the Condor*, 35(1), 10–14. https://doi.org/10.2307/1363459

Aguirre, J. C. (2019, April 25). Australia is deadly serious about killing millions of cats. *New York Times*. https://nyti.ms/2LCGHhs

Aguirre-Hernández, E., & Muñoz-Ocotero, V. (2015). El chile como alimento. *Ciencia*, 5, 16–23.

Ahmed, S., Zaabi, R. A., Soorae, P. S., Shah, J. N., Hammadi, E. A., Pusey, R., & Dhaheri, S. A. (2016). Rediscovering the Arabian sand cat (*Felis margarita harrisoni*) after a gap of 10 years using camera traps in the Western Region of Abu Dhabi, United Arab Emirates. *European Journal of Wildlife Research*, 62(5), 627–631. https://doi.org/10.1007/s10344-016-1035-8

Alberts, E. C. (2020, September 29). 500 years of species loss: Humans drive defaunation across Neotropics. *Mongabay Environmental News*. https://news .mongabay.com/2020/09/500-years-of-species-loss-humans-drive-defaunat ion-across-neotropics

Alcover, J. A., Sans, A., & Palmer, M. (1998). The extent of extinctions of mammals on islands. *Journal of Biogeography*, 25(5), 913–918. https://doi.org /10.1046/j.1365-2699.1998.00246.x

Algeo, T. J., Heckel, P. H., Maynard, J. B., Blakey, R., Rowe, H., Pratt, B. R., & Holmden, C. (2008). Modern and ancient epeiric seas and the super-estuarine circulation model of marine anoxia. In *Dynamics of epeiric seas: Sedimentological, paleontological and geochemical perspectives* (pp. 7–38). Geological Association of Canada.

Allan, J. R., Possingham, H. P., Atkinson, S., Waldron, A., Di Marco, M., Butchart, S. H. M., Adams, V. M., Kissling, W. D., Worsdell, T., Sandbrook, C., Gibbon,

G. E. M., Kumar, K., Mehta, P., Maron, M., Williams, B. A., Jones, K. R., Wintle, B. A., Reside, A. E., & Watson, J. (2022). The minimum land area requiring conservation attention to safeguard biodiversity. *Science*, 376(6597), 1094–1101. https://doi.org/10.1126/science.abl9127

Allison, S. D., & Martiny, J. B. H. (2008). Resistance, resilience, and redundancy in microbial communities. *Proceedings of the National Academy of Sciences of the United States of America*, 105(S1), 11512–11519. https://doi.org/10.1073/pnas .0801925105

Almond, R. E. A., Grooten, M., Juffe Bignoli, D., & Petersen, T. (Eds.). (2022). *Living Planet Report 2022—Building a nature-positive society*. WWF.

Alpine Garden Society. (2023). Navigating the encyclopaedia. *Alpine Garden Society*. http://encyclopaedia.alpinegardensociety.net/plants/encyclopaedia/home#top

Altizer, S., Hobson, K. A., Davis, A. K., De Roode, J. C., & Wassenaar, L. I. (2015). Do healthy monarchs migrate farther? Tracking natal origins of parasitized vs. uninfected monarch butterflies overwintering in Mexico. *PLOS ONE*, 10(11), e0141371. https://doi.org/10.1371/journal.pone.0141371

Altizer, S., Oberhauser, K. S., & Brower, L. P. (2000). Associations between host migration and the prevalence of a protozoan parasite in natural populations of adult monarch butterflies. *Ecological Entomology*, 25(2), 125–139. https://doi .org/10.1046/j.1365-2311.2000.00246.x

Álvarez, W., Kauffman, E. G., Surlyk, F., Alvarez, L. W., Asaro, F., & Michel, H. V. (1984). Impact theory of mass extinctions and the invertebrate fossil record. *Science*, 223(4641), 1135–1141. https://doi.org/10.1126/science.223.4641.1135

Andreoni, K. J., Wagnon, C. J., Bestelmeyer, B. T., & Schooley, R. L. (2021). Exotic oryx interact with shrub encroachment in the Chihuahuan Desert. *Journal of Arid Environments*, 184, 104302. https://doi.org/10.1016/j.jaridenv.2020.104302

Animal Welfare Institute. (2023). *Bird trade*. https://awionline.org/content/bird -trade

Anthony, L., & Spence, G. (2013). *The Last Rhinos: My Battle to Save One of the World's Greatest Creatures*. Macmillan.

Armitage, K. B. (2014). *Marmot Biology: Sociality, Individual Fitness, and Population Dynamics*. Cambridge University Press.

Arnett, E. B., Huso, M. M., Schirmacher, M. R., & Hayes, J. P. (2010). Altering turbine speed reduces bat mortality at wind-energy facilities. *Frontiers in Ecology and the Environment*, 9(4), 209–214. https://doi.org/10.1890/100103

Arnold, C. (2019). Save our parasites! *New Scientist*, 244, 36–39.

Ascensão, F., Fahrig, L., Clevenger, A. P., Corlett, R. T., Jaeger, J. A., Laurance, W. F., & Pereira, H. M. (2018). Environmental challenges for the Belt and

Road Initiative. *Nature Sustainability, 1*(5), 206–209. https://doi.org/10.1038
/s41893-018-0059-3

Austin, K. (2010). The "Hamburger Connection" as ecologically unequal
exchange: A cross-national investigation of beef exports and deforestation
in less-developed countries. *Rural Sociology, 75*(2), 270–299. https://doi.org/10
.1111/j.1549-0831.2010.00017.x

Bachman, S. P., Lughadha, E. N., & Rivers, M. (2018). Quantifying progress
toward a conservation assessment for all plants. *Conservation Biology, 32*(3),
516–524. https://doi.org/10.1111/cobi.13071

Bakan, J. (2020). *The new corporation: How "good" corporations are bad for
democracy.* Vintage.

Baker, B. W. (2003). Beaver (*Castor canadensis*) in heavily browsed environments.
Lutra, 46(2), 173–181. http://landscouncil.org/documents/beaver_project
/articles/baker_2003_beaver_in_heavily_browsed_env.pdf

Baker, D. M., Freeman, C., Wong, J. C. Y., Fogel, M. L., & Knowlton, N. (2018).
Climate change promotes parasitism in a coral symbiosis. *The ISME Journal,
12*(3), 921–930. https://doi.org/10.1038/s41396-018-0046-8

Bakker, V. J., Smith, D. R., Copeland, H. E., Brandt, J., Wolstenholme, R., Burnett,
J., Kirkland, S., & Finkelstein, M. E. (2016). Effects of lead exposure, flock behavior, and management actions on the survival of California condors
(*Gymnogyps californianus*). *Ecohealth, 14*(S1), 92–105. https://doi.org/10.1007
/s10393-015-1096-2

Balmer, J. (2015). White-nose syndrome has almost completely wiped out some
North American bat colonies. *Science.* https://doi.org/10.1126/science.aaa7819

Baptista-Rosas, R. C., Hinojosa-Corona, A., & Riquelme, M. (2007). Ecological
niche modeling of *Coccidioides* spp. in western North American deserts.
Annals of the New York Academy of Sciences, 1111(1), 35–46. https://doi.org/10
.1196/annals.1406.003

Barnosky, A. D. (2014). *Dodging Extinction: Power, Food, Money, and the Future of
Life on Earth.* Univ of California Press.

Barnosky, A. D., Matzke, N. J., Tomiya, S., Wogan, G. O. U., Swartz, B. K.,
Quental, T. B., Marshall, C. R., McGuire, J. L., Lindsey, E., Maguire, K. C.,
Mersey, B., & Ferrer, E. A. (2011). Has the Earth's sixth mass extinction
already arrived? *Nature, 471*(7336), 51–57. https://doi.org/10.1038
/nature09678

Barrett, S., Lenton, T. M., Millner, A., Tavoni, A., Carpenter, S. R., Anderies,
J. M., Chapin, F. S., Crépin, A., Daily, G. C., Ehrlich, P. R., Folke, C., Galaz, V.,
Hughes, T. P., Kautsky, N., Lambin, É. F., Naylor, R. L., Nyborg, K., Polasky, S.,

Scheffer, M., . . . De Zeeuw, A. (2014). Climate engineering reconsidered. *Nature Climate Change, 4*(7), 527–529. https://doi.org/10.1038/nclimate2278

Barrionuevo, J. S., Aguayo, R., & Lavilla, E. O. (2008). First record of chytridiomycosis in Bolivia (*Rhinella quechua*; Anura: Bufonidae). *Diseases of aquatic organisms, 82,* 161–163. https://doi.org/10.3354/dao01980

Bauchner, H., Rivara, F. P., Bonow, R. O., Bressler, N. M., Disis, M. L., Heckers, S., Josephson, S. A., Kibbe, M. R., Piccirillo, J. F., Redberg, R. F., Rhee, J. S., & Robinson, J. K. (2017). Death by gun violence—A public health crisis. *JAMA, 318*(18), 1763. https://doi.org/10.1001/jama.2017.16446

Bauer, D. M., & Wing, I. S. (2010). Economic Consequences of pollinator declines: A synthesis. *Agricultural and Resource Economics Review, 39*(3), 368–383. https://doi.org/10.1017/s1068280500007371

Baums, I. B., Johnson, M. E., Devlin-Durante, M., & Miller, M. W. (2010). Host population genetic structure and zooxanthellae diversity of two reef-building coral species along the Florida Reef Tract and wider Caribbean. *Coral Reefs, 29*(4), 835–842. https://doi.org/10.1007/s00338-010-0645-y

BGCI (Botanic Gardens Conservation International). (2022, April 5). ThreatSearch online database. *BGCI.* www.bgci.org/threat_search.php.

Beard, P. H. (1965). *The End of the Game: The Last Word from Paradise: A Pictorial Documentation of the Origins, History & Prospects of the Big Game in Africa.* Chronicle Books.

Beattie, A. (2016). Species loss: Diverse takes on biodiversity. *Nature, 537,* 617–617.

Beattie, A. (2017, January 3). The futile war between conservationists and farmers: Conservation and production biodiversity. *Millennium Alliance for Humanity and Biosphere.* https://mahb.stanford.edu/blog/conservation-production-biodiversity

Beattie, A. J., Barthlott, W., Elisabetsky, E., Farrel, R., Teck Kheng, T., & Prance, I. (2005). New products and industries from biodiversity. *Ecosystems and human well-being: Current state and trends: Findings of the Condition and Trends Working Group, 1,* 271.

Beaune, D. (2015). What would happen to the trees and lianas if apes disappeared? *Oryx, 49*(3), 442–446. https://doi.org/10.1017/s0030605314000878

Beck, M. W., Brumbaugh, R. D., Airoldi, L., Carranza, A., Coen, L. D., Crawford, C., Defeo, O., Edgar, G. J., Hancock, B., Kay, M. C., Lenihan, H. S., Luckenbach, M. W., Toropova, C., Zhang, G., & Guo, X. (2011). Oyster reefs at risk and recommendations for conservation, restoration, and management. *BioScience, 61*(2), 107–116. https://doi.org/10.1525/bio.2011.61.2.5

Bednaršek, N., Harvey, C. J., Kaplan, I. C., Feely, R. A., & Možina, J. (2016). Pteropods on the edge: Cumulative effects of ocean acidification, warming, and deoxygenation. *Progress in Oceanography, 145*, 1–24. https://doi.org/10.1016/j.pocean.2016.04.002

Beebe, W. (1906). *The Bird: Its Form and Function.* H. Holt.

Beehler, B. M. (2019, December 19). How to solve America's bird crisis. *The Hill.* https://thehill.com/changing-america/sustainability/465203-our-nations-declining-birdlife-can-be-brought-back-heres-how

Beier, P. (1995). Dispersal of juvenile cougars in fragmented habitat. *Journal of Wildlife Management, 59*(2), 228. https://doi.org/10.2307/3808935

Beier, P., & Noss, R. F. (1998). Do habitat corridors provide connectivity? *Conservation Biology, 12*(6), 1241–1252. https://doi.org/10.1111/j.1523-1739.1998.98036.x

Beissinger, S. R. (2000). Ecological mechanisms of extinction. *Proceedings of the National Academy of Sciences of the United States of America, 97*(22), 11688–11689. https://doi.org/10.1073/pnas.97.22.11688

Beissinger, S. R., & McCullough, D. R. (2002). *Population Viability Analysis.* University of Chicago Press.

Bellard, C., Bertelsmeier, C., Leadley, P., Thuiller, W., & Courchamp, F. (2012). Impacts of climate change on the future of biodiversity. *Ecology Letters, 15*(4), 365–377. https://doi.org/10.1111/j.1461-0248.2011.01736.x

Bennett, E. L. (2014). Legal ivory trade in a corrupt world and its impact on African elephant populations. *Conservation Biology, 29*(1), 54–60. https://doi.org/10.1111/cobi.12377

Bennett, N., Roth, R., Klain, S. C., Chan, K. M. A., Christie, P., Clark, D. A., Cullman, G., Curran, D., Durbin, T. J., Epstein, G., Greenberg, A., Nelson, M. P., Sandlos, J., Stedman, R. C., Teel, T. L., Thomas, R., Veríssimo, D., & Wyborn, C. (2017). Conservation social science: Understanding and integrating human dimensions to improve conservation. *Biological Conservation, 205*, 93–108. https://doi.org/10.1016/j.biocon.2016.10.006

Berg, C. C. (1972). Olmedieae, Brosimeae (Moraceae). *Flora Neotropica, 7*, 168–172. https://www.jstor.org/stable/4393672

Betts, H. C., Puttick, M. N., Clark, J. W., Williams, T. A., Donoghue, P. C. J., & Pisani, D. (2018). Integrated genomic and fossil evidence illuminates life's early evolution and eukaryote origin. *Nature Ecology and Evolution, 2*(10), 1556–1562. https://doi.org/10.1038/s41559-018-0644-x

Bhattacharya, S. (2017). Deadly new wheat disease threatens Europe's crops. *Nature, 542*(7640), 145–146. https://doi.org/10.1038/nature.2017.21424

Black, R. (March 5, 2012). A Dinosaur's pterosaur lunch. *Smithsonian Magazine.* https://www.smithsonianmag.com/science-nature/a-dinosaurs-pterosaur-lunch-114418691/

Blackburn, T. M., Bellard, C., & Ricciardi, A. (2019). Alien versus native species as drivers of recent extinctions. *Frontiers in Ecology and the Environment*, 17(4), 203-207. https://doi.org/10.1002/fee.2020

Blackburn, T. M., Cassey, P., Duncan, R. P., Evans, K. L., & Gaston, K. J. (2004). Avian extinction and mammalian introductions on oceanic islands. *Science*, 305(5692), 1955-1958. https://doi.org/10.1126/science.1101617

Bodsworth, F. (1952). *Last of the Curlews*. Counterpoint.

Boggs, C. L., Holdren, C. E., Kulahci, I. G., Bonebrake, T. C., Inouye, B. D., Fay, J. P., Mcmillan, A., Williams, E. H., & Ehrlich, P. R. (2006). Delayed population explosion of an introduced butterfly. *Journal of Animal Ecology*, 75(2), 466-475. https://doi.org/10.1111/j.1365-2656.2006.01067.x

Böhm, M., Collen, B., Baillie, J., Bowles, P., Chanson, J., Cox, N. A., Hammerson, G. A., Hoffmann, M., Livingstone, S. R., Ram, M., Rhodin, A. G. J., Stuart, S. N., Van Dijk, P. P., Young, B. E., Afuang, L. E., Aghasyan, A., García, A., Aguilar, C., Ajtić, R., . . . Guo, P. (2013). The conservation status of the world's reptiles. *Biological Conservation*, 157, 372-385. https://doi.org/10.1016/j.biocon.2012.07.015

Bond, D. P., & Grasby, S. E. (2017). On the causes of mass extinctions. *Palaeogeography, Palaeoclimatology, Palaeoecology*, 478, 3-29. https://doi.org/10.1016/j.palaeo.2016.11.005

Bonebrake, T. C., & Cooper, D. S. (2014). A Hollywood drama of butterfly extirpation and persistence over a century of urbanization. *Journal of Insect Conservation*, 18(4), 683-692. https://doi.org/10.1007/s10841-014-9675-z

Booth, J., Gaston, K. J., Evans, K. L., & Armsworth, P. R. (2011). The value of species rarity in biodiversity recreation: A birdwatching example. *Biological Conservation*, 144(11), 2728-2732. https://doi.org/10.1016/j.biocon.2011.02.018

Bosak, T., Knoll, A. H., & Petroff, A. P. (2013). The meaning of stromatolites. *Annual Review of Earth and Planetary Sciences*, 41(1), 21-44. https://doi.org/10.1146/annurev-earth-042711-105327

Bowman, M. J. (2013). A tale of two CITES: Divergent perspectives upon the effectiveness of the wildlife trade convention. *Review of European, Comparative and International Environmental Law*, 22(3), 228-238. https://doi.org/10.1111/reel.12049

Boyle, J. H., Dalgleish, H. J., & Puzey, J. R. (2019). Monarch butterfly and milkweed declines substantially predate the use of genetically modified crops. *Proceedings of the National Academy of Sciences of the United States of America*, 116(8), 3006-3011. https://doi.org/10.1073/pnas.1811437116

Boyles, J. G., Cryan, P. M., McCracken, G. F., & Kunz, T. (2011). Economic importance of bats in agriculture. *Science, 332*(6025), 41–42. https://doi.org/10.1126/science.1201366

Bradshaw, C. J. A., Ehrlich, P. R., Beattie, A. J., Ceballos, G., Crist, E., Diamond, J., Dirzo, R., Ehrlich, A. H., Harte, J., Harte, M. E., Pyke, G. H., Raven, P. H., Ripple, W. J., Saltré, F., Turnbull, C., Wackernagel, M., & Blumstein, D. T. (2021). Underestimating the challenges of avoiding a ghastly future. *Frontiers in Conservation Science, 1*, 1–10. https://doi.org/10.3389/fcosc.2020.615419

Brannen, P. (2017). *The Ends of the World: Volcanic Apocalypses, Lethal Oceans and Our Quest to Understand Earth's Past Mass Extinctions.* Simon and Schuster.

Brashares, J. S., Arcese, P., Sam, M. K., Coppolillo, P., Sinclair, A. R. E., & Balmford, A. (2004). Bushmeat hunting, wildlife declines, and fish supply in West Africa. *Science, 306*(5699), 1180–1183. https://doi.org/10.1126/science.1102425

Bratman, G. N., Hamilton, J. P., Hahn, K. S., Daily, G. C., & Gross, J. J. (2015). Nature experience reduces rumination and subgenual prefrontal cortex activation. *Proceedings of the National Academy of Sciences of the United States of America, 112*(28), 8567–8572. https://doi.org/10.1073/pnas.1510459112

Breed, G. A., Matthews, C. J. D., Marcoux, M., Higdon, J. W., LeBlanc, B., Petersen, S. D., Orr, J., Reinhart, N. R., & Ferguson, S. H. (2017). Sustained disruption of narwhal habitat use and behavior in the presence of Arctic killer whales. *Proceedings of the National Academy of Sciences of the United States of America, 114*(10), 2628–2633. https://doi.org/10.1073/pnas.1611707114

Breedlove, D. E., & Ehrlich, P. R. (1968). Plant-herbivore coevolution: Lupines and lycaenids. *Science, 162*(3854), 671–672. https://doi.org/10.1126/science.162.3854.671

Brennan, A., Naidoo, R., Greenstreet, L., Mehrabi, Z., Ramankutty, N., & Kremen, C. (2022). Functional connectivity of the world's protected areas. *Science, 376*(6597), 1101–1104. https://doi.org/10.1126/science.abl8974

Brewis, H. (2019, December 28). Nearly 500 million animals killed in Australian bushfires. *Evening Standard.* https://www.standard.co.uk/news/world/australian-bushfires-new-south-wales-koalas-sydney-a4322071.html

Bridgewater, P. (2016). The Anthropocene biosphere: Do threatened species, Red Lists, and protected areas have a future role in nature conservation? *Biodiversity and Conservation, 25*(3), 603–607. https://doi.org/10.1007/s10531-016-1062-5

Britannica. 2015. Great Northern Expedition. *Encyclopedia Britannica.* https://www.britannica.com/event/Great-Northern-Expedition.

Brooke, M. L. (2019). Rat eradication in the Pitcairn Islands, South Pacific: A 25-year perspective. *Proceedings of the International Conference on Island Invasives*, 95–99.

Brower, L. P., Castilleja, G., Peralta, A., López-García, J., Bojórquez-Tapia, L. A., Diaz, S. E., Melgarejo, D., & Missrie, M. (2002). Quantitative changes in forest quality in a principal overwintering area of the Monarch Butterfly in Mexico, 1971–1999. *Conservation Biology, 16*(2), 346–359. https://doi.org/10.1046/j.1523 -1739.2002.00572.x

Brower, L. P., Taylor, O. R., Williams, E. H., Slayback, D. A., Zubieta, R. R., & Isabel, M. (2011). Decline of monarch butterflies overwintering in Mexico: Is the migratory phenomenon at risk? *Insect Conservation and Diversity, 5*(2), 95–100. https://doi.org/10.1111/j.1752-4598.2011.00142.x

Brown, K. (2019, July 5). Heart of Ecuador's Yasuni, home to uncontacted tribes, opens for oil drilling. *Mongabay Environmental News*. https://news.mongabay .com/2019/07/heart-of-ecuadors-yasuni-home-to-uncontacted-tribes-opens -for-oil-drilling

Brussaard, L., De Ruiter, P., & Brown, G. G. (2007). Soil biodiversity for agricultural sustainability. *Agriculture, Ecosystems & Environment, 121*(3), 233–244. https://doi .org/10.1016/j.agee.2006.12.013

Burkhard, B., Crossman, N. D., Nedkov, S., Petz, K., & Alkemade, R. (2013). Mapping and modelling ecosystem services for science, policy and practice. *Ecosystem Services, 4*, 1–3. https://doi.org/10.1016/j.ecoser.2013.04.005

Burkhead, N. M. (2012). Extinction rates in North American freshwater fishes, 1900–2010. *BioScience, 62*(9), 798–808. https://doi.org/10.1525/bio.2012.62.9.5

Burkle, L. A., Marlin, J. C., & Knight, T. M. (2013). Plant-pollinator interactions over 120 years: Loss of species, co-occurrence, and function. *Science, 339*(6127), 1611–1615. https://doi.org/10.1126/science.1232728

Burnett, L. J., Sorenson, K. J., Brandt, J., Sandhaus, E. A., Ciani, D., Clark, M., David, C., Theule, J., Kasielke, S., & Risebrough, R. W. (2013). Eggshell thinning and depressed hatching success of California condors reintroduced to central California. *Ornithological Applications Prev the Condor, 115*(3), 477–491. https://doi.org/10.1525/cond.2013.110150

Butchart, S. H. M., Walpole, M., Collen, B., Van Strien, A., Scharlemann, J. P. W., Almond, R., Baillie, J., Bomhard, B., Brown, C., Bruno, J. P., Carpenter, K. E., Carr, G. M., Chanson, J., Chenery, A. M., Csirke, J., Davidson, N. C., Dentener, F., Foster, M., Galli, A., . . . Watson, R. (2010). Global biodiversity: Indicators of recent declines. *Science, 328*(5982), 1164–1168. https://doi.org/10.1126 /science.1187512

CABI (Centre for Agriculture and Bioscience International). (2021, November 16). *Phytophthora infestans* (Phytophthora blight). *CABI Digital Library*. https://www.cabidigitallibrary.org/doi/10.1079/cabicompendium .40970#toimpact

Cahill, J. A., Heintzman, P. D., Harris, K., Teasdale, M. D., Kapp, J. D., Soares, A. E. R., Stirling, I., Bradley, D. G., Edwards, C. J., Graim, K., Kisleika, A. A., Malev, A. V., Monaghan, N. T., Green, E., & Shapiro, B. (2018). Genomic evidence of widespread admixture from polar bears into brown bears during the last ice age. *Molecular Biology and Evolution, 35*(5), 1120–1129. https://doi .org/10.1093/molbev/msy018

Cain, A. J., & Sheppard, P. M. (1954). Natural selection in *Cepaea. Genetics, 39*(1), 89–116. https://doi.org/10.1093/genetics/39.1.89

Camero, K. (2019, August 5). This tiny insect could be delivering toxic pesticides to honey bees and other beneficial bugs. Excreted honeydew contains high doses of neonicotinoids. *Scienceshots*. https://www.science.org/content /article/tiny-insect-could-be-delivering-toxic-pesticides-honey-bees-and -other-beneficial-bugs

Carlson, C. J., Hopkins, S. R., Bell, K. C., Doña, J., Godfrey, S. S., Kwak, M. L., Lafferty, K. D., Moir, M. L., Speer, K. A., Strona, G., Torchin, M. E., & Wood, C. L. (2020). A global parasite conservation plan. *Biological Conservation, 250*, 108596. https://doi.org/10.1016/j.biocon.2020.108596

Carlton, J. T., Vermeij, G. J., Lindberg, D. R., Carlton, D. A., & Dubley, E. C. (1991). The first historical extinction of a marine invertebrate in an ocean basin: The demise of the eelgrass limpet *Lottia alveus. The Biological Bulletin, 180*(1), 72–80. https://doi.org/10.2307/1542430

Carrington, D. (2021, October 29). Warning of "ecological Armageddon" after dramatic plunge in insect numbers. *The Guardian*. https://www.theguardian .com/environment/2017/oct/18/warning-of-ecological-armageddon-after -dramatic-plunge-in-insect-numbers

Carwardine, J., Wilson, K. A., Ceballos, G., Ehrlich, P. R., Naidoo, R., Iwamura, T., Hajkowicz, S., & Possingham, H. P. (2008). Cost-effective priorities for global mammal conservation. *Proceedings of the National Academy of Sciences of the United States of America, 105*(32), 11446–11450. https://doi.org/10.1073/pnas .0707157105

Cavicchioli, R., Ripple, W. J., Timmis, K. N., Azam, F., Bakken, L. R., Baylis, M., Behrenfeld, M. J., Boetius, A., Boyd, P. W., Classen, A. T., Crowther, T. W., Danovaro, R., Foreman, C. M., Huisman, J., Hutchins, D. A., Jansson, J. K., Karl, D. M., Koskella, B., Welch, D., . . . Webster, N. S. (2019). Scientists' warning to humanity: microorganisms and climate change. *Nature Reviews Microbiology*, 17(9), 569–586. https://doi.org/10.1038/s41579-019-0222-5

Ceballos, G. (ed.). 2011. *Animales Amenazados de América: el Reto de su Sobrevivencia*. Telmex, México D. F.

Ceballos, G., Díaz Pardo, E., Martínez Estévez, L., & Espinoza, H. (2016). *Los peces dulceacuícolas de México en Peligro de Extinción*. Fondo de Cultura Económica—UNAM. México D. F.

Ceballos, G., Ehrlich, A. H., & Ehrlich, P. R. (2015a). *The Annihilation of Nature: Human Extinction of Birds and Mammals*. Johns Hopkins University Press.

Ceballos, G., & Ehrlich, P. R. (2002). Mammal population losses and the extinction crisis. *Science*, 296(5569), 904–907. https://doi.org/10.1126/science.1069349

Ceballos, G., & Ehrlich, P. R. (2006). Global mammal distributions, biodiversity hotspots, and conservation. *Proceedings of the National Academy of Sciences of the United States of America*, 103(51), 19374–19379. https://doi.org/10.1073/pnas.0609334103

Ceballos, G., & Ehrlich, P. R. (2009). Discoveries of new mammal species and their implications for conservation and ecosystem services. *Proceedings of the National Academy of Sciences of the United States of America*, 106(10), 3841–3846. https://doi.org/10.1073/pnas.0812419106

Ceballos, G., & Ehrlich, P. R. (2023). Mutilation of the tree of life via mass extinction of animal genera. *Proceedings of the National Academy of Sciences of the United States of America*, 120(39). https://doi.org/10.1073/pnas.2306987120

Ceballos, G., Ehrlich, P. R., Barnosky, A. D., García, A., Pringle, R. M., & Palmer, T. M. (2015b). Accelerated modern human–induced species losses: Entering the sixth mass extinction. *Science Advances*, 1(5). https://doi.org/10.1126/sciadv.1400253

Ceballos, G., Ehrlich, P. R., & Dirzo, R. (2017). Biological annihilation via the ongoing sixth mass extinction signaled by vertebrate population losses and declines. *Proceedings of the National Academy of Sciences of the United States of America*, 114(30). https://doi.org/10.1073/pnas.1704949114

Ceballos, G., Ehrlich, P. R., Pacheco, J., Valverde-Zúñiga, N., & Daily, G. C. (2019). Conservation in human-dominated landscapes: Lessons from the distribution of the Central American squirrel monkey. *Biological Conservation*, 237, 41–49. https://doi.org/10.1016/j.biocon.2019.06.008

Ceballos, G., Ehrlich, P. R., & Raven, P. H. (2020). Vertebrates on the brink as indicators of biological annihilation and the sixth mass extinction. *Proceedings of the National Academy of Sciences of the United States of America*, 117(24), 13596–13602. https://doi.org/10.1073/pnas.1922686117

Ceballos, G., & Navarro, D. (1991). Diversity and conservation of Mexican mammals. In M. A. Mares & D. J. Schmidly (Eds.), *Latin American Mammalogy: History, Diversity and Conservation* (pp. 167–198). University of Oklahoma Press.

Ceballos, G., & Pacheco, J. (2021). *Guía de o de los mamíferos de la Reserva de la biosfera Janos, Chihuahua*. Universidad Nacional Autónoma de México.

Cerqueira, A. E. S., Hammer, T. J., Moran, N. A., Santana, W. C., Kasuya, M. C. M., & Da Silva, C. C. (2021). Extinction of anciently associated gut bacterial symbionts in a clade of stingless bees. *The ISME Journal, 15*(9), 2813–2816. https://doi.org/10.1038/s41396-021-01000-1

Challenger, D. W., Nash, H. C., & Waterman, C. (2019). *Pangolins: Science, Society and Conservation*. Academic Press.

Chaplin-Kramer, R., Dombeck, E., Gerber, J., Knuth, K. A., Mueller, N. D., Mueller, M., Ziv, G., & Klein, A. (2014). Global malnutrition overlaps with pollinator-dependent micronutrient production. *Proceedings of the Royal Society B: Biological Sciences, 281*(1794), 20141799. https://doi.org/10.1098/rspb.2014.1799

Chaplin-Kramer, R., Sharp, R., Weil, C., Bennett, E. M., Pascual, U., Arkema, K. K., Brauman, K. A., Bryant, B. P., Guerry, A. D., Haddad, N. M., Hamann, M., Hamel, P., Johnson, J. A., Mandle, L., Pereira, H. M., Polasky, S., Ruckelshaus, M., Shaw, M. R., Silver, J. M., . . . Daily, G. C. (2019). Global modeling of nature's contributions to people. *Science, 366*(6462), 255–258. https://doi.org/10.1126/science.aaw3372

Chapman, C. A., & Peres, C. A. (2001). Primate conservation in the new millennium: The role of scientists. *Evolutionary Anthropology: Issues, News, and Reviews, 10*(1), 16–33. https://doi.org/10.1002/1520 -6505(2001)10:1%3C16::AID-EVAN1010%3E3.0.CO;2-O

Chase, M. J., Schlossberg, S., Griffin, C. R., Bouché, P., Djene, S. W., Elkan, P. W., Ferreira, S. M., Grossman, F., Kohi, E. M., Landen, K., Omondi, P., Peltier, A., Selier, J., & Sutcliffe, R. (2016). Continent-wide survey reveals massive decline in African savannah elephants. *PeerJ, 4*, e2354. https://doi.org/10.7717 /peerj.2354

Chatzky, A., & McBride, J. (2020, January 28). China's massive Belt and Road initiative. *Council on Foreign Relations*. https://www.cfr.org/backgrounder /chinas-massive-belt-and-road-initiative

Cheek, M., Lughadha, E. N., Kirk, P. M., Lindon, H., Carretero, J., Looney, B. P., Douglas, B., Haelewaters, D., Gaya, E., Llewellyn, T., Ainsworth, A. M., Gafforov, Y., Hyde, K. D., Crous, P. W., Hughes, M., Walker, B. E., Forzza, R. C., Wong, K. M., & Niskanen, T. (2020). New scientific discoveries: Plants and fungi. *Plants, People, Planet, 2*(5), 371–388. https://doi.org/10.1002/ppp3.10148

Chen, J., Oliveri, P., Gao, F., Dornbos, S. Q., Li, C., Bottjer, D. J., & Davidson, E. H. (2002). Precambrian animal life: Probable developmental and adult cnidarian forms from Southwest China. *Developmental Biology, 248*(1), 182–196. https://doi.org/10.1006/dbio.2002.0714

Chiarenza, A. A., Farnsworth, A., Mannion, P. D., Lunt, D. J., Valdes, P. J., Morgan, J., & Allison, P. A. (2020). Asteroid impact, not volcanism, caused the end-Cretaceous dinosaur extinction. *Proceedings of the National Academy of Sciences of the United States of America, 117*(29), 17084–17093. https://doi.org/10.1073/pnas.2006087117

Chiba, S., & Cowie, R. H. (2016). Evolution and extinction of land snails on oceanic islands. *Annual Review of Ecology, Evolution, and Systematics, 47*(1), 123–141. https://doi.org/10.1146/annurev-ecolsys-112414-054331

Chivian, E., Roberts, C. M., & Bernstein, A. (2003). The threat to cone snails. *Science, 302*(5644), 391. https://doi.org/10.1126/science.302.5644.391b

Chiyo, P. I., Obanda, V., & Korir, D. K. (2015). Illegal tusk harvest and the decline of tusk size in the African elephant. *Ecology and Evolution, 5*(22), 5216–5229. https://doi.org/10.1002/ece3.1769

Cibois, A., Beadell, J. S., Graves, G. R., Pasquet, É., Slikas, B., Sonsthagen, S. A., Thibault, J., & Fleischer, R. C. (2011). Charting the course of reed-warblers across the Pacific islands. *Journal of Biogeography, 38*(10), 1963–1975. https://doi.org/10.1111/j.1365-2699.2011.02542.x

Cinner, J. E., Graham, N. a. J., Huchery, C., & MacNeil, M. A. (2012). Global effects of local human population density and distance to markets on the condition of coral reef fisheries. *Conservation Biology, 27*(3), 453–458. https://doi.org/10.1111/j.1523-1739.2012.01933.x

Civeyrel, L., & Simberloff, D. (1996). A tale of two snails: Is the cure worse than the disease? *Biodiversity and Conservation, 5*(10), 1231–1252. https://doi.org/10.1007/bf00051574

Clark, J. G., & Linares-Matás, G. (2023). Seasonal resource categorisation and behavioral adaptation among chimpanzees: Implications for early hominin carnivory. *PubMed, 101*, 1–35. https://doi.org/10.4436/jass.10006

Clarke, M. F., Taylor, R. S., Dare, A. J., Grey, M. J., & Oldland, J. M. (2007). Challenges in managing miners. *Victorian Naturalist, 124*, 102.

Clulow, J., Upton, R., Trudeau, V. L., & Clulow, S. (2019). Amphibian assisted reproductive technologies: Moving from technology to application. In *Advances in Experimental Medicine and Biology* (pp. 413–463). https://doi.org/10.1007/978-3-030-23633-5_14

Coghlan, A. (2018, January 25). Chimps are now dying of the common cold and they are all at risk. *New Scientist*. https://www.newscientist.com/article

/2159509-chimps-are-now-dying-of-the-common-cold-and-they-are-all-at
-risk

Colborn, T., Dumanoski, D., & Myers, J. P. (1996). *Our Stolen Future*. Dutton.
http://ci.nii.ac.jp/ncid/BA32464581

Colwell, R. K., Brehm, G., Cardelús, C. L., Gilman, A., & Longino, J. T. (2008).
Global warming, elevational range shifts, and lowland biotic attrition in the
wet tropics. *Science, 322*(5899), 258–261. https://doi.org/10.1126/science
.1162547

Conniff, R. (2019, July 23). Why isn't publicly funded conservation on private
land more accountable? *Yale Environment, 360*. https://e360.yale.edu/features
/why-isnt-publicly-funded-conservation-on-private-land-more-accountable

Connor, E. F., Hafernik, J. E., Levy, J. M., Moore, V. L., & Rickman, J. K. (2002).
Insect conservation in an urban biodiversity hotspot: The San Francisco Bay
Area. *Journal of Insect Conservation, 6*(4), 247–259. https://doi.org/10.1023
/a:1024426727504

Conrad, K. F., Warren, M. S., Fox, R., Parsons, M., & Woiwod, I. P. (2006). Rapid
declines of common, widespread British moths provide evidence of an insect
biodiversity crisis. *Biological Conservation, 132*(3), 279–291. https://doi.org/10
.1016/j.biocon.2006.04.020

Cooke, L. R., Schepers, H., Hermansen, A., Bain, R. A., Bradshaw, N. J., Ritchie,
F., Shaw, D., Evenhuis, A., Kessel, G., Wander, J., Andersson, B., Hansen, J. G.,
Hannukkala, A., Nærstad, R., & Nielsen, B. (2011). Epidemiology and
integrated control of potato late blight in Europe. *Potato Research, 54*(2), 183–
222. https://doi.org/10.1007/s11540-011-9187-0

Corbet, S. A., Williams, I. H., & Osborne, J. L. (1991). Bees and the pollination of
crops and wild flowers in the European community. *Bee World, 72*(2), 47–59.
https://doi.org/10.1080/0005772x.1991.11099079

Corlett, R. T. (2020). Safeguarding our future by protecting biodiversity. *Plant
Diversity, 42*(4), 221–228. https://doi.org/10.1016/j.pld.2020.04.002

Cortés-Capano, G., Toivonen, T., Soutullo, Á., & Di Minin, E. (2019). The
emergence of private land conservation in scientific literature: A review.
Biological Conservation, 237, 191–199. https://doi.org/10.1016/j.biocon.2019.07
.010

Cousins, I. T., Johansson, J. H., Salter, M., Sha, B., & Scheringer, M. (2022).
Outside the safe operating space of a new planetary boundary for per- and
polyfluoroalkyl substances (PFAS). *Environmental Science & Technology,
56*(16), 11172–11179. https://doi.org/10.1021/acs.est.2c02765

Cowie, S. (2020, February 3). The jungle metropolis: how sprawling Manaus is
eating into the Amazon. *The Guardian*. https://www.theguardian.com/cities

/2019/jul/23/the-jungle-metropolis-how-sprawling-manaus-is-eating-into
-the-amazon

Cox, N. A., Young, B. E., Bowles, P., Fernández, M., Marin, J., Rapacciuolo, G., Böhm, M., Brooks, T. M., Hedges, S. B., Hilton-Taylor, C., Hoffmann, M., Jenkins, R. K. B., Tognelli, M. F., Alexander, G. J., Allison, A., Ananjeva, N. B., Auliya, M., Ávila, L. J., Chapple, D. G., . . . Xie, Y. (2022). A global reptile assessment highlights shared conservation needs of tetrapods. *Nature*, 605(7909), 285-290. https://doi.org/10.1038/s41586-022-04664-7

Crall, J. D., Switzer, C. M., Oppenheimer, R. L., Versypt, A. N. F., Dey, B., Brown, A. N., Eyster, M., Guérin, C., Pierce, N. E., Combes, S. A., & De Bivort, B. L. (2018). Neonicotinoid exposure disrupts bumblebee nest behavior, social networks, and thermoregulation. *Science*, 362(6415), 683-686. https://doi.org/10.1126/science.aat1598

Craigie, I. D., Baillie, J., Balmford, A., Carbone, C., Collen, B., Green, R. E., & Hutton, J. (2010). Large mammal population declines in Africa's protected areas. *Biological Conservation*, 143(9), 2221-2228. https://doi.org/10.1016/j.biocon.2010.06.007

Criss, D. (2019, January 7). The monarch butterfly population in California has plummeted 86% in one year. *CNN*. https://edition.cnn.com/2019/01/07/us/monarch-butterflies-decline-trnd/index.html

Crist, E., Ripple, W. J., Ehrlich, P. R., Rees, W. E., & Wolf, C. (2022). Scientists' warning on population. *Science of the Total Environment*, 845, 157166. https://doi.org/10.1016/j.scitotenv.2022.157166

Cronk, Q. (2016). Plant extinctions take time. *Science*, 353(6298), 446-447. https://doi.org/10.1126/science.aag1794

Crooks, K. R., & Soulé, M. E. (1999). Mesopredator release and avifaunal extinctions in a fragmented system. *Nature*, 400(6744), 563-566. https://doi.org/10.1038/23028

Crump, M. (2000). *In Search of the Golden Toad*. University of Chicago Press.

Crump, M. L., Hensley, F. R., & Clark, K. L. 1992. Apparent decline of the Golden toad: Underground or extinct? *Copeia*, 1992(2), 413-420. https://www.jstor.org/stable/1446201

Curnutt, J. L., & Pimm, S. L. (2002). How many bird species in Hawaii and the Central Pacific before first contact? *Studies in Avian Biology*, 22:15-30. https://www.researchgate.net/publication/210289376_How_many_bird_species_in_Hawai'i_and_the_Central_Pacific_before_first_contact

Curry, J., Byrne, D., & Schmidt, O. (2002). Intensive cultivation can drastically reduce earthworm populations in arable land. *European Journal of Soil Biology*, 38(2), 127-130. https://doi.org/10.1016/s1164-5563(02)01132-9

Curtis, T. (2006). Microbial ecologists: It's time to "go large." *Nature Reviews Microbiology*, 4(7), 488. https://doi.org/10.1038/nrmicro1455

Dahl, A. L. (2009). The financial crisis and consumer citizenship. *Making a Difference: Putting Consumer Citizenship into Action. Proceedings of the Sixth International Conference of the Consumer Citizenship Network, Berlin, Germany* 2009, 20.

Daily, G. C. (1997). *Nature's Services*. Island Press.

Daily, G. C. (1999). Developing a scientific basis for managing Earth's life support systems. *Conservation Ecology*, 3(2). https://doi.org/10.5751/es-00140-030214

Daily, G. C., Ceballos, G., Pacheco, J., Suzán, G., & Sánchez-Azofeifa, A. (2003). Countryside biogeography of neotropical mammals: Conservation opportunities in agricultural landscapes of Costa Rica. *Conservation Biology*, 17(6), 1814–1826. https://doi.org/10.1111/j.1523-1739.2003.00298.x

Daily, G. C., Ehrlich, P. R., & Haddad, N. M. (1993). Double keystone bird in a keystone species complex. *Proceedings of the National Academy of Sciences of the United States of America*, 90(2), 592–594. https://doi.org/10.1073/pnas.90.2.592

Daily, G. C., Ehrlich, P. R., & Sánchez-Azofeifa, G. A. (2001). Countryside biogeography: Use of human-dominated habitats by the avifauna of southern Costa Rica. *Ecological Applications*, 11(1), 1–13. https://doi.org/10.1890/1051-0761(2001)011

Dallaire, R. 2007. *Shake Hands with the Devil: The Failure of Humanity in Rwanda*. Da Capo Press.

Daniel, T. C., Muhar, A., Aznar, O., Boyd, J., Chan, K. M. A., Costanza, R., Flint, C. G., Gobster, P. H., Grêt-Regamey, A., Penker, M., Ribe, R. G., & Spierenburg, M. (2012). Reply to Kirchhoff: Cultural values and ecosystem services. *Proceedings of the National Academy of Sciences of the United States of America*, 109(46). https://doi.org/10.1073/pnas.1213520109

Darwin, C. (1859). *The Origin of Species*. Reprint. Modern Library.

Darwin, C. (1892). *The Formation of Vegetable Mould, through the Action of Worms, with Observations on their Habits*. J. Murray.

Dasgupta, P. (2021). *The Economics of Biodiversity: The Dasgupta Review*. HM Treasury.

Davis, J., Mengersen, K., Abram, N. K., Ancrenaz, M., Wells, J. A., & Meijaard, E. (2013). It's not just conflict that motivates killing of orangutans. *PLOS ONE*, 8(10), e75373. https://doi.org/10.1371/journal.pone.0075373

Delsink, A., Kirkpatrick, J. F., Van Altena, J., Bertschinger, H., Ferreira, S. M., & Slotow, R. (2013). Lack of spatial and behavioral responses to immunocontraception application in African elephants (*Loxodonta africana*).

Journal of Zoo and Wildlife Medicine, 44(4S), S52–S74. https://doi.org/10.1638
/1042-7260-44.4s.s52

Delsink, A., Van Altena, J. J., Grobler, D., Bertschinger, H., Kirkpatrick, J. F., & Slotow, R. (2007). Implementing immunocontraception in free-ranging African elephants at Makalali Conservancy. *Journal of the South African Veterinary Association, 78*(1), 25–30. https://doi.org/10.4102/jsava.v78i1.282

Dennis, B. R., & Kemp, W. P. (2016). How hives collapse: Allee effects, ecological resilience, and the honey bee. *PLOS ONE, 11*(2), e0150055. https://doi.org/10.1371/journal.pone.0150055

DePalma, R. A., Smit, J., Burnham, D. A., Kuiper, K. F., Manning, P. L., Oleinik, A. E., Larson, P. L., Maurrasse, F. J., Vellekoop, J., Richards, M. A., Gurche, L., & Álvarez, W. (2019). A seismically induced onshore surge deposit at the KPg boundary, North Dakota. *Proceedings of the National Academy of Sciences of the United States of America, 116*(17), 8190–8199. https://doi.org/10.1073/pnas.1817407116

Department of the Interior. (1973). *Endangered Species Act of 1973 as amended.* US Fish and Wildlife Service.

Deutz, A., et al., Paulson Institute, Nature Conservancy, & Cornell Atkinson Center for Sustainability. (2021, November 11). *Financing Nature: Closing the Global Biodiversity Financing Gap.* Paulson Institute. https://www.paulson institute.org/conservation/financing-nature-report

De Vos, J. M., Joppa, L., Gittleman, J. L., Stephens, P. R., & Pimm, S. L. (2014). Estimating the normal background rate of species extinction. *Conservation Biology, 29*(2), 452–462. https://doi.org/10.1111/cobi.12380

De Waal, F. B., & Lanting, F. 2023. *Bonobo: The Forgotten Ape.* University of California Press.

Diamond, J. M. (1982). Rediscovery of the Yellow-Fronted Gardener Bowerbird. *Science, 216*(4544), 431–434. https://doi.org/10.1126/science.216.4544.431

Diamond, J. M. (1972). *Avifauna of the Eastern Highlands of New Guinea.* Nuttall Ornithological Club.

Diamond, J. (1985). In quest of the wild and weird. *Discover, 6*(3): 34–42.

Diamond, J. (2001). *Guns, Germs and Steel: The Fates of Human Societies.* W. W. Norton.

Diamond, J. M. (2001). Dammed experiments!, *Science, 294*(5548), 1847–1848. https://doi.org/10.1126/science.1067012

Diamond, J. M. (2005). *Collapse: How Societies Choose to Fail or Succeed.* Viking.

Diamond, J. M. (2011). *Collapse: How Societies Choose to Fail or Succeed: Revised Edition.* Penguin.

Díaz, S., Settele, J., Brondízio, E., Ngo, H., Guèze, M., Agard, J., Arneth, A., Balvanera, P., Brauman, K., & Butchart, S. (2019). *The Global Assessment Report on Biodiversity and Ecosystem Services: Summary for Policy Makers.* Intergovernmental Science-Policy Platform on Biodiversity and Ecosystem Services.

Dibartolomeis, M. J., Kegley, S. E., Mineau, P., Radford, R., & Klein, K. (2019). An assessment of acute insecticide toxicity loading (AITL) of chemical pesticides used on agricultural land in the United States. *PLOS ONE, 14*(8), e0220029. https://doi.org/10.1371/journal.pone.0220029

Dickman, C. R. (1996). *Overview of the Impact of Feral Cats on Australian Native Fauna.* Australian Nature Conservation Agency.

Dickson, P., & Adams, W. M. (2009). Science and uncertainty in South Africa's elephant culling debate. *Environment and Planning C-Government and Policy, 27*(1), 110–123. https://doi.org/10.1068/c0792j

Diffendorfer, J. E., Thogmartin, W. E., Drum, R. G., & Schultz, C. B. (2020). Editorial: North American monarch butterfly ecology and conservation. *Frontiers in Ecology and Evolution, 8.* https://doi.org/10.3389/fevo.2020.576281

Di Minin, E., & Toivonen, T. (2015). Global protected area expansion: Creating more than paper parks. *BioScience, 65*(7), 637–638. https://doi.org/10.1093/biosci/biv064

Dirzo, R. 1994. *Mexican Diversity of Flora.* CEMEX and Agrupación Sierra Madre.

Dirzo, R., Ceballos, G., & Ehrlich, P. R. (2022). Circling the drain: The extinction crisis and the future of humanity. *Philosophical Transactions of the Royal Society B, 377*(1857). https://doi.org/10.1098/rstb.2021.0378

Dirzo, R., & Miranda, Á. (1990). Contemporary neotropical defaunation and forest structure, function, and diversity—A sequel to John Terborgh. *Conservation Biology, 4*(4), 444–447. https://doi.org/10.1111/j.1523-1739.1990.tb00320.x

Dirzo, R., & Raven, P. H. (2003). Global state of biodiversity and loss. *Annual Review of Environment and Resources, 28*(1), 137–167. https://doi.org/10.1146/annurev.energy.28.050302.105532

Dirzo, R., Young, H. S., Galetti, M., Ceballos, G., Isaac, N. J. B., & Collen, B. (2014). Defaunation in the Anthropocene. *Science, 345*(6195), 401–406. https://doi.org/10.1126/science.1251817

Di Silveri, R. (2023). *The Return of the Bison.* Mountaineer Books.

Dodd, M. S., Papineau, D., Grenne, T., Slack, J. F., Rittner, M., Pirajno, F., O'Neil, J., & Little, C. T. S. (2017). Evidence for early life in Earth's oldest hydrothermal vent precipitates. *Nature, 543*(7643), 60–64. https://doi.org/10.1038/nature21377

Doherty, T. S., Glen, A. S., Nimmo, D. G., Ritchie, E. G., & Dickman, C. R. (2016). Invasive predators and global biodiversity loss. *Proceedings of the National Academy of Sciences of the United States of America, 113*(40), 11261–11265. https://doi.org/10.1073/pnas.1602480113

Druce, H., Mackey, R. L., & Slotow, R. (2011). How immunocontraception can contribute to elephant management in small, enclosed reserves: Munyawana Population as a case study. *PLOS ONE, 6*(12), e27952. https://doi.org/10.1371/journal.pone.0027952

Duda, T. F., Kohn, A. J., & Matheny, A. M. (2009). Cryptic species differentiated in *Conus ebraeus*, a widespread tropical marine gastropod. *The Biological Bulletin, 217*(3), 292–305. https://doi.org/10.1086/bblv217n3p292

Dudgeon, D., Arthington, A., Gessner, M. O., Kawabata, Z., Knowler, D., Lévêque, C., Naiman, R. J., Prieur-Richard, A., Soto, D., Stiassny, M. L. J., & Sullivan, C. A. (2006). Freshwater biodiversity: importance, threats, status and conservation challenges. *Biological Reviews, 81*(2), 163–182. https://doi.org/10.1017/s1464793105006950

Dutertre, S., & Lewis, R. J. (2013). Cone snail biology, bioprospecting and conservation. In *HAL (Le Centre pour la Communication Scientifique Directe)*. https://hal.archives-ouvertes.fr/hal-02306901

Dyer, J. E. (1994). Land use pattern, forest migration, and global warming. *Landscape and Urban Planning, 29*(2-3), 77–83. https://doi.org/10.1016/0169-2046(94)90019-1

Edwards, C. J., Suchard, M. A., Lemey, P., Welch, J. J., Barnes, I., Fulton, T. L., Barnett, R., O'Connell, T. C., Coxon, P., Monaghan, N. T., Valdiosera, C., Lorenzen, E. D., Willerslev, E., Baryshnikov, G., Rambaut, A., Thomas, M. G., Bradley, D. G., & Shapiro, B. (2011). Ancient hybridization and an Irish origin for the modern polar bear matriline. *Current Biology, 21*(15), 1251–1258. https://doi.org/10.1016/j.cub.2011.05.058

Efferth, T., Banerjee, M., Paul, N. W., Abdelfatah, S., Arend, J., Elhassan, G., Hamdoun, S., Hamm, R., Hong, C., Kadioglu, O., Naß, J., Ochwang'i, D. O., Ooko, E., Özenver, N., Saeed, M. E., Schneider, M., Seo, E. J., Wu, C. F., Yan, G., . . . Titinchi, S. J. (2016). Biopiracy of natural products and good bioprospecting practice. *Phytomedicine, 23*(2), 166–173. https://doi.org/10.1016/j.phymed.2015.12.006

Eggert, L. S., Rasner, C. A., & Woodruff, D. S. (2002). The evolution and phylogeography of the African elephant inferred from mitochondrial DNA sequence and nuclear microsatellite markers. *Proceedings: Biological Sciences, 269*(1504), 1993–2006. http://www.jstor.org/stable/3558824

Ehrlich, A. H., & Ehrlich, R. P. (2014, December 30). Collapse of avian biodiversity in the Pacific. *Millennium Alliance for Humanity and the Biosphere.* https://mahb.stanford.edu/wp-content/uploads/2014/12/MAHBBlog _CollapseOfAvianBiodiversity_PEhrlichAEhrlich_Dec2014.pdf

Ehrlich, P. R. (1958). The integumental anatomy of the monarch butterfly *Danaus plexippus* L. (Lepidoptera: Danaiidae). *The University of Kansas Science Bulletin,* 38:1315–1349.

Ehrlich, P. R. (1961). Intrinsic barriers to dispersal in checkerspot butterfly. *Science,* 134:108–109.

Ehrlich, P. R. (1995). The scale of the human enterprise and biodiversity loss. In *Extinction Rates* (pp. 214–226). https://doi.org/10.1093/oso/9780198548294.003 .0014

Ehrlich, P. R. (2023). *Life: A Journey Through Science and Politics.* Yale University Press.

Ehrlich, P. R., Breedlove, D. E., Brussard, P. F., & Sharp, M. (1972). Weather and the "regulation" of subalpine populations. *Ecology,* 53(2), 243–247. https://doi .org/10.2307/1934077

Ehrlich, P. R., & Ehrlich, A. H. (1981). *Extinction: The Causes and Consequences of the Disappearance of Species.* Random House.

Ehrlich, P. R., & Ehrlich, A. H. (2013). Can a collapse of global civilization be avoided? *Proceedings of the Royal Society B: Biological Sciences,* 280(1754), 20122845. https://doi.org/10.1098/rspb.2012.2845

Ehrlich, P. R., & Ehrlich, A. H. (2022). Returning to "normal"? Evolutionary roots of the human prospect. *BioScience,* 72(8), 778–788. https://doi.org/10.1093 /biosci/biac044

Ehrlich, P. R., Ehrlich, A. H., & Holdren, J. P. (1977). *Ecoscience: Population, Resources, Environment.* W. H. Freeman and Co.

Ehrlich, P. R., & Hanski, I., eds. (2004). *On the Wings of Checkerspots: A Model System for Population Biology.* Oxford University Press.

Ehrlich, P. R., & Harte, J. (2018). Pessimism on the food front. *Sustainability,* 10(4), 1120. https://doi.org/10.3390/su10041120

Ehrlich, P. R., & Holdren, J. P. (1971). Impact of population growth. *Science,* 171(3977), 1212–1217. https://doi.org/10.1126/science.171.3977.1212

Ehrlich, P. R., Kareiva, P., & Daily, G. C. (2012). Securing natural capital and expanding equity to rescale civilization. *Nature,* 486(7401), 68–73. https://doi .org/10.1038/nature11157

Ehrlich, P. R., & Mooney, H. A. (1983). Extinction, substitution, and ecosystem services. *BioScience,* 33(4), 248–254. https://doi.org/10.2307/1309037

Ehrlich, P. R., & Raven, P. H. (1964). Butterflies and plants: A study in coevolution. *Evolution, 18*(4), 586. https://doi.org/10.2307/2406212

Eilers, E. J., Kremen, C., Greenleaf, S. S., Garber, A. K., & Klein, A. (2011). Contribution of pollinator-mediated crops to nutrients in the human food supply. *PLOS ONE, 6*(6), e21363. https://doi.org/10.1371/journal.pone.0021363

Eilperin, J. (2019, May 10). How the West Virginia coal industry changed federal endangered species policy. *West Virginia Gazette Mail.* https://www.wvgazettemail.com/how-the-west-virginia-coal-industry-changed-federal-endangered-species-policy/image_3bfccef5-74ee-5121-bfe3-00e4cddae9c2.html

Eilperin, J. (2023, April 8). Trump administration working toward renewed drilling in Arctic National Wildlife Refuge. *Washington Post.* https://www.washingtonpost.com/politics/trump-administration-working-toward-renewed-drilling-in-arctic-national-wildlife-refuge/2017/09/15/bfa5765e-97ea-11e7-87fc-c3f7ee4035c9_story.html

Erwin, D. H. (1998). The end and the beginning: Recoveries from mass extinctions. *Trends in Ecology and Evolution, 13*(9), 344–349. https://doi.org/10.1016/s0169-5347(98)01436-0

Erwin, T. L. (1982). Tropical forests: Their richness in coleoptera and other arthropod species. *Coleopterists Bulletin, 36*(1), 74–75. http://entomology.si.edu/StaffPages/Erwin/T's%20updated%20pub%20PDFs%20Jan2014/065_1982_TropicalForests_30MILLION.pdf

Espinosa, C. (2013). The riddle of leaving the oil in the soil—Ecuador's Yasuní-ITT project from a discourse perspective. *Forest Policy and Economics, 36,* 27–36. https://doi.org/10.1016/j.forpol.2012.07.012

Esterman, I. (2023, May 9). Death of last female Yangtze softshell turtle signals end for 'god' turtle. *Mongabay Environmental News.* https://news.mongabay.com/2023/05/death-of-last-female-yangtze-softshell-turtle-signals-end-for-god-turtle

Estes, J. A. (2006). *Whales, Whaling, and Ocean Ecosystems.* University of California Press.

Estes, J. A., Burdin, A. M., & Doak, D. F. (2015). Sea otters, kelp forests, and the extinction of Steller's sea cow. *Proceedings of the National Academy of Sciences of the United States of America, 113*(4), 880–885. https://doi.org/10.1073/pnas.1502552112

Estrada, A., Garber, P. A., Rylands, A. B., Roos, C., Fernández-Duque, E., Di Fiore, A., Nekaris, K. a. I., Nijman, V., Heymann, E. W., Lambert, J. E., Rovero, F., Barelli, C., Setchell, J. M., Gillespie, T. R., Mittermeier, R. A., Arregoitia, L. D. V., De Guinea, M., Gouveia, S. F., Dobrovolski, R., . . . Li, B. (2017).

Impending extinction crisis of the world's primates: Why primates matter. *Science Advances, 3*(1). https://doi.org/10.1126/sciadv.1600946

Evrensel, A., & Ceylan, M. E. (2015). The gut-brain axis: The missing link in depression. *Clinical Psychopharmacology and Neuroscience: The Official Scientific Journal of the Korean College of Neuropsychopharmacology, 13*(3), 239–244. https://doi.org/10.9758/cpn.2015.13.3.239

Fa, J. E., Peres, C. A., & Meeuwig, J. J. (2002). Bushmeat exploitation in tropical forests: An intercontinental comparison. *Conservation Biology, 16*(1), 232–237. https://doi.org/10.1046/j.1523-1739.2002.00275.x

Facchinelli, F., Pappalardo, S. E., Codato, D., Diantini, A., Della Fera, G., Crescini, E., & Marchi, D. M. (2019). Unburnable and unleakable carbon in Western Amazon: Using VIIRS Nightfire data to map gas flaring and policy compliance in the Yasuní Biosphere Reserve. *Sustainability, 12*(1), 58. https://doi.org/10.3390/su12010058

Fanning, J. C., Tyler, M. J., & Shearman, D. J. C. (1982). Converting a stomach to a uterus: The microscopic structure of the stomach of the Gastric brooding frog *Rheobatrachus silus. Gastroenterology, 82*(1), 62–70. https://doi.org/10.1016/0016-5085(82)90124-x

FAO (Food and Agriculture Organization of the United Nations). (2016). *State of the World's Forests 2016. Forests and Agriculture: Land-Use Challenges and Opportunities.* Food and Agriculture Organization of the United Nations.

FAO (Food and Agriculture Organization of the United Nations). (2019). *State of World's Biodiversity for Food and Agriculture.* Food and Agriculture Organization of the United Nations.

Farhadinia, M. S., Maheshwari, A., Nawaz, M. A., Ambarlı, H., Gritsina, M. A., Koshkin, M. A., Rosen, T., Hinsley, A., & Macdonald, D. W. (2019). Belt and Road Initiative may create new supplies for illegal wildlife trade in large carnivores. *Nature Ecology and Evolution, 3*(9), 1267–1268. https://doi.org/10.1038/s41559-019-0963-6

Fattorini, S. (2011). Insect extinction by urbanization: A long term study in Rome. *Biological Conservation, 144*(1), 370–375. https://doi.org/10.1016/j.biocon.2010.09.014

Festa-Bianchet, M. (2016). When does selective hunting select, how can we tell, and what should we do about it? *Mammal Review, 47*(1), 76–81. https://doi.org/10.1111/mam.12078

Field, C. B., Behrenfeld, M. J., Randerson, J. T., & Falkowski, P. G. (1998). Primary production of the biosphere: Integrating terrestrial and oceanic components. *Science, 281*(5374), 237–240. https://doi.org/10.1126/science.281.5374.237

Fierer, N. (2019). Earthworms' place on Earth. *Science, 366*(6464), 425–426. https://doi.org/10.1126/science.aaz5670

Fierer, N., Ladau, J., Clemente, J. C., Leff, J. W., Owens, S. M., Pollard, K. S., Knight, R., Gilbert, J. A., & McCulley, R. L. (2013). Reconstructing the microbial diversity and function of pre-agricultural tallgrass prairie soils in the United States. *Science, 342*(6158), 621–624. https://doi.org/10.1126/science.1243768

Fillios, M., Crowther, M. S., & Letnic, M. (2012). The impact of the dingo on the thylacine in Holocene Australia. *World Archaeology, 44*(1), 118–134. https://doi.org/10.1080/00438243.2012.646112

Finlayson, C. (2018). Ducks unlimited (DU). In Finlayson, C. M., Everard, M., Irvine, K., McInnes, R. J., Middleton, B. A., Van Dam, A. A., & Davidson, N. C. (Eds.), *The Wetland Book I: Structure and Function, Management and Methods* (pp. 659–663). Springer.

Finn, C., Grattarola, F., & Pincheira-Donoso, D. (2023). More losers than winners: Investigating Anthropocene defaunation through the diversity of population trends. *Biological Reviews, 98*(5), 1732–1748. https://doi.org/10.1111/brv.12974

Foissner, W. (2006). Biogeography and dispersal of micro-organisms: A review emphasizing protists. *Acta Protozoologica, 45*(45), 111–136. http://www1.nencki.gov.pl/pdf/ap/ap921.pdf

Forister, M. L., Cousens, B., Harrison, J. G., Anderson, K., Thorne, J. H., Waetjen, D., Nice, C. C., De Parsia, M., Hladik, M. L., Meese, R. J., Van Vliet, H., & Shapiro, A. M. (2016). Increasing neonicotinoid use and the declining butterfly fauna of lowland California. *Biology Letters, 12*(8), 20160475. https://doi.org/10.1098/rsbl.2016.0475

Forister, M. L., Halsch, C. A., Nice, C. C., Fordyce, J. A., Dilts, T. E., Oliver, J. C., Prudic, K. L., Shapiro, A. M., Wilson, J. K., & Glassberg, J. (2021). Fewer butterflies seen by community scientists across the warming and drying landscapes of the American West. *Science, 371*(6533), 1042–1045. https://doi.org/10.1126/science.abe5585

Forister, M. L., Pelton, E., & Black, S. (2019). Declines in insect abundance and diversity: We know enough to act now. *Conservation Science and Practice, 1*(8). https://doi.org/10.1111/csp2.80

Fortin, D., Beyer, H. L., Boyce, M. S., Smith, D. W., Duchesne, T., & Mao, J. S. (2005). Wolves influence elk movements: Behavior shapes a trophic cascade in Yellowstone National Park. *Ecology, 86*(5), 1320–1330. http://www.jstor.org/stable/3450894

Fossey, D. (1983). *Gorillas in the Mist.* Houghton Mifflin.

Foster, J. A., & Neufeld, K. M. (2013). Gut-brain axis: How the microbiome influences anxiety and depression. *Trends in Neurosciences, 36*(5), 305-312. https://doi.org/10.1016/j.tins.2013.01.005

Fox, R. (2012). The decline of moths in Great Britain: A review of possible causes. *Insect Conservation and Diversity, 6*(1), 5-19. https://doi.org/10.1111/j.1752-4598.2012.00186.x

Franklin, J. F. (1993). Preserving biodiversity: Species, ecosystems, or landscapes? *Ecological Applications, 3*(2), 202-205. https://doi.org/10.2307/1941820.

French, H. (2014). *China's Second Continent: How a Million Migrants Are Building a New Empire in Africa.* Knopf Editorial.

Friedman, W. E. (2009). The meaning of Darwin's "abominable mystery." *American Journal of Botany, 96*(1), 5-21. https://doi.org/10.3732/ajb.0800150

Fuller, E. (1987). *Extinct Birds.* Facts on File.

Fuller, E. (1999). *The Great Auk.* Bunker Hill Publishing.

Fuller, E. (2000). *Extinct Birds.* Oxford University Press.

Fuller, E. (2001). *Extinct Birds.* Comstock Publishing.

Fulton, G. R. (2017). The Bramble Cay melomys: The first mammalian extinction due to human-induced climate change. *Pacific Conservation Biology, 23*(1), 1. https://doi.org/10.1071/pcv23n1_ed

Furuichi, T. (2011). Female contributions to the peaceful nature of bonobo society. *Evolutionary Anthropology, 20*(4), 131-142. https://doi.org/10.1002/evan.20308

Futuyma, D. J. (1986). *Evolutionary biology.* Sinauer Associates.

Galetti, M., & Dirzo, R. (2013). Ecological and evolutionary consequences of living in a defaunated world. *Biological Conservation, 163*, 1-6. https://doi.org/10.1016/j.biocon.2013.04.020

Garibaldi, L. A., Aizen, M. A., Klein, A., Cunningham, S. A., & Harder, L. D. (2011). Global growth and stability of agricultural yield decrease with pollinator dependence. *Proceedings of the National Academy of Sciences of the United States of America, 108*(14), 5909-5914. https://doi.org/10.1073/pnas.1012431108

Gaston, K. J., Cox, D. T., Canavelli, S. B., García, D., Hughes, B., Maas, B., Martínez, D., Ogada, D., & Inger, R. (2018). Population abundance and ecosystem service provision: The case of birds. *BioScience, 68*(4), 264-272. https://doi.org/10.1093/biosci/biy005

Gaston, K. J., & Fuller, R. A. (2008). Commonness, population depletion and conservation biology. *Trends in Ecology and Evolution, 23*(1), 14-19. https://doi.org/10.1016/j.tree.2007.11.001

Gelabert, P., Sandoval-Velasco, M., Serres, A., De Manuel, M., Renom, P., Margaryan, A., Stiller, J., De-Dios, T., Fang, Q., Feng, S., Mañosa, S., Pacheco, G., Ferrando-Bernal, M., Shi, G., Hao, F., Chen, X., Petersen, B., Olsen, R. A., Navarro, A., . . . Lalueza-Fox, C. (2020). Evolutionary history, genomic adaptation to toxic diet, and extinction of the Carolina parakeet. *Current Biology, 30*(1), 108–114.e5. https://doi.org/10.1016/j.cub.2019.10.066

Gepts, P. (2006). Plant genetic resources conservation and utilization: The accomplishments and future of a societal insurance policy. *Crop Science, 46*(5), 2278–2292. https://doi.org/10.2135/cropsci2006.03.0169gas

Ghazoul, J. (2015, January 6). Earthworms under threat. *Mongabay Environmental News.* https://news.mongabay.com/2015/01/earthworms-under-threat

Gill, J. L. (2013). Ecological impacts of the late Quaternary megaherbivore extinctions. *New Phytologist, 201*(4), 1163–1169. https://doi.org/10.1111/nph.12576

Gill, R. A. (1990). The value of honeybee pollination to society. In *VI International Symposium on Pollination, 288,* 62–68.

Gill, V. (2023). Sold for a song. *BBC News.* https://www.bbc.co.uk/news/resources/idt-sh/sold_for_a_song

Gillson, L., & Lindsay, K. (2003). Ivory and ecology—Changing perspectives on elephant management and the international trade in ivory. *Environmental Science & Policy, 6*(5), 411–419. https://doi.org/10.1016/s1462-9011(03)00078-9

Gleick, P. (2023). *The Three Ages of Water: Prehistoric Past, Imperiled Present, and a Hope for the Future.* Public Affairs.

Glidden, C. K., Nova, N., Kain, M. P., Lagerstrom, K. M., Skinner, E. B., Mandle, L., Sokolow, S. H., Plowright, R. K., Dirzo, R., De Leo, G. A., & Mordecai, E. A. (2021). Human-mediated impacts on biodiversity and the consequences for zoonotic disease spillover. *Current Biology, 31*(19), R1342–R1361. https://doi.org/10.1016/j.cub.2021.08.070

Goldfarb, B. (2016). Can captive breeding save Mexico's vaquita? *Science, 353*(6300), 633–634. https://doi.org/10.1126/science.353.6300.633

González, A. V., Estévez, L. M., Villeda, M. E. Á., & Ceballos, G. (2018). The extinction of the Catarina pupfish *Megupsilon aporus* and the implications for the conservation of freshwater fish in Mexico. *Oryx, 54*(2), 154–160. https://doi.org/10.1017/s003060531800056x

González-Suárez, M., Gómez, A., & Revilla, E. (2013). Which intrinsic traits predict vulnerability to extinction depends on the actual threatening processes. *Ecosphere, 4*(6), 1–16. https://doi.org/10.1890/es12-00380.1

Gonzalez-Voyer, A., González-Suárez, M., Vilà, C., & Revilla, E. (2016). Larger brain size indirectly increases vulnerability to extinction in mammals. *Evolution, 70*(6), 1364–1375. https://doi.org/10.1111/evo.12943 5

Gore, A. C., Chappell, V. A., Fenton, S. E., Flaws, J. A., Nadal, Á., Prins, G. S., Toppari, J., & Zoeller, R. T. (2015). EDC-2: The Endocrine Society's second scientific statement on endocrine-disrupting chemicals. *Endocrine Reviews, 36*(6), E1–E150. https://doi.org/10.1210/er.2015-1010

Goulson, D. (2019). The insect apocalypse, and why it matters. *Current Biology, 29*(19), R967–R971. https://doi.org/10.1016/j.cub.2019.06.069

Gowler, C. D., Leon, K. E., Hunter, M. D., & De Roode, J. C. (2015). Secondary defense chemicals in milkweed reduce parasite infection in monarch butterflies, *Danaus plexippus. Journal of Chemical Ecology, 41*(6), 520–523. https://doi.org/10.1007/s10886-015-0586-6

Green, J. J., & Elisseeff, J. H. (2016). Mimicking biological functionality with polymers for biomedical applications. *Nature, 540*(7633), 386–394. https://doi.org/10.1038/nature21005

Grenham, S., Clarke, G., Cryan, J. F., & Dinan, T. G. (2011). Brain–gut–microbe communication in health and disease. *Frontiers in Physiology, 2.* https://doi.org/10.3389/fphys.2011.00094

Grey, M. J., Clarke, M. F., & Loyn, R. (1998). Influence of the Noisy Miner *Manorina melanocephala* on avian diversity and abundance in remnant Grey Box woodland. *Pacific Conservation Biology, 4*(1), 55. https://doi.org/10.1071/pc980055

Gruber, T., & Norscia, I. (2016). A comparison between bonobos and chimpanzees: A review and update. *Evolutionary Anthropology, 25*(5), 239–252. https://doi.org/10.1002/evan.21501

Gu, T., Wu, H., Yang, F., Gaubert, P., Heighton, S. P., Fu, Y., Lan, K., Luo, S., Zhang, H., Hu, J., & Yu, L. (2023). Genomic analysis reveals a cryptic pangolin species. *Proceedings of the National Academy of Sciences of the United States of America, 120*(40). https://doi.org/10.1073/pnas.2304096120

Guo, Y., Cao, Q., Hong, Z. S., Tan, Y., Chen, S. D., Jin, H., Tan, K. S., Wang, D. Y., & Yan, Y. (2020). The origin, transmission and clinical therapies on coronavirus disease 2019 (COVID-19) outbreak—An update on the status. *Military Medical Research, 7*(1). https://doi.org/10.1186/s40779-020-00240-0

Habel, J. C., Samways, M. J., & Schmitt, T. (2019a). Mitigating the precipitous decline of terrestrial European insects: Requirements for a new strategy. *Biodiversity and Conservation, 28*(6), 1343–1360. https://doi.org/10.1007/s10531-019-01741-8

Habel, J. C., Segerer, A. H., Ulrich, W., Torchyk, O., Weisser, W. W., & Schmitt, T. (2016). Butterfly community shifts over two centuries. *Conservation Biology, 30*(4), 754–762. https://doi.org/10.1111/cobi.12656

Habel, J. C., Ulrich, W., & Schmitt, T. (2019b). Butterflies in corridors: Quality matters for specialists. *Insect Conservation and Diversity*, 13(1), 91–98. https://doi.org/10.1111/icad.12386

Haberl, H., Erb, K., & Krausmann, F. (2014). Human appropriation of net primary production: Patterns, trends, and planetary boundaries. *Annual Review of Environment and Resources*, 39(1), 363–391. https://doi.org/10.1146/annurev-environ-121912-094620

Habibullah, M. S., Din, B. H., Tan, S., & Zahid, H. (2021). Impact of climate change on biodiversity loss: Global evidence. *Environmental Science and Pollution Research*, 29(1), 1073–1086. https://doi.org/10.1007/s11356-021-15702-8

Haddad, N. M. (2019). *The Last Butterflies: A Scientist's Quest to Save a Rare and Vanishing Creature*. Princeton University Press.

Haddad, N. M., & Baum, K. A. (1999). An experimental test of corridor effects on butterfly densities. *Ecological Applications*, 9(2), 623. https://doi.org/10.2307/2641149

Haddad, N. M., & Tewksbury, J. J. (2005). Low-quality habitat corridors as movement conduits for two butterfly species. *Ecological Applications*, 15(1), 250–257. https://doi.org/10.1890/03-5327

Haig, S. M., Murphy, S., Matthews, J., Arisméndi, I., & Safeeq, M. (2019). Climate-altered wetlands challenge waterbird use and migratory connectivity in arid landscapes. *Scientific Reports*, 9(1). https://doi.org/10.1038/s41598-019-41135-y

Hailer, F., Kutschera, V. E., Hallström, B. M., Klassert, D., Fain, S. R., Leonard, J. A., Árnason, Ú., & Janke, A. (2012). Nuclear genomic sequences reveal that polar bears are an old and distinct bear lineage. *Science*, 336(6079), 344–347. https://doi.org/10.1126/science.1216424

Halliday, T. (1978). *Vanishing Birds: Their Natural History and Conservation*. Holt, Rinehart and Winston.

Hallmann, C. A., Sorg, M., Jongejans, E., Siepel, H., Hofland, N., Schwan, H., Stenmans, W., Müller, A., Sumser, H., Hörren, T., Goulson, D., & De Kroon, H. (2017). More than 75 percent decline over 27 years in total flying insect biomass in protected areas. *PLOS ONE*, 12(10), e0185809. https://doi.org/10.1371/journal.pone.0185809

Hallock, P., Lidz, B. H., Cockey-Burkhard, E. M., & Donnelly, K. B. (2003). Foraminifera as bioindicators in coral reef assessment and monitoring: the FORAM Index. *Environmental Monitoring and Assessment*, 81, 221–238. https://doi.org/10.1023/A:1021337310386

Hamilton, A. J., Basset, Y., Benke, K. K., Grimbacher, P. S., Miller, S. E., Novotný, V., Samuelson, G. A., Stork, N. E., Weiblen, G. D., & Yen, J. D. L. (2010). Quantifying uncertainty in estimation of tropical arthropod species richness. *The American Naturalist, 176*(1), 90–95. https://doi.org/10.1086/652998

Hammarlund, E. U., Dahl, T. W., Harper, D. a. T., Bond, D. P., Nielsen, A. T., Bjerrum, C. J., Schovsbo, N. H., Schönlaub, H. P., Zalasiewicz, J., & Canfield, D. E. (2012). A sulfidic driver for the end-Ordovician mass extinction. *Earth and Planetary Science Letters, 331-332*, 128–139. https://doi.org/10.1016/j.epsl .2012.02.024

Hanes, S. (2006). *What Follows Genocide?* Pulitzer Center. https://pulitzercenter .org/stories/what-follows-genocide

Hannibal, M. E. (2016). *Citizen Scientist: Searching for Heroes and Hope in an Age of Extinction.* The Experiment.

Hansen, M. C., Potapov, P., Moore, R., Hancher, M., Turubanova, S., Tyukavina, A., Thau, D., Stehman, S. V., Goetz, S. J., Loveland, T. R., Kommareddy, A., Egorov, A., Chini, L. P., Justice, C. O., & Townshend, J. R. G. (2013). High-resolution global maps of 21st-century forest cover change. *Science, 342*(6160), 850–853. https://doi.org/10.1126/science.1244693

Hanski, I. (2002). In the midst of ecology, conservation, and competing interests in the society. *Annales Zoologici Fennici, 39*(3), 183–186. https://helda.helsinki .fi/bitstream/1975/7535/3/inthemid.pdf

Hanski, I., & Ovaskainen, O. (2002). Extinction debt at extinction threshold. *Conservation Biology, 16*(3), 666–673. https://doi.org/10.1046/j.1523-1739.2002 .00342.x

Harvey, A. L. (2008). Natural products in drug discovery. *Drug Discovery Today, 13*(19-20), 894–901. https://doi.org/10.1016/j.drudis.2008.07.004

Hastings, R. A., & Beattie, A. J. (2006). Stop the bullying in the corridors: Can including shrubs make your revegetation more Noisy Miner free? *Ecological Management & Restoration, 7*(2), 105–112. https://doi.org/10.1111/j.1442-8903 .2006.00264.x

Haverkort, A. J., Boonekamp, P., Hutten, R., Jacobsen, E., Lotz, L., Kessel, G., Visser, R., & Van Der Vossen, E. (2008). Societal costs of late blight in potato and prospects of durable resistance through cisgenic modification. *Potato Research, 51*(1), 47–57. https://doi.org/10.1007/s11540-008-9089-y

Hayes, M. A. (2013). Bats killed in large numbers at United States wind energy facilities. *BioScience, 63*(12), 975–979. https://doi.org/10.1525/bio.2013.63.12.10

Hayes, T. B., Collins, A., Lee, M., Mendoza, M. A., Noriega, N., Stuart, A., & Vonk, A. (2002). Hermaphroditic, demasculinized frogs after exposure to the herbicide atrazine at low ecologically relevant doses. *Proceedings of the National Academy of Sciences of the United States of America, 99*(8), 5476–5480. https://doi.org/10.1073/pnas.082121499

He, F., Zarfl, C., Bremerich, V., David, J. N. W., Hogan, Z., Kalinkat, G., Tockner, K., & Jähnig, S. C. (2019). The global decline of freshwater megafauna. *Global Change Biology, 25*(11), 3883–3892. https://doi.org/10.1111/gcb.14753

Hernández-Yáñez, H., Kim, S. Y., & Che-Castaldo, J. (2022). Demographic and life history traits explain patterns in species vulnerability to extinction. *PLOS ONE, 17*(2), e0263504. https://doi.org/10.1371/journal.pone.0263504

Herrmann, E., Hernández-Lloreda, M. V., Call, J., Hare, B., & Tomasello, M. (2009). The structure of individual differences in the cognitive abilities of children and chimpanzees. *Psychological Science, 21*(1), 102–110. https://doi.org/10.1177/0956797609356511

Hesman Saey, T. (2016, January 8). Body's bacteria don't outnumber human cells so much after all. *ScienceNews, 189.* https://www.sciencenews.org/article/bodys-bacteria-dont-outnumber-human-cells-so-much-after-all

Hess, G. (1996). Disease in metapopulation models: Implications for conservation. *Ecology, 77*(5), 1617–1632. https://doi.org/10.2307/2265556

Hickey, J. R., Nackoney, J., Nibbelink, N. P., Blake, S., Bonyenge, A., Coxe, S., Dupain, J., Emetshu, M., Furuichi, T., Grossmann, F., Guislain, P., Hart, J., Hashimoto, C., Ikembelo, B., Ilambu, O., Inogwabini, B., Liengola, I., Lokasola, A. L., Lushimba, A., . . . Kühl, H. S. (2013). Human proximity and habitat fragmentation are key drivers of the rangewide bonobo distribution. *Biodiversity and Conservation, 22*(13-14), 3085–3104. https://doi.org/10.1007/s10531-013-0572-7

Hill, K. D. (1993). The Endangered Species Act: What do we mean by species? *Boston College Environmental Affairs Law Review, 20*(2), 239. https://lawdigital commons.bc.edu/cgi/viewcontent.cgi?article=1424&context=ealr

Hoegh-Guldberg, O., Mumby, P. J., Hooten, A. J., Steneck, R. S., Greenfield, P. F., Gomez, E. D., Harvell, C. D., Sale, P. F., Edwards, A. J., Caldeira, K., Knowlton, N., Eakin, C. M., Iglesias-Prieto, R., Muthiga, N. A., Bradbury, R., Dubi, A. M., & Hatziolos, M. E. (2007). Coral reefs under rapid climate change and ocean acidification. *Science, 318*(5857), 1737–1742. https://doi.org/10.1126/science.1152509

Holdren, C. E., & Ehrlich, P. R. (1981). Long range dispersal in checkerspot butterflies: Transplant experiments with *Euphydryas gillettii. Oecologia, 50*(1), 125–129. https://doi.org/10.1007/bf00378805

Holdren, J. P., & Ehrlich, P. R. (1974). Human population and the global environment. *American Scientist, 62*(3), 282–292. https://pubmed.ncbi.nlm.nih.gov/4832978

Hooper, D. U., Adair, E. C., Cardinale, B. J., Byrnes, J. E. K., Hungate, B. A., Matulich, K. L., Gonzalez, A., Duffy, J. E., Gamfeldt, L., & O'Connor, M. I.

(2012). A global synthesis reveals biodiversity loss as a major driver of ecosystem change. *Nature, 486*(7401), 105–108. https://doi.org/10.1038/nature11118

Hornaday, W. T. (1889). *The extermination of the American bison*. Reprint. Createspace Independent Publishing Platform.

Horz, H., Rich, V. I., Avrahami, S., & Bohannan, B. J. M. (2005). Methane-oxidizing bacteria in a California upland grassland soil: Diversity and response to simulated global change. *Applied and Environmental Microbiology, 71*(5), 2642–2652. https://doi.org/10.1128/aem.71.5.2642-2652.2005

Hoyt, J. R., Kilpatrick, A. M., & Langwig, K. E. (2021). Ecology and impacts of white-nose syndrome on bats. *Nature Reviews Microbiology, 19*(3), 196–210. https://doi.org/10.1038/s41579-020-00493-5

Hug, L., Baker, B. J., Anantharaman, K., Brown, C. T., Probst, A. J., Castelle, C. J., Butterfield, C. N., Hernsdorf, A. W., Amano, Y., Ise, K., Suzuki, Y., Dudek, N., Relman, D. A., Finstad, K., Amundson, R., Thomas, B. C., & Banfield, J. F. (2016). A new view of the tree of life. *Nature Microbiology, 1*(5), 16048. https://doi.org/10.1038/nmicrobiol.2016.48

Hughes, J. B., Daily, G. C., & Ehrlich, P. R. (1997). Population diversity: Its extent and extinction. *Science, 278*(5338), 689–692. https://doi.org/10.1126/science.278.5338.689

Hughes, J. B., Daily, G. C., & Ehrlich, P. R. (2000). Conservation of insect diversity: A habitat approach. *Conservation Biology, 14*(6), 1788–1797. https://doi.org/10.1111/j.1523-1739.2000.99187.x

Hughes, K. (2021). *The World's Forgotten Fishes*. World Wildlife Fund (WWF).

Hughes, T. P., Graham, N. a. J., Jackson, J. B. C., Mumby, P. J., & Steneck, R. S. (2010). Rising to the challenge of sustaining coral reef resilience. *Trends in Ecology and Evolution, 25*(11), 633–642. https://doi.org/10.1016/j.tree.2010.07.011

Humphrey, C. (2023, April 28). An extremely rare, revered reptile is on the brink of extinction after last female dies. *Time*. https://time.com/6275373/giant-yangtze-softshell-turtle-female-dies

Humphreys, A. M., Govaerts, R., Ficinski, S. Z., Lughadha, E. N., & Vorontsova, M. S. (2019). Global dataset shows geography and life form predict modern plant extinction and rediscovery. *Nature Ecology and Evolution, 3*(7), 1043–1047. https://doi.org/10.1038/s41559-019-0906-2

Hunter, C. M., Caswell, H., Runge, M. C., Regehr, E. V., Amstrup, S. C., & Stirling, I. (2010). Climate change threatens polar bear populations: a stochastic demographic analysis. *Ecology, 91*(10), 2883–2897. https://doi.org/10.1890/09-1641.1

Hutchings, J. A., & Reynolds, J. D. (2004). Marine fish population collapses: Consequences for recovery and extinction risk. *BioScience, 54*(4), 297. https://doi.org/10.1641/0006-3568(2004)054

Hutchinson, G. E. (1959). Homage to Santa Rosalia or why are there so many kinds of animals? *The American Naturalist, 93*(870), 145–159. http://www.jstor.org/stable/2458768

Hvenegaard, G. T., Butler, J. R., & Krystofiak, D. K. (1989). Economic values of bird-watching at Point Pelee National Park, Canada. *Wildlife Society Bulletin, 17*, 526–531. http://www.jstor.org/stable/3782724

Inamine, H., Ellner, S. P., Springer, J. P., & Agrawal, A. A. (2016). Linking the continental migratory cycle of the monarch butterfly to understand its population decline. *Oikos, 125*(8), 1081–1091. https://doi.org/10.1111/oik.03196

iNaturalist. (2023). *Connect with Nature. Explore and Share Your Observations from the Natural World.* iNaturalist. https://www.inaturalist.org.

IPBES (Intergovernmental Science-Policy Platform on Biodiversity and Ecosystem Services). (2018). *Summary for policymakers of the thematic assessment report on land degradation and restoration of the Intergovernmental Science-Policy Platform on Biodiversity and Ecosystem Services.* IPBES Secretariat.

Isaac, N. J. B., & Cowlishaw, G. (2004). How species respond to multiple extinction threats. *Proceedings of the Royal Society B: Biological Sciences, 271*(1544), 1135–1141. https://doi.org/10.1098/rspb.2004.2724

IUCN (International Union for the Conservation of Nature). (2021). *Seventy Five Years of Experience. 75 Years of Vision and Impact.* IUCN. https://www.iucn.org/about/iucn-a-brief-history

IUCN (International Union for the Conservation of Nature). (2022). *The IUCN Red List of Threatened Species.* IUCN. https://www.iucnredlist.org/

IUCN (International Union for the Conservation of Nature). (2023, December 11). *Freshwater Fish Highlight Escalating Climate Impacts on Species - IUCN Red List.* Press release, IUCN.

Ivantsov, A. Y., Zakrevskaya, M. A., & Nagovitsyn, A. (2019). Morphology of integuments of the Precambrian animals, Proarticulata. *Invertebrate Zoology, 16*(1), 19–26. https://doi.org/10.15298/invertzool.16.1.03

Jablonski, D. (1986). Background and mass extinctions: The alternation of macroevolutionary regimes. *Science, 231*(4734), 129–133. https://doi.org/10.1126/science.231.4734.129

Jablonski, D. (2005). Mass extinctions and macroevolution. *Paleobiology, 31*(sp5), 192–210. https://doi.org/10.1666/0094-8373(2005)031

Jäger, K. (2017, January 30). Dwindling earthworm numbers. *dw.com*. https://
www.dw.com/en/earthworm-numbers-dwindle-threatening-soil-health/a
-37325923

Janzen, D. H. (2001). Latent extinction—The living dead. In *Encyclopedia of Biodiversity*, 4, 590–598. https://doi.org/10.1016/b0-12-226865-2/00173-5

Jaramillo-Legorreta, A., Bonilla-Garzón, A., Cardenas-Hinojosa, G., Nieto, E., Taylor, B. L., Mesnick, S., Henry, A., Sánchez-Alós, L., Van Sull, F., Booth, C., & Thomas, L. (2023). *Survey report for vaquita research*. Unpublished report. Ciudad de Mexico.

Jenniskens, P. 2019. Tunguska eyewitness accounts, injuries and casualties. *Icarus 327*, 4–18.

Johnson, K. A., Burbidge, A. A., & McKenzie, N. L. (1989). Australian Macropodoidea: Causes of decline and future research and management. In Grigg, G. C., Jarman, P. J., & Hume, I. D. (Eds.), *Kangaroos, Wallabies and Rat-Kangaroos* (pp. 641–657). Australian Mammal Society.

Jones, C. G., Lawton, J. H., & Shachak, M. (1994). Organisms as ecosystem engineers. *Oikos*, 69(3), 373. https://doi.org/10.2307/3545850

Jones, D. S., Martini, A. M., Fike, D. A., & Kaiho, K. (2017). A volcanic trigger for the Late Ordovician mass extinction? Mercury data from south China and Laurentia. *Geology*, 45(7), 631–634. https://doi.org/10.1130/g38940.1

Jørgensen, D. (2013). Reintroduction and de-extinction. *BioScience*, 63(9), 719–720. https://doi.org/10.1093/bioscience/63.9.719

Jouffray, J., Wedding, L. M., Norström, A. V., Donovan, M. K., Williams, G. J., Crowder, L. B., Erickson, A. S., Friedlander, A. M., Graham, N. a. J., Gove, J. M., Kappel, C. V., Kittinger, J. N., Lecky, J., Oleson, K. L., Selkoe, K. A., White, C., Williams, I. D., & Nyström, M. (2019). Parsing human and biophysical drivers of coral reef regimes. *Proceedings of the Royal Society B: Biological Sciences*, 286(1896), 20182544. https://doi.org/10.1098/rspb.2018.2544

Karesh, W. B., Dobson, A., Lloyd-Smith, J. O., Liu, J., Dixon, M., Bennett, M. J., Aldrich, S., Harrington, T., Formenty, P., Loh, E. H., Machalaba, C., Thomas, M. J., & Heymann, D. L. (2012). Ecology of zoonoses: natural and unnatural histories. *The Lancet*, 380(9857), 1936–1945. https://doi.org/10.1016/s0140 -6736(12)61678-x

Karp, D. S., Mendenhall, C. D., Sandi, R. F., Chaumont, N., Ehrlich, P. R., Hadly, E. A., & Daily, G. C. (2013). Forest bolsters bird abundance, pest control and coffee yield. *Ecology Letters*, 16(11), 1339–1347. https://doi.org/10.1111/ele.12173

Kawai, T., Faulkes, Z., & Scholtz, G. (2015). *Freshwater Crayfish: A Global Overview* (1st ed.). CRC Press. https://doi.org/10.1201/b18723

Keating, J. (2014, December 4). China's new execution rules could lead to an organ shortage. *Slate Magazine.* https://slate.com/news-and-politics/2014/12/china-says-it-will-stop-harvesting-organs-from-executed-prisoners.html

Kellogg, S. K., Fink, L. S., & Brower, L. P. (2003). Parasitism of native luna moths, *Actias luna* (L.)(Lepidoptera: Saturniidae) by the introduced *Compsilura concinnata* (Meigen)(Diptera: Tachinidae) in central Virginia, and their hyperparasitism by trigonalid wasps (Hymenoptera: Trigonalidae). *Environmental Entomology, 32*(5), 1019–1027. https://doi.org/10.1603/0046-225x-32.5.1019

Kelly, B. P., Whiteley, A. R., & Tallmon, D. A. (2010). The Arctic melting pot. *Nature, 468*(7326), 891. https://doi.org/10.1038/468891a

Kent, A. G., Dupont, C. L., Yooseph, S., & Martiny, A. C. (2016). Global biogeography of *Prochlorococcus* genome diversity in the surface ocean. *The ISME Journal, 10*(8), 1856–1865. https://doi.org/10.1038/ismej.2015.265

Kepler, C. B., Irvine, G. W., DeCapita, M. E., & Weinrich, J. (1996). The conservation management of Kirtland's Warbler *Dendroica kirtlandii. Bird Conservation International, 6*(1), 11–22. https://doi.org/10.1017/s0959270900001271

Kerley, G. I. H., & Shrader, A. M. (2007). Elephant contraception: Silver bullet or a potentially bitter pill? *South African Journal of Science, 103,* 181–182. http://www.scielo.org.za/pdf/sajs/v103n5-6/05.pdf

Kesler, K. K. (2019). *The Direct and Indirect Effects of Site Suitability on Eastern Monarch Butterfly Migratory Populations* [PhD thesis]. University of North Carolina at Greensboro. https://libres.uncg.edu/ir/uncg/f/Kesler_uncg_0154M_12768.pdf

Kidd, K. A., Paterson, M., Rennie, M. D., Podemski, C. L., Findlay, D. L., Blanchfield, P. J., & Liber, K. (2014). Direct and indirect responses of a freshwater food web to a potent synthetic oestrogen. *Philosophical Transactions of the Royal Society B, 369*(1656), 20130578. https://doi.org/10.1098/rstb.2013.0578

King, D., & Finch, D. M. (2013). *The Effects of Climate Change on Terrestrial Birds of North America.* US Department of Agriculture, Forest Service, Climate Change Resource Center.

King, M. C., & Wilson, A. C. (1975). Evolution at two levels in humans and chimpanzees. *Science, 188*(4184), 107–116. https://doi.org/10.1126/science.1090005

Kirby, J. S., Stattersfield, A. J., Butchart, S. H. M., Evans, M. I., Grimmett, R. F., Jones, V. R., O'Sullivan, J. B., Tucker, G., & Newton, I. (2008). Key conservation issues for migratory land- and waterbird species on the world's major flyways. *Bird Conservation International, 18*(S1), S49–S73. https://doi.org/10.1017/s0959270908000439

Kirchhoff, T. (2012). Pivotal cultural values of nature cannot be integrated into the ecosystem services framework. *Proceedings of the National Academy of Sciences of the United States of America, 109*(46). https://doi.org/10.1073/pnas.1212409109

Kirkpatrick, J. F., Lyda, R. O., & Frank, K. M. (2011). Contraceptive vaccines for wildlife: A review. *American Journal of Reproductive Immunology, 66*(1), 40–50. https://doi.org/10.1111/j.1600-0897.2011.01003.x

Kleijn, D., Winfree, R., Bartomeus, I., Carvalheiro, L. G., Henry, M., Isaacs, R., Klein, A., Kremen, C., M'Gonigle, L. K., Rader, R., Ricketts, T. H., Williams, N. M., Adamson, N. L., Ascher, J. S., Báldi, A., Batáry, P., Benjamin, F., Biesmeijer, J. C., Blitzer, E. J., . . . Potts, S. G. (2015). Delivery of crop pollination services is an insufficient argument for wild pollinator conservation. *Nature Communications, 6*(1). https://doi.org/10.1038/ncomms8414

Klein, A., Vaissière, B., Cane, J. H., Steffan-Dewenter, I., Cunningham, S. A., Kremen, C., & Tscharntke, T. (2006). Importance of pollinators in changing landscapes for world crops. *Proceedings of the Royal Society B: Biological Sciences, 274*(1608), 303–313. https://doi.org/10.1098/rspb.2006.3721

Klein, B. C. (1989). Effects of forest fragmentation on dung and carrion beetle communities in Central Amazonia. *Ecology, 70*(6), 1715–1725. https://doi.org/10.2307/1938106

Klem, D. (2009). Preventing bird—window collisions. *The Wilson Journal of Ornithology, 121*(2), 314–321. http://www.jstor.org/stable/20616902

Klok, C., Zorn, M. I., Koolhaas, J. E., Eijsackers, H., & Van Gestel, C. A. (2006). Does reproductive plasticity in *Lumbricus rubellus* improve the recovery of populations in frequently inundated river floodplains? *Soil Biology & Biochemistry, 38*(3), 611–618. https://doi.org/10.1016/j.soilbio.2005.06.013

Knapp, W. M., Frances, A., Noss, R. F., Naczi, R. F. C., Weakley, A. S., Gann, G. D., Baldwin, B. G., Miller, J. S., McIntyre, P. J., Mishler, B. D., Moore, G., Olmstead, R. G., Strong, A. W., Gluesenkamp, D., & Kennedy, K. (2020). Regional records improve data quality in determining plant extinction rates. *Nature Ecology and Evolution, 4*(4), 512–514. https://doi.org/10.1038/s41559-020-1146-1

Knoll, A. H., & Nowak, M. A. (2017). The timetable of evolution. *Science Advances, 3*(5). https://doi.org/10.1126/sciadv.1603076

Koch, H., Frickel, J., Valiadi, M., & Becks, L. (2014). Why rapid, adaptive evolution matters for community dynamics. *Frontiers in Ecology and Evolution, 2.* https://doi.org/10.3389/fevo.2014.00017

Kolding, J., Béné, C., & Bavinck, M. (2014). Small-scale fisheries: importance, vulnerability, and deficient knowledge. In *Governance of Marine Fisheries and*

Biodiversity Conservation: Interaction and Coevolution, 317–331. Wiley-Blackwell.

Kotchen, M. J., & Burger, N. (2007). Should we drill in the Arctic National Wildlife Refuge? An economic perspective. *Energy Policy, 35*(9), 4720–4729. https://doi.org/10.1016/j.enpol.2007.04.007

Kottasova, I. (2014, September 30). WWF: World has lost more than half its wildlife in 40 years. *CNN.* https://www.cnn.com/2014/09/30/business/wild-life-decline-wwf/index.html

Kouakou, C. Y., Boesch, C., & Kuehl, H. (2009). Estimating chimpanzee population size with nest counts: Validating methods in Taï National Park. *American Journal of Primatology, 71*(6), 447–457. https://doi.org/10.1002/ajp.20673

Kremen, C., & Miles, A. (2012). Ecosystem services in biologically diversified versus conventional farming systems: Benefits, externalities, and trade-offs. *Ecology and Society, 17*(4). https://doi.org/10.5751/es-05035-170440

Krishnaveni, K., Sudarshan, S., Vimaladevi, S., Vijayarahavan, V., & Alamelu, V. (2021). An impact of invasive alien species in aquatic ecosystem. *Biotica Research Today, 3:*1123–1126. https://www.biospub.com/index.php/biorestoday/article/download/1331/955

Kristensen, H. M., & McKinzie, M. G. (2015). Nuclear arsenals: Current developments, trends and capabilities. *International Review of the Red Cross, 97*(899), 563–599. https://doi.org/10.1017/s1816383116000308

Kroeber, T. 1961. *Ishi in Two Worlds.* University of California Press.

Kutt, A. S., & Woinarski, J. C. Z. (2007). The effects of grazing and fire on vegetation and the vertebrate assemblage in a tropical savanna woodland in north-eastern Australia. *Journal of Tropical Ecology, 23*(1), 95–106. https://doi.org/10.1017/s0266467406003579

Kutz, S., & Tomaselli, M. (2019). "Two-eyed seeing" supports wildlife health. *Science, 364*(6446), 1135–1137. https://doi.org/10.1126/science.aau6170

LaBastille, A. (1983). Drastic decline in Guatemala's giant pied-billed grebe population. *Environmental Conservation, 10*(4), 346–348. https://doi.org/10.1017/s0376892900013072

Lah, K., & Moya, A. (2016, May 23). Aquatic "cocaine": Fish bladders are latest Mexican smuggling commodity. *CNN.* http://cnn.it/2bBIoYH

Laidre, K. L., Stirling, I., Estes, J. A., Kochnev, A., & Roberts, J. J. (2018). Historical and potential future importance of large whales as food for polar bears. *Frontiers in Ecology and the Environment, 16*(9), 515–524. https://doi.org/10.1002/fee.1963

LaJeunesse, T. C., Smith, R. T., Finney, J. C., & Oxenford, H. A. (2009). Outbreak and persistence of opportunistic symbiotic dinoflagellates during the 2005

Caribbean mass coral 'bleaching' event. *Proceedings of the Royal Society B: Biological Sciences, 276*(1676), 4139–4148. https://doi.org/10.1098/rspb.2009 .1405

Landrigan, P. J., Collins, T. J., & Myers, J. P. (2020). Controlling toxic exposures. In Myers, S. & Frumkin, H. (Eds.), *Planetary Health. Protecting Health to Protect Ourselves* (pp. 359–386). Island Press.

Langpap, C., & Kerkvliet, J. (2012). Endangered species conservation on private land: Assessing the effectiveness of habitat conservation plans. *Journal of Environmental Economics and Management, 64*(1), 1–15. https://doi.org/10.1016/j .jeem.2012.02.002

Larkin, A. A., & Martiny, A. C. (2017). Microdiversity shapes the traits, niche space, and biogeography of microbial taxa. *Environmental Microbiology Reports, 9*(2), 55–70. https://doi.org/10.1111/1758-2229.12523

Larsen, B. B., Miller, E. C., Rhodes, M. K., & Wiens, J. J. (2017). Inordinate fondness multiplied and redistributed: The number of species on Earth and the new pie of life. *The Quarterly Review of Biology, 92*(3), 229–265. https://doi .org/10.1086/693564

Lazarus, S. (2019, December 26). Can tech save the kakapo, New Zealand's 'gorgeous, hilarious' parrot? Call to Earth. *CNN.* https://www.cnn.com/2019 /12/26/world/kakapo-conservation-scn-c2e-intl-hnk/index.html

Leach, W. R. (2013). *Butterfly People: An American Encounter with the Beauty of the World.* Random House.

Leather, S. R. (2017). "Ecological Armageddon"—More evidence for the drastic decline in insect numbers. *Annals of Applied Biology, 172*(1), 1–3. https://doi .org/10.1111/aab.12410

Lebreton, L., Slat, B., Ferrari, F., Sainte-Rose, B., Aitken, J., Marthouse, R., Hajbane, S., Cunsolo, S., Schwarz, A., Levivier, A., Noble, K., Debeljak, P., Maral, H., Schoeneich-Argent, R., Brambini, R., & Reisser, J. (2018). Evidence that the Great Pacific Garbage Patch is rapidly accumulating plastic. *Scientific Reports, 8*(1). https://doi.org/10.1038/s41598-018-22939-w

Legge, S., Murphy, B. P., McGregor, H., Woinarski, J. C. Z., Augusteyn, J., Ballard, G., Baseler, M., Buckmaster, T., Dickman, C. R., Doherty, T. S., Edwards, G., Eyre, T. J., Fancourt, B. A., Ferguson, D. J., Forsyth, D. M., Geary, W. L., Gentle, M., Gillespie, G. R., Greenwood, L., . . . Zewe, F. (2017). Enumerating a continental-scale threat: How many feral cats are in Australia? *Biological Conservation, 206*, 293–303. https://doi.org/10.1016/j .biocon.2016.11.032

Lemoine, N. P. (2015). Climate change may alter breeding ground distributions of Eastern Migratory Monarchs (*Danaus plexippus*) via range expansion of

Asclepias host plants. *PLOS ONE, 10*(2), e0118614. https://doi.org/10.1371
/journal.pone.0118614

Leroy, B., Paschetta, M., Canard, A., Bakkenes, M., Isaia, M., & Ysnel, F.
(2013). First assessment of effects of global change on threatened spiders:
Potential impacts on *Dolomedes plantarius* (*Clerck*) and its conservation
plans. *Biological Conservation, 161*, 155–163. https://doi.org/10.1016/j.biocon
.2013.03.022

Lewis, J. (1978). Game domestication for animal production in Kenya: Shade
behaviour and factors affecting the herding of eland, oryx, buffalo and zebu
cattle. *The Journal of Agricultural Science, 90*(3), 587–595. https://doi.org/10
.1017/s0021859600056124

Lewis, T. (2018, February, 18). Wings of desire: Why is an obsessive British
collector risking jail to kill rare butterflies? What's the appeal in this one-
sided pursuit? *Esquire*. https://www.esquire.com/uk/life/a16811279/butterfly
-killer-uk-phillip-cullen

Liem, D. S. (1973). A new genus of frog of the family *Leptodactylidae* from S. E.
Queensland, Australia. *Memoirs of the Queensland Museum, 16*, 459–470.
https://biostor.org/reference/153354

Ligon, B. L. (2004). Penicillin: Its discovery and early development. *Seminars in
Pediatric Infectious Diseases, 15*(1), 52–57.

Lindsey, E. W., & Altizer, S. (2008). Sex differences in immune defenses and
response to parasitism in monarch butterflies. *Evolutionary Ecology, 23*(4),
607–620. https://doi.org/10.1007/s10682-008-9258-0

Linnaeus, C. 1758. *Systema naturae*. Vol. 1. Laurentii Salvii.

Liu, J., Hull, V., Yang, W., Viña, A., Chen, X., Ouyand, Z., & Zhang, H. (2016).
Pandas and People: Coupling Human and Natural Systems for Sustainability.
Oxford University Press.

Locey, K. J., & Lennon, J. T. (2016). Scaling laws predict global microbial
diversity. *Proceedings of the National Academy of Sciences of the United States of
America, 113*(21), 5970–5975. https://doi.org/10.1073/pnas.1521291113

Locke, H., & Dearden, P. (2005). Rethinking protected area categories and the
new paradigm. *Environmental Conservation, 32*(1), 1–10. https://doi.org/10.1017
/s0376892905001852

Lockwood, J. A., Latchininsky, A. V., & Sergeev, M. G. (2012). *Grasshoppers and
Grassland Health: Managing Grasshopper Outbreaks without Risking
Environmental Disaster*. Springer Science & Business Media.

López-Bao, J. V. (2019). Protect giraffes from wildlife trade. *Science, 364*(6442),
744. https://doi.org/10.1126/science.aax4485

Losey, J. E., & Vaughan, M. (2006). The economic value of ecological services provided by insects. *BioScience, 56*(4), 311. https://doi.org/10.1641/0006 -3568(2006)56

Losos, J. B. (2019). *Deadly Cats Down Under.* American Association for the Advancement of Science.

Loss, S. R., Will, T., & Marra, P. P. (2013a). Corrigendum: The impact of free-ranging domestic cats on wildlife of the United States. *Nature Communications, 4,* 2691. https://doi.org/10.1038/ncomms3961

Loss, S. R., Will, T., & Marra, P. P. (2013b). The impact of free-ranging domestic cats on wildlife of the United States. *Nature Communications, 4,*1396.

Lötters, S., Plewnia, A., Catenazzi, A., Neam, K., Acosta-Galvis, A. R., Vela, Y. A., Allen, J. P., Segundo, J. O. A., De Lourdes Almendáriz Cabezas, A., Barboza, G. A., Alves-Silva, K. R., Anganoy-Criollo, M., Ortiz, E., Lojano, J. D. A., Arteaga, A., Ballestas, O., Moscoso, D. B., Barros-Castañeda, J. D., Batista, A., . . . La Marca, E. (2023). Ongoing harlequin toad declines suggest the amphibian extinction crisis is still an emergency. *Communications Earth & Environment, 4*(1). https://doi.org/10.1038/s43247-023-01069-w

Low, T. (2013). Considering corridors. *Wildlife Australia, 50,* 4.

Lowery, C. M., & Fraass, A. (2019). Morphospace expansion paces taxonomic diversification after end Cretaceous mass extinction. *Nature Ecology and Evolution, 3*(6), 900–904. https://doi.org/10.1038/s41559-019-0835-0

Lu, C., Warchol, K. M., & Callahan, R. A. (2014). Sub-lethal exposure to neonicotinoids impaired honey bees winterization before proceeding to colony collapse disorder. *Bulletin of Insectology, 67*(1), 125–130. http:// pesticidetruths.com/wp-content/uploads/2014/08/Reference-PCP-Bees-2014 -03-27-Chensheng-Lu-Lunatic-Report-On-Sub-Lethal-That-Impaired-Honey -Bees-Italy.pdf

Lughadha, E. N., Bachman, S. P., Leão, T., Forest, F., Halley, J. M., Moat, J., Acedo, C., Bacon, K. L., Brewer, R. F., Gâteblé, G., Gonçalves, S. C., Govaerts, R., Hollings-worth, P., Krisai-Greilhüber, I., De Lírio, E. J., Moore, P. G., Negrão, R., Onana, J. M., Rajaovelona, L., . . . Walker, B. E. (2020). Extinction risk and threats to plants and fungi. *Plants, People, Planet, 2*(5), 389–408. https://doi.org/10.1002 /ppp3.10146

Lundholm, C. (1997). DDE-induced eggshell thinning in birds: Effects of p,p′-DDE on the calcium and prostaglandin metabolism of the eggshell gland. *Comparative Biochemistry and Physiology Part C: Pharmacology, Toxicology and Endocrinology, 118*(2), 113–128. https://doi.org/10.1016/s0742-8413(97) 00105-9

Lyson, T. R., Miller, I. M., Bercovici, A., Weissenburger, K., Fuentes, A., Clyde, W. C., Hagadorn, J. W., Butrim, M. J., Johnson, K. R., Fleming, R. W., Barclay, R. S., Maccracken, S. A., Lloyd, B. D., Wilson, G. P., Krause, D. W., & Chester, S. G. B. (2019). Exceptional continental record of biotic recovery after the Cretaceous–Paleogene mass extinction. *Science, 366*(6468), 977–983. https://doi.org/10.1126/science.aay2268

Maceda-Veiga, A., Domínguez-Domínguez, O., Escribano-Alacid, J., & Lyons, J. D. (2014). The aquarium hobby: Can sinners become saints in freshwater fish conservation? *Fish and Fisheries, 17*(3), 860–874. https://doi.org/10.1111/faf.12097

Mäder, P., Fließbach, A., Dubois, D., Gunst, L., Fried, P. M., & Niggli, U. (2002). Soil fertility and biodiversity in organic farming. *Science, 296*(5573), 1694–1697. https://doi.org/10.1126/science.1071148.

Mahon, M. B., Campbell, K. U., & Crist, T. O. (2019). Experimental effects of white-tailed deer and an invasive shrub on forest ant communities. *Oecologia, 191*(3), 633–644. https://doi.org/10.1007/s00442-019-04516-8

Mahon, M. B., & Crist, T. O. (2019). Invasive earthworm and soil litter response to the experimental removal of white-tailed deer and an invasive shrub. *Ecology, 100*(5). https://doi.org/10.1002/ecy.2688

Maisels, F., Strindberg, S., Blake, S., Wittemyer, G., Hart, J., Williamson, E. A., Aba'a, R., Abitsi, G., Ambahe, R. D., Amsini, F., Bakabana, P. C., Hicks, T. C., Bayogo, R. E., Bechem, M., Beyers, R., Bezangoye, A. N., Boundja, P., Bout, N., Akou, M. E., . . . Warren, Y. (2013). Devastating decline of forest elephants in Central Africa. *PLOS ONE, 8*(3), e59469. https://doi.org/10.1371/journal.pone.0059469

Mak, G. J. K., & Song, W. (2018). Transnational norms and governing illegal wildlife trade in China and Japan: Elephant ivory and related products under CITES. *Cambridge Review of International Affairs, 31*, 373–391.

Malnick, S., Fisher, D., Somin, M., & Neuman, M. G. (2021). Treating the metabolic syndrome by fecal transplantation—Current status. *Biology, 10*(5), 447. https://doi.org/10.3390/biology10050447

Mandrak, N. E., Pratt, T. C., & Reid, S. M. (2014). *Evaluating the Current Status of Deepwater Ciscoes (Coregonus spp.) in the Canadian Waters of Lake Huron, 2002–2012, with Emphasis on Shortjaw Cisco (C. zenithicus).* Canadian Science Advisory Secretariat.

Mariner, J. C., House, J. A., Mebus, C. A., Sollod, A. E., Chibeu, D., Jones, B., Roeder, P. L., Admassu, B., & Van't Klooster, G. G. M. (2012). Rinderpest eradication: Appropriate technology and social innovations. *Science, 337*(6100), 1309–1312. https://doi.org/10.1126/science.1223805

Marshall, B. M., Strine, C. T., & Hughes, A. C. (2020). Thousands of reptile species threatened by under-regulated global trade. *Nature Communications*, 11(1). https://doi.org/10.1038/s41467-020-18523-4

Martel, A., Blooi, M., Adriaensen, C., Van Rooij, P., Beukema, W., Fisher, M. C., Farrer, R. A., Schmidt, B. R., Tobler, U., Goka, K., Lips, K. R., Muletz, C. R., Zamudio, K. R., Bosch, J., Lötters, S., Wombwell, E., Garner, T. W. J., Cunningham, A. A., Sluijs, A. S. D., . . . Pasmans, F. (2014). Recent introduction of a chytrid fungus endangers Western Palearctic salamanders. *Science*, 346(6209), 630–631. https://doi.org/10.1126/science.1258268

Martel, A., Sluijs, A. S. D., Blooi, M., Bert, W., Ducatelle, R., Fisher, M. C., Woeltjes, A. G. W., Bosman, W., Chiers, K., Bossuyt, F., & Pasmans, F. (2013). *Batrachochytrium salamandrivorans* sp. nov. causes lethal chytridiomycosis in amphibians. *Proceedings of the National Academy of Sciences of the United States of America*, 110(38), 15325–15329. https://doi.org/10.1073/pnas.1307356110

Martin, R. B. (2000). When CITES works and when it does not. In *Endangered Species, Threatened Convention: The Past, Present and Future of CITES, the Convention on International Trade in Endangered Species of Wild Fauna and Flora*. Routledge.

Martin, P. & Klein, R. G., eds. (1984). *Quaternary Extinctions: A Prehistoric Revolution*. University of Arizona Press.

Martínez, E., & Ramos, C. H. (1989). *Lacandoniaceae* (Triuridales): Una nueva familia de México. *Annals Missouri Botanical Garden*, 76, 128–135.

Martinez-Alier, J., Nnimmo, B., & Bond, P. (2013, August 17). Yasuni ITT is dead. Blame President Correa. *Environmental Justice Organization, Liabilities and Trade*. http://www.ejolt.org/2013/08/yasuni-itt-is-dead-blame-president-correa

Martínez-Estévez, L., Balvanera, P., Pacheco, J., & Ceballos, G. (2013). Prairie dog decline reduces the supply of ecosystem services and leads to desertification of semiarid grasslands. *PLOS ONE*, 8(10), e75229. https://doi.org/10.1371/journal.pone.0075229

Martiny, J. B. H., Bohannan, B. J. M., Brown, J. H., Colwell, R. K., Fuhrman, J. A., Green, J. L., Horner-Devine, M. C., Kane, M. D., Krumins, J. A., Kuske, C. R., Morin, P. J., Naeem, S., Øvreås, L., Reysenbach, A., Smith, V. H., & Staley, J. T. (2006). Microbial biogeography: Putting microorganisms on the map. *Nature Reviews Microbiology*, 4(2), 102–112. https://doi.org/10.1038/nrmicro1341

Martiny, J. B. H., Jones, S., Lennon, J. T., & Martiny, A. C. (2015). Microbiomes in light of traits: A phylogenetic perspective. *Science*, 350(6261). https://doi.org/10.1126/science.aac9323

Mattsson, K., Johnson, E. V., Malmendal, A., Linse, S., Hansson, L., & Cedervall, T. (2017). Brain damage and behavioural disorders in fish induced by plastic nanoparticles delivered through the food chain. *Scientific Reports, 7*(1). https://doi.org/10.1038/s41598-017-10813-0

Matulich, K. L., & Martiny, J. B. H. (2015). Microbial composition alters the response of litter decomposition to environmental change. *Ecology, 96*(1), 154–163. https://doi.org/10.1890/14-0357.1

Maxwell, S., Fuller, R. A., Brooks, T. M., & Watson, J. (2016). Biodiversity: The ravages of guns, nets and bulldozers. *Nature, 536*(7615), 143–145. https://doi.org/10.1038/536143a

McCallum, H., & Dobson, A. (2002). Disease, habitat fragmentation and conservation. *Proceedings of the Royal Society B: Biological Sciences, 269*(1504), 2041–2049. https://doi.org/10.1098/rspb.2002.2079

McCarthy, M. 2016. *The Moth Snowstorm: Nature and Joy.* New York Review of Books.

McCauley, D. J., DeSalles, P. A., Young, H. S., Dunbar, R. B., Dirzo, R., Mills, M. M., & Micheli, F. (2012). From wing to wing: The persistence of long ecological interaction chains in less-disturbed ecosystems. *Scientific Reports, 2*(1). https://doi.org/10.1038/srep00409

McGuire, J. L., & Shipley, B. R. (2022). Dynamic priorities for conserving species. *Science, 376*(6597), 1048–1049. https://doi.org/10.1126/science.abq0788

McKelvey, K. S., Copeland, J. P., Schwartz, M. K., Littell, J. S., Aubry, K. B., Squires, J. R., Parks, S. A., Elsner, M. M., & Mauger, G. (2011). Climate change predicted to shift wolverine distributions, connectivity, and dispersal corridors. *Ecological Applications, 21*(8), 2882–2897. https://doi.org/10.1890/10-2206.1

McKittrick, E. (2019, March 1). The shells of wild sea butterflies are already dissolving. *Hakay Magazine.* https://hakaimagazine.com/news/the-shells-of-wild-sea-butterflies-are-already-dissolving

McLennan, M. R., & Hockings, K. J. (2015). The aggressive apes? Causes and contexts of great ape attacks on local persons. In *Springer eBooks* (pp. 373–394). https://doi.org/10.1007/978-3-319-22246-2_18

McMenamin, A. J., & Gensersch, E. (2015). Honey bee colony losses and associated viruses. *Current Opinion in Insect Science, 8*, 121–129. https://doi.org/10.1016/j.cois.2015.01.015

Mearns, E. A. (1907). *Mammals of the Mexican Boundary of the United States: A Descriptive Catalogue of the Species of Mammals Occurring in that Region; with a General Summary of the Natural History, and a List of Trees.* Vol. 56. US Government Printing Office.

Mee, A., Rideout, B. A., Hamber, J. A., Todd, J. N., Austin, G., Clark, M., & Wallace, M. P. (2007). Junk ingestion and nestling mortality in a reintroduced population of California Condors *Gymnogyps californianus*. *Bird Conservation International, 17*(2), 119-130. https://doi.org/10.1017/s095927090700069x

Menon, V., & Tiwari, S. K. (2019). Population status of Asian elephants *Elephas maximus* and key threats. *International Zoo Yearbook, 53*, 17-30.

Meyerson, L. A., Carlton, J. T., Simberloff, D., & Lodge, D. M. (2019). The growing peril of biological invasions. *Frontiers in Ecology and the Environment, 17*(4), 191. https://doi.org/10.1002/fee.2036

Miao, Z., Wang, Q., Cui, X., Conrad, K., Ji, W., Zhang, W., Zhou, X., & MacMillan, D. C. (2022). The dynamics of the illegal ivory trade and the need for stronger global governance. *Journal of International Wildlife Law & Policy, 25*(1), 84-96. https://doi.org/10.1080/13880292.2022.2077393

Milardi, M., Green, A. J., Mancini, M., Trotti, P., Kiljunen, M., Torniainen, J., & Castaldelli, G. (2022). Invasive catfish in northern Italy and their impacts on waterbirds. *NeoBiota, 72*, 109-128. https://doi.org/10.3897/neobiota.72.80500

Milcu, A. I., Hanspach, J., Abson, D. J., & Fischer, J. (2013). Cultural ecosystem services: A literature review and prospects for future research. *Ecology and Society, 18*(3). https://doi.org/10.5751/es-05790-180344

Millennium Ecosystem Assessment. (2005). *Ecosystems and Human Well-Being: Biodiversity Synthesis*. World Resources Institute.

Milman, O. (2022). *The Insect Crisis. The Fall of the Tiny Empires that Run the World*. W. W. Norton.

Miransari, M. (2013). Soil microbes and the availability of soil nutrients. *Acta physiologiae plantarum, 35*, 3075-3084.

Mohanty, S. R., Bodelier, P. L. E., & Conrad, R. (2007). Effect of temperature on composition of the methanotrophic community in rice field and forest soil. *FEMS Microbiology Ecology, 62*(1), 24-31. https://doi.org/10.1111/j.1574-6941 .2007.00370.x

Molina, E., Arenillas, I., & Arz, J. A. (1998). Mass extinction in planktic foraminifera at the Cretaceous/Tertiary boundary in subtropical and temperate latitudes. *Bulletin de la Société géologique de France, 169*, 351-363.

Monosson, E. (2023). *Blight: Fungi and the coming pandemic*. W. W. Norton.

Montgomery, S. (2016). *The soul of an octopus: A surprising exploration into the wonder of consciousness*. Simon and Schuster.

Moodley, Y., Russo, I. M., Dalton, D. L., Kotzé, A., Muya, S., Haubensak, P., Bálint, B., Munimanda, G. K., Deimel, C., Setzer, A., Dicks, K., Herzig-Straschil, B., Kalthoff, D. C., Siegismund, H. R., Robovský, J., O'Donoghue, P., & Bruford, M. W. (2017). Extinctions, genetic erosion and conservation options for the black rhinoceros (*Diceros bicornis*). *Scientific Reports, 7*(1). https://doi.org/10.1038/srep41417

Mora, C., Tittensor, D. P., Adl, S. M., Simpson, A. G. B., & Worm, B. (2011). How many species are there on Earth and in the ocean? *PLOS Biology, 9*(8), e1001127. https://doi.org/10.1371/journal.pbio.1001127.

Morat, P., Jaffré, T., Tronchet, F., Munzinger, J., Pillon, Y., Veillon, J., Chalopin, M., Birnbaum, P., Rigault, F., Dagostini, G., Tinel, J., & Lowry, P. P. (2012). The taxonomic reference base "FLORICAL" and characteristic of the native vascular flora of New Caledonia. *Adansonia, 34*(2), 179–221. https://doi.org/10.5252/a2012n2a1

Morell, V. (2008). Into the wild: Reintroduced animals face daunting odds. *Science, 320*(5877), 742–743. https://doi.org/10.1126/science.320.5877.742

Moritz, R. F. A., & Erler, S. (2016). Lost colonies found in a data mine: Global honey trade but not pests or pesticides as a major cause of regional honeybee colony declines. *Agriculture, Ecosystems & Environment, 216*, 44–50. https://doi.org/10.1016/j.agee.2015.09.027

Morong, T. (1981). The flora of the Desert of Atacama. *Bulletin of the Torrey Botanical Club, 18*, 39–48.

Morris, J., Puttick, M. N., Clark, J. W., Edwards, D., Kenrick, P., Pressel, S., Wellman, C. H., Yang, Z., Schneider, H., & Donoghue, P. C. J. (2018). The timescale of early land plant evolution. *Proceedings of the National Academy of Sciences of the United States of America, 115*(10). https://doi.org/10.1073/pnas.1719588115

Morris, S. C. (1989). Burgess shale faunas and the Cambrian explosion. *Science, 246*(4928), 339–346. http://www.jstor.org/stable/1703956

Morris, S. S., Neidecker-Gonzales, O., Carletto, C., Munguía, M., Medina, J. M., & Wodon, Q. (2002). Hurricane Mitch and the livelihoods of the rural poor in Honduras. *World Development, 30*(1), 49–60. https://doi.org/10.1016/s0305-750x(01)00091-2

Morse, I. (2020). Rare plant threatened by quarry. *New Scientist, 247*(3296), 14. https://doi.org/10.1016/s0262-4079(20)31437-8

Mulawka, E. J. (2014). *The Cockatoos: A Complete Guide to the 21 Species.* McFarland.

Mumby, P. J. (2009). Phase shifts and the stability of macroalgal communities on Caribbean coral reefs. *Coral Reefs, 28*(3), 761–773. https://doi.org/10.1007/s00338-009-0506-8

Murphy, A. E., Sageman, B. B., & Hollander, D. J. (2000). Eutrophication by decoupling of the marine biogeochemical cycles of C, N, and P: A mechanism for the Late Devonian mass extinction. *Geology, 28*(5), 427–430. https://doi.org/10.1130/0091-7613(2000)28%3C427:EBDOTM%3E2.0.CO;2

Murphy, D. D., & Weiland, P. (2010). The route to best science in implementation of the Endangered Species Act's consultation mandate: The benefits of structured effects analysis. *Environmental Management, 47*(2), 161–172. https://doi.org/10.1007/s00267-010-9597-9

Murphy, D. D., & Weiland, P. (2016). Guidance on the use of best available science under the U.S. Endangered Species Act. *Environmental Management, 58*(1), 1–14. https://doi.org/10.1007/s00267-016-0697-z

Myers, N., Mittermeier, R. A., Mittermeier, C. G., Fonseca, G., & Kent, J. (2000). Biodiversity hotspots for conservation priorities. *Nature, 403*(6772), 853–858. https://doi.org/10.1038/35002501

Naeem, S. (1998). Species redundancy and ecosystem reliability. *Conservation Biology, 12*(1), 39–45. https://doi.org/10.1046/j.1523-1739.1998.96379.x

Nair, P. (2012). Woese and fox: Life, rearranged. *Proceedings of the National Academy of Sciences of the United States of America, 109*(4), 1019–1021. https://doi.org/10.1073/pnas.1120749109

Narat, V., Pennec, F., Simmen, B., Ngawolo, J. C. B., & Krief, S. (2015). Bonobo habituation in a forest-savanna mosaic habitat: Influence of ape species, habitat type, and sociocultural context. *Primates, 56*(4), 339–349. https://doi.org/10.1007/s10329-015-0476-0

NASA. (2023). Basics | Stars—NASA universe exploration. *NASA.* https://universe.nasa.gov/stars/basics

Nash, S. R. (2019). Frontline lessons from Vietnam's battle to save biodiversity. *BioScience, 69*(6), 411–417. https://doi.org/10.1093/biosci/biz033

National Geographic. (2022, September 27). The economics of the illicit ivory trade. *National Geographic.* https://education.nationalgeographic.org/resource/economics-illicit-ivory-trade

Nelson, M. B., Martiny, A. C., & Martiny, J. B. H. (2016). Global biogeography of microbial nitrogen-cycling traits in soil. *Proceedings of the National Academy of Sciences of the United States of America, 113*(29), 8033–8040. https://doi.org/10.1073/pnas.1601070113

Nemergut, D. R., Costello, E. K., Hamady, M., Lozupone, C., Jiang, L., Schmidt, S. K., Fierer, N., Townsend, A. R., Cleveland, C. C., Stanish, L. F., & Knight, R. (2011). Global patterns in the biogeography of bacterial taxa. *Environmental Microbiology, 13*(1), 135–144. https://doi.org/10.1111/j.1462-2920.2010.02315.x

Nicholls, H. (2006). Digging for dodo. *Nature, 443*(7108), 138–140. https://doi.org/10.1038/443138a

Nicol, S., Bowie, A. R., Jarman, S., Lannuzel, D., Meiners, K. M., & Van Der Merwe, P. (2010). Southern Ocean iron fertilization by baleen whales and Antarctic krill. *Fish and Fisheries, 11*(2), 203–209. https://doi.org/10.1111/j.1467-2979.2010.00356.x

Nicolaï, A., & Ansart, A. (2017). Conservation at a slow pace: Terrestrial gastropods facing fast-changing climate. *Conservation Physiology, 5*(1). https://doi.org/10.1093/conphys/cox007

Nicole, W. (2015). Pollinator power: nutrition security benefits of an ecosystem service. *Environmental Health Perspectives, 123,* A210.

Noël, M., Loseto, L. L., & Stern, G. A. (2018). Legacy contaminants in the Eastern Beaufort Sea beluga whales (*Delphinapterus leucas*): Are temporal trends reflecting regulations? *Arctic Science.* https://doi.org/10.1139/as-2017-0049

Nowak, D. J., Crane, D. E., & Stevens, J. C. (2006). Air pollution removal by urban trees and shrubs in the United States. *Urban Forestry & Urban Greening, 4*(3-4), 115-123. https://doi.org/10.1016/j.ufug.2006.01.007

Nuwer, R. (2021, May 3). Illegal trade in pangolins keeps growing as criminal networks expand. *Animals.* https://www.nationalgeographic.com/animals/article/pangolin-scale-trade-shipments-growing

Nuwer, R. L. (2018). *Poached: Inside the Dark World of Wildlife Trafficking.* Hachette UK.

Oberhauser, K. S., Elmquist, D., Perilla-Lopez, J. M., Gebhard, I., Lukens, L., & Stireman, J. O. (2017). Tachinid fly (*Diptera: Tachinidae*) parasitoids of *Danaus plexippus* (*Lepidoptera: Nymphalidae*). *Annals of the Entomological Society of America, 110*(6), 536-543. https://doi.org/10.1093/aesa/sax048

Ødegaard, F. (2000). How many species of arthropods? Erwin's estimate revised. *Biological Journal of the Linnean Society, 71*(4), 583-597. https://doi.org/10.1006/bijl.2000.0468

Ogutu-Ohwayo, R., Hecky, R. E., Cohen, A. S., & Kaufman, L. (1997). Human impacts on the African Great Lakes. *Environmental Biology of Fishes, 50*(2), 117-131. https://doi.org/10.1023/a:1007320932349

Ollerton, J., Price, V., Armbruster, W. S., Memmott, J., Watts, S., Waser, N. M., Totland, Ø., Goulson, D., Alarcón, R., Stout, J. C., & Tarrant, S. (2012). Overplaying the role of honey bees as pollinators: A comment on Aebi and Neumann (2011). *Trends in Ecology and Evolution, 27*(3), 141-142. https://doi.org/10.1016/j.tree.2011.12.001

Oreskes, N. & Conway, E. M. (2023). *The Big Myth: How American Business Taught Us to Loathe Government and Love the Free Market.* Bloomsbury USA.

Orford, K. A., Vaughan, I. P., & Memmott, J. (2015). The forgotten flies: The importance of non-syrphid *Diptera* as pollinators. *Proceedings of the Royal Society B: Biological Sciences, 282*(1805), 20142934. https://doi.org/10.1098/rspb.2014.2934

Oster, R. M. (2015). *Great Britain in the Age of Sail: Scarce Resources, Ruthless Actions and Consequences.* Air Command and Staff College.

Ovaskainen, O. (2002). Long-term persistence of species and the SLOSS problem. *Journal of Theoretical Biology, 218*(4), 419–433. https://doi.org/10.1016/s0022-5193(02)93089-4

Paez, S., Kraus, R. H. S., Shapiro, B., Gilbert, M., Jarvis, E. D., Al-Ajli, F. O. M., Ceballos, G., Crawford, A. J., Fédrigo, O., Johnson, R. N., Johnson, W. E., Marqués-Bonet, T., Morin, P. A., Mueller, R., Ryder, O. A., Teeling, E. C., & Venkatesh, B. (2022). Reference genomes for conservation. *Science, 377*(6604), 364–366. https://doi.org/10.1126/science.abm8127

Pagano, A. M., Durner, G. M., Rode, K. D., Atwood, T. C., Atkinson, S. N., Peacock, E., Costa, D. P., Owen, M. A., & Williams, T. M. (2018). High-energy, high-fat lifestyle challenges an Arctic apex predator, the polar bear. *Science, 359*(6375), 568–572. https://doi.org/10.1126/science.aan8677

Paltridge, R., Gibson, D. F., & Edwards, G. (1997). Diet of the feral cat (*Felis catus*) in Central Australia. *Wildlife Research, 24*(1), 67. https://doi.org/10.1071/wr96023

Palumbi, S. R., Barshis, D. J., Traylor-Knowles, N., & Bay, R. A. (2014). Mechanisms of reef coral resistance to future climate change. *Science, 344*(6186), 895–898. https://doi.org/10.1126/science.1251336

Pan, Y., Birdsey, R. A., Fang, J., Houghton, R. A., Kauppi, P. E., Kurz, W. A., Phillips, O. L., Shvidenko, A., Lewis, S. L., Canadell, J. G., Ciais, P., Jackson, R. B., Pacala, S. W., McGuire, A. D., Piao, S., Rautiainen, A., Sitch, S., & Hayes, D. J. (2011). A large and persistent carbon sink in the world's forests. *Science, 333*(6045), 988–993. https://doi.org/10.1126/science.1201609

Pandolfi, J. M., Connolly, S. R., Marshall, D. J., & Cohen, A. L. (2011). Projecting coral reef futures under global warming and ocean acidification. *Science, 333*(6041), 418–422. https://doi.org/10.1126/science.1204794

Parker, M. (2018). *Shut Down the Business School.* University of Chicago Press Economics Books.

Pearce, F. (2019). Rivers in the sky. *New Scientist, 244*, 40–43.

Pecenka, J. R., & Lundgren, J. G. (2015). Non-target effects of clothianidin on monarch butterflies. *The Science of Nature, 102*(3–4). https://doi.org/10.1007/s00114-015-1270-y

Pellegrini, A. F. A., Pringle, R. M., Govender, N., & Hedin, L. O. (2016). Woody plant biomass and carbon exchange depend on elephant-fire interactions across a productivity gradient in African savanna. *Journal of Ecology, 105*(1), 111–121. https://doi.org/10.1111/1365-2745.12668

Perdok, A. A., De Boer, W. F., & Stout, T. A. E. (2007). Prospects for managing African elephant population growth by immunocontraception: A review. *Pachyderm, 42*, 95–105. https://edepot.wur.nl/31543

Perry, W. J. (2015). *My Journey to the Nuclear Brink.* Stanford Security Studies.

Peters, S. E., & Foote, M. (2002). Determinants of extinction in the fossil record. *Nature, 416*(6879), 420-424. https://doi.org/10.1038/416420a

Petipas, R. H., & Brody, A. K. (2014). Termites and ungulates affect arbuscular mycorrhizal richness and infectivity in a semiarid savanna. *Botany, 92*(3), 233-240. https://doi.org/10.1139/cjb-2013-0223

Philippot, L., Bru, D., Saby, N. P., Čuhel, J., Arrouays, D., Šimek, M., & Hallin, S. (2009). Spatial patterns of bacterial taxa in nature reflect ecological traits of deep branches of the 16S rRNA bacterial tree. *Environmental Microbiology, 11*(12), 3096-3104. https://doi.org/10.1111/j.1462-2920.2009.02014.x

Phillips, H. R. P., Guerra, C. A., Bartz, M. L. C., Briones, M. J. I., Brown, G. G., Crowther, T. W., Ferlian, O., Gongalsky, K. B., Van Den Hoogen, J., Krebs, J., Orgiazzi, A., Routh, D., Schwarz, B., Bach, E. M., Bennett, J. M., Brose, U., Decaëns, T., König-Ries, B., Loreau, M., . . . Nuutinen, V. (2019). Global distribution of earthworm diversity. *Science, 366*(6464), 480-485. https://doi.org/10.1126/science.aax4851

Pietsch, C., Petsios, E., & Bottjer, D. J. (2016). Sudden and extreme hyperthermals, low-oxygen, and sediment influx drove community phase shifts following the end-Permian mass extinction. *Palaeogeography, Palaeoclimatology, Palaeoecology, 451*, 183-196. https://doi.org/10.1016/j.palaeo.2016.02.056

Pimiento, C., & Pyenson, N. D. (2021). When sharks nearly disappeared. *Science, 372*(6546), 1036-1037. https://doi.org/10.1126/science.abj2088

Pimm, S. L., & Jenkins, C. N. (2010). Extinctions and the practice of preventing them. *Conservation Biology for All, 1*(9), 181-198.

Pimm, S. L., Jenkins, C. N., Abell, R., Brooks, T. M., Gittleman, J. L., Joppa, L., Raven, P. H., Roberts, C. M., & Sexton, J. O. (2014). The biodiversity of species and their rates of extinction, distribution, and protection. *Science, 344*(6187). https://doi.org/10.1126/science.1246752

Pimm, S. L., & Joppa, L. (2015). How many plant species are there, where are they, and at what rate are they going extinct? *Annals of the Missouri Botanical Garden, 100*(3), 170-176. https://doi.org/10.3417/2012018

Pimm, S. L., Raven, P. H., Peterson, A. M., Şekercioğlu, Ç. H., & Ehrlich, P. R. (2006). Human impacts on the rates of recent, present, and future bird extinctions. *Proceedings of the National Academy of Sciences of the United States of America, 103*(29), 10941-10946. https://doi.org/10.1073/pnas.0604181103

Pleasants, J. M., & Oberhauser, K. S. (2012). Milkweed loss in agricultural fields because of herbicide use: Effect on the monarch butterfly population. *Insect Conservation and Diversity, 6*(2), 135-144. https://doi.org/10.1111/j.1752-4598.2012.00196.x

Pongracz, J., Paetkau, D., Branigan, M., & Richardson, E. (2017). Recent hybridization between a polar bear and grizzly bears in the Canadian Arctic. *Arctic, 70*(2), 151. https://doi.org/10.14430/arctic4643

Poorter, L., Van Der Sande, M. T., Thompson, J., Arets, E., Alarcón, A., Álvarez-Sánchez, F. J., Ascarrunz, N., Balvanera, P., Barajas-Guzmán, G., Boit, A., Bongers, F., Carvalho, F. A., Casanoves, F., Cornejo-Tenorio, G., Costa, F. R. C., Castilho, C. V., Duivenvoorden, J. F., Dutrieux, L., Enquist, B. J., . . . Peña-Claros, M. (2015). Diversity enhances carbon storage in tropical forests. *Global Ecology and Biogeography, 24*(11), 1314-1328. https://doi.org/10.1111/geb.12364

Popovich, N., Albeck-Ripka, L., & Pierre-Louis, K. (2021, October 6). The Trump administration is reversing nearly 100 environmental rules. Here's the full list. *New York Times.* https://www.nytimes.com/interactive/2020/climate /trump-environment-rollbacks.html

Potts, S. G., Imperatriz-Fonseca, V. L., Ngo, H. T., Aizen, M. A., Biesmeijer, J. C., Breeze, T. D., Dicks, L. V., Garibaldi, L. A., Hill, R., Settele, J., & Vanbergen, A. J. (2016). Safeguarding pollinators and their values to human well-being. *Nature, 540*(7632), 220-229. https://doi.org/10.1038/nature20588

Prowse, T. A. A., Johnson, C. N., Bradshaw, C. J. A., & Brook, B. W. (2014). An ecological regime shift resulting from disrupted predator–prey interactions in Holocene Australia. *Ecology, 95*(3), 693-702. https://doi.org/10.1890/13-0746.1

Pyron, R. A. (2017, April 9). We don't need to save endangered species. Extinction is part of evolution. *Washington Post.* http://wapo.st/2kgcRSk

Queensland Dung Beetle Project. (2002). *Improving Sustainable Land Management Systems in Queensland Using Beetles: Final Report of the 2001-2002 Queensland Dung Beetle Project.* Queensland Government. http://www.dungbeetle.com.au /queenslandreport.pdf

Quétier, F., Regnery, B., & Levrel, H. (2014). No net loss of biodiversity or paper offsets? A critical review of the French no net loss policy. *Environmental Science & Policy, 38*, 120-131. https://doi.org/10.1016/j.envsci.2013.11.009

Ramette, A., & Tiedje, J. M. (2006). Biogeography: An emerging cornerstone for understanding prokaryotic diversity, ecology, and evolution. *Microbial Ecology, 53*(2), 197-207. https://doi.org/10.1007/s00248-005-5010-2

Rates, S. M. K. (2001). Plants as source of drugs. *Toxicon, 39*(5), 603-613. https://doi.org/10.1016/s0041-0101(00)00154-9

Raunkiaer, C. (1934). *The Life Forms of Plants and Statistical Plant Geography.* Oxford University Press.

Raup, D. M. (1986). Biological extinction in earth history. *Science, 231*, 1528-1533.

Raup, D. M., & Sepkoski, J. J. (1982). Mass extinctions in the marine fossil record. *Science, 215*(4539), 1501-1503. https://doi.org/10.1126/science.215.4539.1501

Raven, P. H., Gereau, R. E., Phillipson, P. B., Chatelain, C., Jenkins, C. N., & Ulloa, C. U. (2020). The distribution of biodiversity richness in the tropics. *Science Advances*, 6(37). https://doi.org/10.1126/sciadv.abc6228

Ravilious, K. (2019). Hidden hotspots. *New Scientist*, 243(3246), 40–41. https://doi.org/10.1016/s0262-4079(19)31681-1

Redford, K. H. (1992). The empty forest. *BioScience*, 42(6), 412–422. https://doi.org/10.2307/1311860

Reed, H. E., & Martiny, J. B. H. (2012). Microbial composition affects the functioning of estuarine sediments. *The ISME Journal*, 7(4), 868–879. https://doi.org/10.1038/ismej.2012.154

Reeve, R. (2014). *Policing International Trade in Endangered Species: The CITES Treaty and Compliance*. Routledge.

Regehr, E. V., Laidre, K. L., Akçakaya, H. R., Amstrup, S. C., Atwood, T. C., Lunn, N. J., Obbard, M. E., Stern, H. L., Thiemann, G. W., & Wiig, Ø. (2016). Conservation status of polar bears (*Ursus maritimus*) in relation to projected sea-ice declines. *Biology Letters*, 12(12), 20160556. https://doi.org/10.1098/rsbl.2016.0556

Reynolds, J. H., Knutson, M. G., Newman, K., Silverman, E., & Thompson, W. L. (2016). A road map for designing and implementing a biological monitoring program. *Environmental Monitoring and Assessment*, 188(7). https://doi.org/10.1007/s10661-016-5397-x

Richards, M. A., Álvarez, W., Self, S., Karlstrom, L., Renne, P. R., Manga, M., Sprain, C. J., Smit, J., Vanderkluysen, L., & Gibson, S. A. (2015). Triggering of the largest Deccan eruptions by the Chicxulub impact. *Geological Society of America Bulletin*, 127(11–12), 1507–1520. https://doi.org/10.1130/b31167.1

Ricketts, T. H., Daily, G. C., Ehrlich, P. R., & Fay, J. P. (2001). Countryside biogeography of moths in a fragmented landscape: Biodiversity in native and agricultural habitats. *Conservation Biology*, 15(2), 378–388. https://doi.org/10.1046/j.1523-1739.2001.015002378.x

Ricketts, T. H., Daily, G. C., Ehrlich, P. R., & Michener, C. D. (2004). Economic value of tropical forest to coffee production. *Proceedings of the National Academy of Sciences of the United States of America*, 101(34), 12579–12582. https://doi.org/10.1073/pnas.0405147101

Riddley, M. & Chan, A. (2023). *Viral: The Search for the Origin of COVID-19*. Harper Perennial.

Riegl, B., Purkis, S. J., Al-Cibahy, A. S., Abdel-Moati, M. A., & Hoegh-Guldberg, O. (2011). Present limits to heat-adaptability in corals and population-level responses to climate extremes. *PLOS ONE*, 6(9), e24802. https://doi.org/10.1371/journal.pone.0024802

Rigal, S., Dakos, V., Alonso, H., Auniņš, A., Benkö, Z., Brotóns, L., Chodkiewicz, T., Chylarecki, P., De Carli, E., Del Moral, J. C., Domşa, C., Escandell, V., Fontaine, B., Foppen, R., Gregory, R. D., Harris, S., Herrando, S., Husby, M., Ieronymidou, C., . . . Devictor, V. (2023). Farmland practices are driving bird population decline across Europe. *Proceedings of the National Academy of Sciences of the United States of America, 120*(21). https://doi.org/10.1073/pnas.2216573120

Ripple, W. J., Abernethy, K., Betts, M. G., Chapron, G., Dirzo, R., Galetti, M., Levi, T., Lindsey, P. A., Macdonald, D. W., Machovina, B., Newsome, T. M., Peres, C. A., Wallach, A. D., Wolf, C., & Young, H. S. (2016). Bushmeat hunting and extinction risk to the world's mammals. *Royal Society Open Science, 3*(10), 160498. https://doi.org/10.1098/rsos.160498

Ripple, W. J., Newsome, T. M., Wolf, C., Dirzo, R., Everatt, K. T., Galetti, M., Hayward, M. W., Kerley, G. I. H., Levi, T., Lindsey, P. A., Macdonald, D. W., Malhi, Y., Painter, L. E., Sandom, C. J., Terborgh, J., & Van Valkenburgh, B. (2015). Collapse of the world's largest herbivores. *Science Advances, 1*(4). https://doi.org/10.1126/sciadv.1400103

Ripple, W. J., Wolf, C., Newsome, T. M., Betts, M. G., Ceballos, G., Courchamp, F., Hayward, M. W., Van Valkenburgh, B., Wallach, A. D., & Worm, B. (2019). Are we eating the world's megafauna to extinction? *Conservation Letters, 12*(3). https://doi.org/10.1111/conl.12627

Ripple, W. J., Wolf, C., Newsome, T. M., Galetti, M., Alamgir, M., Crist, E., Mahmoud, M. I., & Laurance, W. F. (2017). World scientists' warning to humanity: A second notice. *BioScience, 67*(12), 1026–1028. https://doi.org/10.1093/biosci/bix125

Robbins, J. (2019, November 14). Climate whiplash: Wild swings in extreme weather are on the rise. *Yale Environment 360*. Yale School of the Environment.

Roberts, D. L., & Solow, A. R. (2003). When did the dodo become extinct? *Nature, 426*(6964), 245. https://doi.org/10.1038/426245a

Rockwell, S. M., Bocetti, C. I., & Marra, P. P. 2012. Carry-over effects of winter climate on spring arrival date and reproductive success in an endangered migratory bird, Kirtland's Warbler (*Setophaga kirtlandii*). *The Auk, 129*, 744–752.

Rodolfo-Metalpa, R., Houlbrèque, F., Tambutté, É., Boisson, F., Baggini, C., Patti, F. P., Jeffree, R. A., Fine, M., Foggo, A., Gattuso, J., & Hall-Spencer, J. M. (2011). Coral and mollusc resistance to ocean acidification adversely affected by warming. *Nature Climate Change, 1*(6), 308–312. https://doi.org/10.1038/nclimate1200

Rodrigues, A. S. L., Pilgrim, J. D., Lamoreux, J. F., Hoffmann, M., & Brooks, T. M. (2006). The value of the IUCN Red List for conservation. *Trends in Ecology and Evolution, 21*(2), 71–76. https://doi.org/10.1016/j.tree.2005.10.010

Root, T. L., Bishop, K. D., Ehrlich, P. R., Schneider, S. H., & Ehrlich, A. H. (2006). Conservation of southeast Asian birds: The role of bird markets and avian flu. *Environ Awareness, 29*, 57–65.

Rose, N., & Burton, P. J. (2009). Using bioclimatic envelopes to identify temporal corridors in support of conservation planning in a changing climate. *Forest Ecology and Management, 258*, S64–S74. https://doi.org/10.1016/j.foreco.2009.07.053

Rosenberg, K. V., Dokter, A. M., Blancher, P. J., Sauer, J. R., Smith, A. C., Smith, P. A., Stanton, J. C., Panjabi, A. O., Helft, L., Parr, M. J., & Marra, P. P. (2019). Decline of the North American avifauna. *Science, 366*(6461), 120–124. https://doi.org/10.1126/science.aaw1313

Rosser, A. M., & Mainka, S. (2002). Overexploitation and species extinctions. *Conservation Biology, 16*(3), 584–586. https://doi.org/10.1046/j.1523-1739.2002.01635.x

Running, S. W., Thornton, P. E., Nemani, R. R., & Glassy, J. M. (2000). Global terrestrial gross and net primary productivity from the Earth observing system. In Sala, O. E., Jackson, R. B., Mooney, H. A., & Howarth, R. W. (Eds.), *Methods in Ecosystem Science* (pp. 44–56), Springer. https://doi.org/10.1007/978-1-4612-1224-9_4

Rutberg, A. T. (2013). Managing wildlife with contraception: Why is it taking so long? *Journal of Zoo and Wildlife Medicine, 44*(4s), S38–S46. https://doi.org/10.1638/1042-7260-44.4s.s38

Rzedowski, J. (1991). Diversidad y orígenes de la flora fanerogámica de México. *Acta Botánica Mexicana, 14*, 3–21.

Sáenz-Romero, C., Rehfeldt, G. E., Duval, P., & Lindig-Cisneros, R. (2012). *Abies religiosa* habitat prediction in climatic change scenarios and implications for monarch butterfly conservation in Mexico. *Forest Ecology and Management, 275*, 98–106. https://doi.org/10.1016/j.foreco.2012.03.004

Safina, C. (2015). *Beyond Words: What Animals Think and Feel*. Henry Holt.

Salih, A. M., Baraibar, M., Mwangi, K., & Artan, G. (2020). Climate change and locust outbreak in East Africa. *Nature Climate Change, 10*(7), 584–585. https://doi.org/10.1038/s41558-020-0835-8

Sallan, L., & Galimberti, A. K. (2015). Body-size reduction in vertebrates following the end-Devonian mass extinction. *Science, 350*(6262), 812–815. https://doi.org/10.1126/science.aac7373

Salzburger, W., Van Bocxlaer, B., & Cohen, A. S. (2014). Ecology and evolution of the African Great Lakes and their faunas. *Annual Review of Ecology, Evolution, and Systematics, 45*(1), 519–545. https://doi.org/10.1146/annurev-ecolsys-120213-091804

San Diego Natural History Museum. (2018, April 19). Museum researchers rediscover animal not seen in 30 years. *ScienceDaily*. www.sciencedaily.com /releases/2018/04/180419172701.htm

Sánchez-Bayo, F., & Wyckhuys, K. A. G. (2019). Worldwide decline of the entomofauna: A review of its drivers. *Biological Conservation, 232*, 8–27. https://doi.org/10.1016/j.biocon.2019.01.020

Sanders, D., Thébault, É., Kehoe, R., & Van Veen, F. J. F. (2018). Trophic redundancy reduces vulnerability to extinction cascades. *Proceedings of the National Academy of Sciences of the United States of America, 115*(10), 2419–2424. https://doi.org/10.1073/pnas.1716825115

Sapolsky, R. M. (2004). *Why Zebras Don't Get Ulcers*. (3rd ed.). Henry Holt and Company.

Sarre, S. D., MacDonald, A. J., Barclay, C., Saunders, G., & Ramsey, D. S. L. (2012). Foxes are now widespread in Tasmania: DNA detection defines the distribution of this rare but invasive carnivore. *Journal of Applied Ecology, 50*(2), 459–468. https://doi.org/10.1111/1365-2664.12011

Satterfield, D. A., Maerz, J. C., & Altizer, S. (2015). Loss of migratory behaviour increases infection risk for a butterfly host. *Proceedings of the Royal Society B: Biological Sciences, 282*(1801), 20141734. https://doi.org/10.1098/rspb.2014.1734

Satterthwaite, D., McGranahan, G., & Tacoli, C. (2010). Urbanization and its implications for food and farming. *Philosophical Transactions of the Royal Society B, 365*(1554), 2809–2820. https://doi.org/10.1098/rstb.2010.0136

Saunders, G., Gentle, M., & Dickman, C. R. (2010). The impacts and management of foxes *Vulpes vulpes* in Australia. *Mammal Review, 40*(3), 181–211. https://doi .org/10.1111/j.1365-2907.2010.00159.x

Saunders, M. E., Janes, J. K., & O'Hanlon, J. C. (2019). Moving on from the insect apocalypse narrative: Engaging with evidence-based insect conservation. *BioScience, 70*(1), 80–89. https://doi.org/10.1093/biosci/biz143

Savage, J. M. (1966). An extraordinary new toad (Bufo) from Costa Rica. *Revista de Biologia Tropical, 14*(2), 153–167.

Schauer, R., Bienhold, C., Ramette, A., & Harder, J. (2009). Bacterial diversity and biogeography in deep-sea surface sediments of the South Atlantic Ocean. *The ISME Journal, 4*(2), 159–170. https://doi.org/10.1038/ismej.2009.106

Schoene, B., Samperton, K. M., Eddy, M. P., Keller, G., Adatte, T., Bowring, S. A., Khadri, S., & Gertsch, B. (2015). U-Pb geochronology of the Deccan Traps and relation to the end-Cretaceous mass extinction. *Science, 347*(6218), 182–184. https://doi.org/10.1126/science.aaa0118

Schopf, J. W., Kitajima, K., Kudryavtsev, A. B., & Valley, J. W. (2017). SIMS analyses of the oldest known assemblage of microfossils document their taxon-correlated carbon isotope compositions. *Proceedings of the National Academy of Sciences of the United States of America, 115*(1), 53–58. https://doi.org /10.1073/pnas.1718063115

Schroeder, J., Martin, J. T., Angulo, D. F., Razo, I. A., Barbosa, J. M., Perea, R., Sebastián-González, E., & Dirzo, R. (2019). Host plant phylogeny and abundance predict root-associated fungal community composition and diversity of mutualists and pathogens. *Journal of Ecology, 107*(4), 1557–1566. https://doi.org/10.1111/1365-2745.13166

Schulte, P., Alegret, L., Arenillas, I., Arz, J. A., Barton, P. J., Bown, P. R., Bralower, T. J., Christeson, G. L., Claeys, P., Cockell, C. S., Collins, G. S., Deutsch, A., Goldin, T., Goto, K., Grajales-Nishimura, J., Grieve, R. A. F., Gulick, S. P. S., Johnson, K. R., Kiessling, W., . . . Willumsen, P. S. (2010). The Chicxulub asteroid impact and mass extinction at the Cretaceous-Paleogene boundary. *Science, 327*(5970), 1214–1218. https://doi.org/10.1126/science.1177265

Schwartz, M. W. (2008). The performance of the Endangered Species Act. *Annual Review of Ecology, Evolution, and Systematics, 39*(1), 279–299. https://doi .org/10.1146/annurev.ecolsys.39.110707.173538

Sekercioglu, C. H. (2017, June 12). Analysis: The economic value of birds. *All About Birds.* https://www.allaboutbirds.org/news/analysis-the-economic -value-of-birds

SEMARNAT. 1994. Norma oficial Mexicana que determina las especies y subespecies de flora y fauna silvestres terrestres y acuáticas, en peligro de extinción, amenazadas, raras y las sujetas a protección especial, y que establece especificaciones para su protección. *Diario Oficial de la Federación, México, Distrito Federal.*

Servick, K. (2017). Desperate rescue for vanishing vaquitas. *Science, 358*(6361), 280. https://doi.org/10.1126/science.358.6361.280

Shannon, G., Slotow, R., Durant, S. M., Sayialel, K., Poole, J. H., Moss, C. J., & McComb, K. (2013). Effects of social disruption in elephants persist decades after culling. *Frontiers in Zoology, 10*(1), 62. https://doi.org/10.1186/1742-9994 -10-62

Sharples, F. E., Husbands, J., Board on Life Sciences; Division on Earth and Life Studies; Committee on Science, Technology, and Law; Policy and Global Affairs; Board on Health Sciences Policy; National Research Council; & Institute of Medicine. (2015). *Potential Risks and Benefits of Gain-of-Function Research: Summary of a Workshop.* National Academies Press.

Shen, S., Crowley, J. L., Wang, Y., Bowring, S. A., Erwin, D. H., Sadler, P. M., Cao, C., Rothman, D. H., Henderson, C. M., Ramezani, J., Zhang, H., Shen, Y., Wang, X., Wang, W., Mu, L., Li, W., Tang, Y., Liu, X., Liu, L., . . . Yugan, J.

(2011). Calibrating the End-Permian mass extinction. *Science, 334*(6061), 1367–1372. https://doi.org/10.1126/science.1213454

Shepherd, C. R., Leupen, B. T., Siriwat, P., & Nijman, V. (2020). International wildlife trade, avian influenza, organised crime and the effectiveness of CITES: The Chinese hwamei as a case study. *Global Ecology and Conservation, 23*, e01185. https://doi.org/10.1016/j.gecco.2020.e01185

Sheppard, P. M. (1952). Natural selection in two colonies of the polymorphic land snail *Cepaea nemoralis. Heredity, 6*(2), 233–238. https://doi.org/10.1038/hdy.1952.23

Sieff, K. (2020, January 5). Climate change is playing havoc with Mexico's monarch butterfly migration. *Washington Post.* https://wapo.st/2MYv823

Simberloff, D., & Cox, J. A. (1987). Consequences and costs of conservation corridors. *Conservation Biology, 1*(1), 63–71. https://doi.org/10.1111/j.1523-1739.1987.tb00010.x

Simmons, B. A., Ray, R., Yang, H., & Gallagher, K. P. (2021). China can help solve the debt and environmental crises. *Science, 371*(6528), 468–470. https://doi.org/10.1126/science.abf4049

Simon, T. P. (2011). Conservation status of North American freshwater crayfish (*Decapoda: Cambaridae*) from the southern United States. *Proceedings of the Indiana Academy of Science, 120*, 71–95.

Sims, D., & Frost, M. (2019). Trade in mislabeled endangered sharks. *Science, 364*(6442), 743–744. https://doi.org/10.1126/science.aax5777

Sinervo, B., Lara Reséndiz, R. A., Miles, D. B., Lovich, J. E., Ennen, J. R., Müller, J., . . . Méndez de la Cruz, F. R. (2017). *Climate Change and Collapsing Thermal Niches of Mexican Endemic Reptiles.* UC Office of the President: UC-Mexico Initiative. https://escholarship.org/uc/item/4xko77hp

Sinervo, B., Mendez-de-la-Cruz, F., Miles, D. B., Heulin, B., Bastiaans, E., Villagran-Santa Cruz, M., Lara-Resendiz, R., Martinez-Mendez, N., Calderon-Espinosa, M. L., Meza-Lazaro, R. N., Gadsden, H., Avila, L. J., Morando, M., De la Riva, I. J., Sepulveda, P. V., Rocha, C. F. D., Ibarguengoytia, N., Puntriano, C. A., Massot, M., & Lepetz, V. (2010). Erosion of lizard diversity by climate change and altered thermal niches. *Science, 328*(5980), 894–899. https://doi.org/10.1126/science.1184695

Singh, B. K., Bardgett, R. D., Smith, P., & Reay, D. (2010). Microorganisms and climate change: Terrestrial feedbacks and mitigation options. *Nature Reviews Microbiology, 8*(11), 779–790. https://doi.org/10.1038/nrmicro2439

Singh, J., Singh, R. P., & Khare, R. (2018). Influence of climate change on Antarctic flora. *Polar Science, 18*, 94–101. https://doi.org/10.1016/j.polar.2018.05.006

Slik, J. W. F., Arroyo-Rodríguez, V., Aiba, S.-I., Alvarez-Loayza, P., Alves, L. F., Ashton, P., Balvanera, P., Bastian, M. L., Bellingham, P. J., van den Berg, E., Bernacci, L., da Conceição Bispo, P., Blanc, L., Böhning-Gaese, K., Boeckx, P., Bongers, F., Boyle, B., Bradford, M., Brearley, F. Q., & Breuer-Ndoundou Hockemba, M. (2015). An estimate of the number of tropical tree species. *Proceedings of the National Academy of Sciences, 112*(24), 7472–7477. https://doi .org/10.1073/pnas.1423147112

Sloan, S., Jenkins, C. N., Joppa, L., Gaveau, D., & Laurance, W. F. (2014). Remaining natural vegetation in the global biodiversity hotspots. *Biological Conservation, 177*, 12–24. https://doi.org/10.1016/j.biocon.2014.05.027

Smil, V. (2015). *Harvesting the Biosphere: What We Have Taken from Nature.* MIT Press.

Snyder, N. F. R., Brown, D. E., & Clark, K. B. (2009). *The Travails of Two Woodpeckers: Ivory-Bills & Imperials.* University of New Mexico Press.

Solan, M., Cardinale, B. J., Downing, A. L., Engelhardt, K. A. M., Ruesink, J. L., & Srivastava, D. S. (2004). Extinction and ecosystem function in the marine benthos. *Science, 306*(5699), 1177–1180. https://doi.org/10.1126/science.1103960

Souza, V., Espinosa-Asuar, L., Escalante, A. E., Eguiarte, L. E., Farmer, J. D., Forney, L. J., Lloret, L., Rodríguez-Martínez, J. M., Soberón, X., Dirzo, R., & Elser, J. J. (2006). An endangered oasis of aquatic microbial biodiversity in the Chihuahuan desert. *Proceedings of the National Academy of Sciences of the United States of America, 103*(17), 6565–6570. https://doi.org/10.1073/pnas .0601434103

Spurgeon, D. J., & Hopkin, S. P. (1996). The effects of metal contamination on earthworm populations around a smelting works: quantifying species effects. *Applied Soil Ecology, 4*(2), 147–160. https://doi.org/10.1016/0929 -1393(96)00109-6

Steadman, D. W. (1989). Extinction of birds in eastern Polynesia: A review of the record, and comparisons with other Pacific Island groups. *Journal of Archaeological Science, 16*(2), 177–205. https://doi.org/10.1016/0305-4403(89) 90065-4

Steadman, D. W. (2006). *Extinction and Biogeography of Tropical Pacific Birds.* University of Chicago Press.

Steffan-Dewenter, I., Potts, S. G., & Packer, L. (2005). Pollinator diversity and crop pollination services are at risk. *Trends in Ecology and Evolution, 20*(12), 651–652. https://doi.org/10.1016/j.tree.2005.09.004

Steinbeck, J. (2000). *Cannery Row* (Penguin Modern Classic Edition). Penguin UK.

Steller, G. W., & Frost, O. W. 1988. *Journal of a Voyage with Bering, 1741–1742.* Stanford University Press.

Stevens, S. (2014). *Indigenous Peoples, National Parks, and Protected Areas: A New Paradigm Linking Conservation, Culture, and Rights.* Univ of Arizona Press.

Stierhoff, K. L., Neuman, M., & Butler, J. L. (2012). On the road to extinction? Population declines of the endangered white abalone, *Haliotis sorenseni.* *Biological Conservation, 152,* 46-52. https://doi.org/10.1016/j.biocon.2012.03.013

Stiles, D. (2015, July 1). Only legal ivory can stop poaching. *Earth Island Journal.* https://www.earthisland.org/journal/index.php/magazine/entry/stiles

Stoddart, D. R., Peake, J., Gordon, C., & Burleigh, R. (1979). Historical records of Indian Ocean giant tortoise populations. *Philosophical Transactions of the Royal Society of London, 286*(1011), 147-161. https://doi.org/10.1098/rstb.1979.0023

Stork, N. E. (2018). How many species of insects and other terrestrial arthropods are there on Earth? *Annual Review of Entomology, 63,* 31-45.

Stuart, S. N., Chanson, J., Cox, N. A., Young, B. E., Rodrigues, A. S. L., Fischman, D. L., & Waller, R. W. (2004). Status and trends of amphibian declines and extinctions worldwide. *Science, 306*(5702), 1783-1786. https://doi.org/10.1126/science.1103538

Stuart-Smith, R. D., Edgar, G., Clausius, E., Oh, E. S., Barrett, N., Emslie, M. J., Bates, A. E., Bax, N. J., Brock, D. J., Cooper, A., Davis, T. R., Day, P. B., Dunic, J. C., Green, A., Hasweera, N., Hicks, J., Holmes, T. H., Jones, B. K., Jordan, A., . . . Mellin, C. (2022). Tracking widespread climate-driven change on temperate and tropical reefs. *Current Biology, 32*(19), 4128-4138.e3. https://doi.org/10.1016/j.cub.2022.07.067

Sukumar, R. (2006). A brief review of the status, distribution and biology of wild Asian elephants *Elephas maximus. International Zoo Yearbook, 40,* 1-8.

Suraci, J. P., Clinchy, M., Zanette, L., & Wilmers, C. C. (2019). Fear of humans as apex predators has landscape-scale impacts from mountain lions to mice. *Ecology Letters, 22*(10), 1578-1586. https://doi.org/10.1111/ele.13344

Sutherland, J. D. (2015). The origin of life—Out of the blue. *Angewandte Chemie International Edition, 55*(1), 104-121. https://doi.org/10.1002/anie.201506585

Suzán, G., Marcé, E., Giermakowski, J. T., Armién, B., Pascale, J. M., Mills, J. N., Ceballos, G., Gómez, A., Aguirre, A. A., Salazar-Bravo, J., Armién, A. G., Parmenter, R., & Yates, T. L. (2008). The effect of habitat fragmentation and species diversity loss on hantavirus prevalence in Panama. *Annals of the New York Academy of Sciences, 1149*(1), 80-83. https://doi.org/10.1196/annals.1428.063

Swaisgood, R. R., Wei, W., & Zhang, Z. (2023). Progress in China's environmental policy in synergy with foundational giant panda conservation program. *BioScience, 73*(8), 592-601. https://doi.org/10.1093/biosci/biad065

Swan, S. H., & Colino, S. (2022). *Count Down: How Our Modern World Is Threatening Sperm Counts, Altering Male and Female Reproductive Development, and Imperiling the Future of the Human Race.* Simon and Schuster.

Taabu-Munyaho, A., Marshall, B. E., Tómasson, T., & Marteinsdóttir, G. (2016). Nile perch and the transformation of Lake Victoria. *African Journal of Aquatic Science, 41*(2), 127–142. https://doi.org/10.2989/16085914.2016 .1157058

Tamburini, G., Bommarco, R., Wanger, T. C., Kremen, C., Van Der Heijden, M. G. A., Liebman, M., & Hallin, S. (2020). Agricultural diversification promotes multiple ecosystem services without compromising yield. *Science Advances, 6*(45). https://doi.org/10.1126/sciadv.aba1715

Tashiro, T., Ishida, A., Hori, M., Igisu, M., Koike, M., Méjean, P., Takahata, N., Sano, Y., & Komiya, T. (2017). Early trace of life from 3.95 Ga sedimentary rocks in Labrador, Canada. *Nature, 549*(7673), 516–518. https://doi.org/10.1038 /nature24019

The Telegraph. (2012). Obituaries: Lawrence Anthony. https://www.telegraph.co .uk/news/obituaries/9131585/Lawrence-Anthony.html

Temming, M. (2019, August 8). Gut bacteria may change the way many drugs work in the body. *Science News.* https://www.sciencenews.org/article/gut -bacteria-may-change-way-many-drugs-work-body

Terborgh, J. 1989. *Where Have All the Birds Gone?* Princeton University Press.

Terborgh, J., Lopez, L., Núñez, P., Rao, M., Shahabuddin, G., Orihuela, G., Riveros, M., Ascanio, R., Adler, G. H., Lambert, T. D., & Balbas, L. (2001). Ecological meltdown in predator-free forest fragments. *Science, 294*(5548), 1923–1926. https://doi.org/10.1126/science.1064397

Thibault, K. M., Ernest, S. K. M., White, E., Brown, J. H., & Goheen, J. R. (2010). Long-term insights into the influence of precipitation on community dynamics in desert rodents. *Journal of Mammalogy, 91*(4), 787–797. https://doi .org/10.1644/09-mamm-s-142.1

Thiele-Bruhn, S., Bloem, J., De Vries, F. T., Kalbitz, K., & Wagg, C. (2012). Linking soil biodiversity and agricultural soil management. *Current Opinion in Environmental Sustainability, 4*(5), 523–528. https://doi.org/10.1016/j.cosust .2012.06.004

Thomas, C. D., Jones, T. H., & Hartley, S. (2019). "Insectageddon": A call for more robust data and rigorous analyses. *Global Change Biology, 25*(6), 1891–1892. https://doi.org/10.1111/gcb.14608

Thomas, J. A. (2016). Butterfly communities under threat. *Science, 353*, 216–218.

Thomas, R. L., Fellowes, M. D. E., & Baker, P. J. (2012). Spatio-temporal variation in predation by urban domestic cats (*Felis catus*) and the acceptability of pos-

sible management actions in the UK. *PLOS ONE, 7*(11), e49369. https://doi.org
/10.1371/journal.pone.0049369

Thomson, K. (1991). *Living Fossil: The Story of the Coelacanth.* W. W. Norton.

Tilman, D., May, R. M., Lehman, C., & Nowak, M. A. (1994). Habitat destruction
and the extinction debt. *Nature, 371*(6492), 65–66. https://doi.org/10.1038
/371065a0

Toibin, C., & Ferriter, D. (2002). *The Irish Famine: A Documentary.* Thomas Dunne
Books.

Toon, O. B., Robock, A., & Turco, R. P. (2014, May 9). Environmental
consequences of nuclear war. *AIP Conference Proceedings, 1596*(1): 65–73.
https://doi.org/10.1063/1.4876320

Tori, G. M., McLeod, S., McKnight, K., Moorman, T., & Reid, F. A. (2002).
Wetland conservation and ducks unlimited: Real world approaches to
multispecies management. *Waterbirds, 25,* 115–121. https://www.jstor.org
/stable/1522457

Trejo, I., & Dirzo, R. (2002). Floristic diversity of Mexican seasonally dry
tropical forests. *Biodiversity and Conservation, 11,* 2063–2084. https://doi.org
/10.1023/A:1020876316013

Turschwell, M. P., Brown, C. J., Pearson, R. M., & Connolly, R. M. (2020). China's
Belt and Road Initiative: Conservation opportunities for threatened marine
species and habitats. *Marine Policy, 112,* 103791. https://doi.org/10.1016/j
.marpol.2019.103791

Tyler, M. J., Shearman, D. J. C., Franco, R., O'Brien, P. E., Seamark, R., & Kelly,
R. W. (1983). Inhibition of gastric acid secretion in the Gastric brooding frog,
Rheobatrachus silus. Science, 220(4597), 609–610. https://doi.org/10.1126
/science.6573024

Union of Concerned Scientists. (1993). *World Scientists' Warning to Humanity.*
Union of Concerned Scientists.

US Fish and Wildlife Service. (1973). *Endangered Species Act.* Department of the
Interior. https://www.fws.gov/law/endangered-species-act

US Fish and Wildlife Service. (1993). *Amendment of Endangered Species Act of 1973.*
Department of the Interior. Citation 16 U.S.C. 1531–1544. https://www.fws.gov
/law/endangered-species-act

US Fish and Wildlife Service. (2011). *FWS-Listed U.S. Species by Taxonomic Group—
Clams.* Department of the Interior. https://ecos.fws.gov/ecp/report/species
-listings-by-tax-group?statusCategory=Listed&groupName=Clams&total=123

US Fish and Wildlife Service. (2019). *Endangered Species Act: Overview.*
Endangered Species. Department of the Interior. https://www.fws.gov/law
/endangered-species-act

US Fish and Wildlife Service. (2022). *FWS-Listed US Species by Taxonomic Group—Insects*. Department of the Interior. https://ecos.fws.gov/ecp/report/species-listings-by-tax-group?statusCategory=Listed&groupName=Insects

US Fish and Wildlife Service. (2023). *FWS Species Listed as Distinct Population Segments (DPS)*. Department of the Interior. https://ecos.fws.gov/ecp/report/dps

Van Aarde, R., Whyte, I. J., & Pimm, S. L. (1999). Culling and the dynamics of the Kruger National Park African elephant population. *Animal Conservation, 2*(4), 287–294. https://doi.org/10.1111/j.1469-1795.1999.tb00075.x

Van Den Hoogen, J., Geisen, S., Routh, D., Ferris, H., Traunspurger, W., Wardle, D. A., De Goede, R., Adams, B. J., Ahmad, W., Andriuzzi, W. S., Bardgett, R. D., Bonkowski, M., Campos-Herrera, R., Cares, J. E., Caruso, T., De Brito Caixeta, L., Chen, X., Costa, S. R., Creamer, R., . . . Crowther, T. W. (2019). Soil nematode abundance and functional group composition at a global scale. *Nature, 572*(7768), 194–198. https://doi.org/10.1038/s41586-019-1418-6

Van Der Heijden, M. G. A., Klironomos, J. N., Ursic, M., Moutoglis, P., Streitwolf-Engel, R., Böller, T., Wiemken, A., & Sanders, I. R. (1998). Mycorrhizal fungal diversity determines plant biodiversity, ecosystem variability and productivity. *Nature, 396*(6706), 69–72. https://doi.org/10.1038/23932

vanEngelsdorp, D., Traynor, K. S., Andree, M., Lichtenberg, E. M., Chen, Y., Saegerman, C., & Cox-Foster, D. (2017). Colony Collapse Disorder (CCD) and bee age impact honey bee pathophysiology. *PLOS ONE, 12*(7), e0179535. https://doi.org/10.1371/journal.pone.0179535

Van Huis A., Van Itterbeeck, J., Klunder, H., Mertens, E., Halloran, A., Muir, G., & Vantomme, P. (2013). *Edible Insects: Future Prospects for Food and Feed Security*. Food and Agriculture Organization of the United Nations.

Van Zwieten, P., Kolding, J., Plank, M. J., Hecky, R. E., Bridgeman, T. B., MacIntyre, S., Seehausen, O., & Silsbe, G. M. (2016). The Nile perch invasion in Lake Victoria: Cause or consequence of the haplochromine decline? *Canadian Journal of Fisheries and Aquatic Sciences, 73*(4), 622–643. https://doi.org/10.1139/cjfas-2015-0130

Vejřík, L., Vejříková, I., Kočvara, L., Blabolil, P., Peterka, J., Sajdlová, Z., Jůza, T., Šmejkal, M., Kolařík, T., Bartoň, D., Kubečka, J., & Čech, M. (2019). The pros and cons of the invasive freshwater apex predator, European catfish *Silurus glanis*, and powerful angling technique for its population control. *Journal of Environmental Management, 241*, 374–382. https://doi.org/10.1016/j.jenvman.2019.04.005

Veron, J., Hoegh-Guldberg, O., Lenton, T. M., Lough, J., Obura, D., Pearce-Kelly, P., Sheppard, C., Spalding, M., Stafford-Smith, M., & Rogers, A. D. (2009). The coral reef crisis: The critical importance of <350ppm CO2. *Marine Pollution Bulletin, 58*(10), 1428–1436. https://doi.org/10.1016/j.marpolbul.2009.09.009

Vethaak, A., & Legler, J. (2021). Microplastics and human health. *Science, 371*(6530), 672–674. https://doi.org/10.1126/science.abe5041

Vice, D. S., Vice, D. L., & Gibbons, J. C. (2005). Multiple predations of wild birds by brown tree snakes (*Boiga irregularis*) on Guam. *Micronesica, 38*(1), 121–124. https://micronesica.org/sites/default/files/6_vice-vice.pdf

Vidal, O., López-García, J., & Rendón-Salinas, E. (2013). Trends in deforestation and forest degradation after a decade of monitoring in the Monarch Butterfly Biosphere Reserve in Mexico. *Conservation Biology, 28*(1), 177–186. https://doi.org/10.1111/cobi.12138

Vidal, O., & Rendón-Salinas, E. (2014). Dynamics and trends of overwintering colonies of the monarch butterfly in Mexico. *Biological Conservation, 180,* 165–175. https://doi.org/10.1016/j.biocon.2014.09.041

Vitousek, P. M., Ehrlich, P. R., Ehrlich, A. H., & Matson, P. A. (1986). Human appropriation of the products of photosynthesis. *BioScience, 36*(6), 368–373. https://doi.org/10.2307/1310258

Vogt, R. C., Dirzo, R., & Soriano, E. G. (1997). *Historia Natural de Region de Los Tuxtlas.* Univerisdad Nacional Autónoma de México, Mexico D.F.

Voigt, M., Wich, S. A., Ancrenaz, M., Meijaard, E., Abram, N. K., Banes, G. L., Campbell-Smith, G., D'Arcy, L., Delgado, R. A., Erman, A., Gaveau, D., Goossens, B., Heinicke, S., Houghton, M., Husson, S. J., Leiman, A., Sánchez, K. L., Makinuddin, N., Marshall, A. J., . . . Kühl, H. S. (2018). Global demand for natural resources eliminated more than 100,000 Bornean orangutans. *Current Biology, 28*(5), 761–769.e5. https://doi.org/10.1016/j.cub.2018.01.053

Von Frisch, K. (1993). *The Dance Language and Orientation of Bees.* Harvard University Press.

Waal, F. B. M., & Lanting, F. (1997). *Bonobo: The Forgotten Ape.* University of California Press.

Wagner, D. L. (2012). Moth decline in the northeastern United States. *News of the Lepidopterists' Society, 54*(2): 52–56. https://nationalmothweek.org/wp-content/uploads/2012/07/Moth-Decline.pdf

Wake, D. B., & Vredenburg, V. T. (2008). Are we in the midst of the sixth mass extinction? A view from the world of amphibians. *Proceedings of the National Academy of Sciences, 105*(supplement_1), 11466–11473. https://doi.org/10.1073/pnas.0801921105

Walter, K. (2017, August 28). Ancient history of Lyme disease in North America revealed with bacterial genomes. *Yale School of Medicine.* https://medicine

.yale.edu/news-article/ancient-history-of-lyme-disease-in-north-america
-revealed-with-bacterial-genomes

Wang, J., Parham, J. F., & Shi, H. (2021). China's turtles need protection in the wild. *Science, 371*(6528), 473. https://doi.org/10.1126/science.abg3541

Wang, S. L., Ball, V. E., Nehring, R. F., Williams, R., & Chau, T. (2017). Impacts of climate change and extreme weather on U.S. agricultural productivity: Evidence and projection. *Agricultural Productivity and Producer Behavior*, 41–75. National Bureau of Economic Research. https://doi.org/10.3386/w23533

Wardle, D. A., Bardgett, R. D., Klironomos, J. N., Setälä, H., Van Der Putten, W. H., & Wall, D. H. (2004). Ecological linkages between aboveground and belowground biota. *Science, 304*(5677), 1629–1633. https://doi.org/10.1126/science.1094875

Warren, M. (2019). Conserving British butterflies: Progress against the odds. *New of The Lepidopterists' Society, 61*(6), 3–6.

Washington, H., & Kopnina, H. (2018). The insanity of endless growth. *Ecological Citizen*, 2. https://www.researchgate.net/publication/323285645_The _insanity_of_endless_growth

Watanabe, J., Matsuoka, H., & Hasegawa, Y. (2018). Pleistocene fossils from Japan show that the recently extinct Spectacled Cormorant (*Phalacrocorax perspicillatus*) was a relict. *The Auk, 135*(4), 895–907. https://doi.org/10.1642 /auk-18-54.1

Watson, K., & Stallins, J. A. (2016). Honey bees and colony collapse disorder: A pluralistic reframing. *Geography Compass, 10*(5), 222–236. https://doi.org/10 .1111/gec3.12266

Watts, J. (2018, February 14). New round of oil drilling goes deeper into Ecuador's Yasuní national park. *The Guardian*. https://www.theguardian.com /environment/2018/jan/10/new-round-of-oil-drilling-goes-deeper-into -ecuadors-yasuni-national-park

Watts, J. (2022, March 27). US-China soy trade war could destroy 13 million hectares of rainforest. *The Guardian*. https://www.theguardian.com /environment/2019/mar/27/us-china-soy-tariff-war-could-destroy-13 -million-hectares-of-amazon-rainforest

Weiss, S. B. (1999). Cars, cows, and checkerspot butterflies: Nitrogen deposition and management of nutrient-poor grasslands for a threatened species. *Conservation Biology, 13*(6), 1476–1486. https://doi.org/10.1046/j.1523-1739.1999 .98468.x

Wenny, D. G., DeVault, T. L., Johnson, M. D., Kelly, D., Şekercioğlu, Ç. H., Tomback, D. F., & Whelan, C. J. (2011). The need to quantify ecosystem services provided by birds. *The Auk, 128*(1), 1–14. https://doi.org/10.1525/auk.2011.10248

Wepprich, T., Adrion, J. R., Ries, L., Wiedmann, J., & Haddad, N. M. (2019). Butterfly abundance declines over 20 years of systematic monitoring in Ohio, USA. *PLOS ONE*, 14(7), e0216270. https://doi.org/10.1371/journal.pone .0216270

West, C., Wolfe, J. D., Wiegardt, A., & Williams-Claussen, T. (2017). Feasibility of California Condor recovery in northern California, USA: Contaminants in surrogate turkey vultures and common ravens. *Ornithological Applications Prev the Condor*, 119(4), 720–731. https://doi.org/10.1650/condor-17-48.1

Wheye, D., & Ehrlich, P. R. (2015). Are there caterpillars on butterfly wings? *News of the Lepidopterists' Society*, 57, 182–185.

Whitaker, R. J., Grogan, D. W., & Taylor, J. W. (2003). Geographic barriers isolate endemic populations of hyperthermophilic archaea. *Science*, 301(5635), 976–978. https://doi.org/10.1126/science.1086909

White, L. C., Mitchell, K. J., & Austin, J. J. (2017). Ancient mitochondrial genomes reveal the demographic history and phylogeography of the extinct, enigmatic thylacine (*Thylacinus cynocephalus*). *Journal of Biogeography*, 45(1), 1–13. https://doi.org/10.1111/jbi.13101

White, L. C., Saltré, F., Bradshaw, C. J. A., & Austin, J. J. (2018). High-quality fossil dates support a synchronous, late Holocene extinction of devils and thylacines in mainland Australia. *Biology Letters*, 14(1), 20170642. https://doi .org/10.1098/rsbl.2017.0642

White, T., Petrovan, S. O., Christie, A. P., Martin, P. A., & Sutherland, W. J. (2022). What is the price of conservation? A review of the status quo and recommendations for improving cost reporting. *BioScience*, 72(5), 461–471. https://doi.org/10.1093/biosci/biac007

Wiens, J. J. (2016). Climate-related local extinctions are already widespread among plant and animal species. *PLOS Biology*, 14(12), e2001104. https://doi .org/10.1371/journal.pbio.2001104

Wilcox, C. (2019, January 9). Lonely George the tree snail dies, and a species goes extinct. *National Geographic*. https://www.nationalgeographic.co.uk /animals/2019/01/lonely-george-the-snail-has-died-marking-the-extinction -of-his-species

Wilcox, C., Van Sebille, E., & Hardesty, B. D. (2015). Threat of plastic pollution to seabirds is global, pervasive, and increasing. *Proceedings of the National Academy of Sciences of the United States of America*, 112(38), 11899–11904. https://doi .org/10.1073/pnas.1502108112

Williams, G. J., Graham, N. A. J., Jouffray, J., Norström, A. V., Nyström, M., Gove, J. M., Heenan, A., & Wedding, L. M. (2019). Coral reef ecology in the

Anthropocene. *Functional Ecology, 33*(6), 1014–1022. https://doi.org/10.1111/1365-2435.13290

Williams, J., ReVelle, C., & Levin, S. A. (2005). Spatial attributes and reserve design models: A review. *Environmental Modeling & Assessment, 10*(3), 163–181. https://doi.org/10.1007/s10666-005-9007-5

Willoughby, J. R., Sundaram, M., Wijayawardena, B. K., Kimble, S. J. A., Ji, Y., Fernández, N. B., Antonides, J. D., Lamb, M. C., Marra, N. J., & DeWoody, J. A. (2015). The reduction of genetic diversity in threatened vertebrates and new recommendations regarding IUCN conservation rankings. *Biological Conservation, 191*, 495–503. https://doi.org/10.1016/j.biocon.2015.07.025

Wilson, E. O. (1988). *Biodiversity*. National Academy of Sciences Press. https://doi.org/10.17226/989.

Wilson, E. O. (2016). *Half-Earth: Our Planet's Fight for Life*. W. W. Norton.

Wittemyer, G., Northrup, J. M., Blanc, J., Douglas-Hamilton, I., Omondi, P., & Burnham, K. P. (2014). Illegal killing for ivory drives global decline in African elephants. *Proceedings of the National Academy of Sciences of the United States of America, 111*(36), 13117–13121. https://doi.org/10.1073/pnas.1403984111

Woinarski, J. C. Z., Burbidge, A. A., & Harrison, P. L. (2015). Ongoing unraveling of a continental fauna: Decline and extinction of Australian mammals since European settlement. *Proceedings of the National Academy of Sciences of the United States of America, 112*(15), 4531–4540. https://doi.org/10.1073/pnas.1417301112

Woinarski, J. C. Z., Murphy, B. P., Legge, S., Garnett, S. T., Lawes, M. J., Comer, S., Dickman, C. R., Doherty, T. S., Edwards, G., Nankivell, A., Paton, D. C., Palmer, R., & Woolley, L. (2017). How many birds are killed by cats in Australia? *Biological Conservation, 214*, 76–87. https://doi.org/10.1016/j.biocon.2017.08.006

Wood, R., Liu, A., Bowyer, F., Wilby, P. R., Dunn, F. S., Kenchington, C. G., Cuthill, J. F. H., Mitchell, E. G., & Penny, A. (2019). Integrated records of environmental change and evolution challenge the Cambrian Explosion. *Nature Ecology and Evolution, 3*(4), 528–538. https://doi.org/10.1038/s41559-019-0821-6

Woolley, L., Geyle, H. M., Murphy, B. P., Legge, S., Palmer, R., Dickman, C. R., Augusteyn, J., Comer, S., Doherty, T. S., Eager, C., Edwards, G., Harley, D., Leiper, I., McDonald, P. J., McGregor, H., Moseby, K. E., Myers, C., Read, J., Riley, J., . . . Woinarski, J. C. Z. (2019). Introduced cats *Felis catus* eating a continental fauna: Inventory and traits of Australian mammal species killed. *Mammal Review, 49*(4), 354–368. https://doi.org/10.1111/mam.12167

World Economic Forum (WEF). (2016, January). *The New Plastics Economy: Rethinking the Future of Plastics*. http://wef.ch/plasticseconomy

Worm, B., Barbier, E. B., Beaumont, N., Duffy, J. E., Folke, C., Halpern, B. S., Jackson, J. B. C., Lotze, H. K., Micheli, F., Palumbi, S. R., Sala, E., Selkoe, K. A., Stachowicz, J. J., & Watson, R. (2006). Impacts of biodiversity loss on ocean ecosystem services. *Science, 314*(5800), 787–790. https://doi.org/10.1126/science.1132294

Worthy, T. H., & Holdaway, R. N. (2002). *The Lost World of the Moa*. University of Indiana Press.

Wynes, S., & Nicholas, K. A. (2017). The climate mitigation gap: Education and government recommendations miss the most effective individual actions. *Environmental Research Letters, 12*(7), 074024. https://doi.org/10.1088/1748-9326/aa7541

Xue, Y., Prado-Martinez, J., Sudmant, P. H., Narasimhan, V. M., Ayub, Q., Szpak, M., Frandsen, P., Chen, Y., Yngvadóttir, B., Cooper, D. N., De Manuel, M., Hernández-Rodríguez, J., Lobón, I., Siegismund, H. R., Pagani, L., Quail, M. A., Hvilsom, C., Mudakikwa, A., Eichler, E. E., . . . Scally, A. (2015). Mountain gorilla genomes reveal the impact of long-term population decline and inbreeding. *Science, 348*(6231), 242–245. https://doi.org/10.1126/science.aaa3952

Yamaura, Y., Kawahara, T., Iida, S., & Ozaki, K. (2008). Relative importance of the area and shape of patches to the diversity of multiple taxa. *Conservation Biology, 22*(6), 1513–1522. https://doi.org/10.1111/j.1523-1739.2008.01024.x

Yashina, S., Gubin, S., Markismovich, S., & Gilichinsky, D. (2012). Regeneration of whole fertile plants from 30,000-y-old fruit tissue buried in Siberian permafrost. *PNAS, 109*, 4008–4013. https://doi.org/10.1073/pnas.1118386109

Yoğurtçuoğlu, B., & Ekmekçi, F. G. (2018). First record of the giant pangasius, *Pangasius sanitwongsei* (*Actinopterygii: Siluriformes: Pangasiidae*), from central Anatolia, Turkey. *Acta Ichthyologica Et Piscatoria, 48*(3), 241–244. https://doi.org/10.3750/aiep/02407

Young, H. S., Dirzo, R., Helgen, K. M., McCauley, D. J., Billeter, S. A., Kosoy, M., Osikowicz, L. M., Salkeld, D. J., Young, T. P., & Dittmar, K. (2014). Declines in large wildlife increase landscape-level prevalence of rodent-borne disease in Africa. *Proceedings of the National Academy of Sciences of the United States of America, 111*(19), 7036–7041. https://doi.org/10.1073/pnas.1404958111

Young, H. S., McCauley, D. J., Galetti, M., & Dirzo, R. (2016). Patterns, causes, and consequences of anthropocene defaunation. *Annual Review of Ecology, Evolution, and Systematics, 47*(1), 333–358. https://doi.org/10.1146/annurev-ecolsys-112414-054142

Youngster, I., Russell, G. H., Pindar, C., Ziv-Baran, T., Sauk, J., & Hohmann, E. (2014a). Oral, capsulized, frozen fecal microbiota transplantation for relapsing *Clostridium difficile* infection. *JAMA, 312*(17), 1772. https://doi.org/10.1001/jama.2014.13875

Youngster, I., Sauk, J., Pindar, C., Wilson, R. L., Kaplan, J. L., Smith, M., Alm, E. J., Gevers, D., Russell, G. H., & Hohmann, E. (2014b). Fecal microbiota transplant for relapsing *Clostridium difficile* infection using a frozen inoculum from unrelated donors: A randomized, open-label, controlled pilot study. *Clinical Infectious Diseases, 58*(11), 1515-1522. https://doi.org/10.1093/cid/ciu135

Zedan, H., Wijnstekers, W., Hepworth, R., Bridgewater, P., & Bandarin, F. (2005). *Biodiversity: Life Insurance for Our Changing World.* CITES. https://cites.org/eng/news/pr/2005/050912_statement.shtml

Zhan, S., Zhang, W., Niitepõld, K., Hsu, J., Haeger, J. F., Zalucki, M. P., Altizer, S., De Roode, J. C., Reppert, S. M., & Kronforst, M. R. (2014). The genetics of monarch butterfly migration and warning colouration. *Nature, 514*(7522), 317-321. https://doi.org/10.1038/nature13812

Zhang, T., Wu, Q., & Zhang, Z. (2020). Probable pangolin origin of SARS-CoV-2 associated with the COVID-19 outbreak. *Current Biology, 30*(7), 1346-1351.e2. https://doi.org/10.1016/j.cub.2020.03.022

Zhang, X., Shu, D., Han, J., Zhang, Z., Liu, J., & Fu, D. (2014). Triggers for the Cambrian Explosion: Hypotheses and problems. *Gondwana Research, 25*(3), 896-909. https://doi.org/10.1016/j.gr.2013.06.001

Zietsman, G. (2019). *Hot Debate on Elephant Culling vs Contraception vs Exports Continues.* Conservation Action Trust. https://www.conservationaction.co.za/recent-news/hot-debate-elephant-culling-vs-contraception-vs-exports-continues

Zimmerman, N., Izard, J., Klatt, C. G., Zhou, J., & Aronson, E. L. (2014). The unseen world: environmental microbial sequencing and identification methods for ecologists. *Frontiers in Ecology and the Environment, 12*(4), 224-231. https://doi.org/10.1890/130055

Zipkin, E. F., Ries, L., Reeves, R., Regetz, J., & Oberhauser, K. S. (2012). Tracking climate impacts on the migratory monarch butterfly. *Global Change Biology, 18*(10), 3039-3049. https://doi.org/10.1111/j.1365-2486.2012.02751.x

Zschokke, S., Armbruster, G. F. J., Ursenbacher, S., & Baur, B. (2011). Genetic differences between the two remaining wild populations of the endangered Indian rhinoceros (*Rhinoceros unicornis*). *Biological Conservation, 144*(11), 2702-2709. https://doi.org/10.1016/j.biocon.2011.07.031

Standard and Scientific Name Index

Page numbers in italics indicate illustrations.

Abominable Snowman (Yeti), xi
Achatina fulica, 121
Achatinella apexfulva, 121
Acipenser sinensis, 95–96
Acrocephalus taiti, 75
Aldabra giant tortoises, 89–90
Alloteropsis semialata, 78
Alouatta seniculus, 176
Amazona ochrocephala, 76
American woodpecker, 71
Andrias davidianus, 85
Anguilla anguilla, 95
Anodorhynchus hyacinthinus, 76
Anopheles mosquitos, 167
Antarctic penguins, xii
antelope, 146, 175, 232
Antiaris toxicaria, 195
Arabian oryx, 232
Arabian sand cat, 186
Ara macao, 76
Ara militaris, 76
Argentinosaurus, 20
Asian giant soft-shell turtle, 89

Ateles paniscus, 175
Atitlán grebe, 69
Atlantic cod, 96
Atlantic halibut, 95
Australian aggressive honeyeater, 226
Australopithecus, 87

Bachman's warbler, xiii, xiv
Batrachochytrium dendrobatidis, 81, 179
Batrachochytrium salamandrivorans, 83
Batrachuperus londongensis, 85
Battus philenor, 102
Bay checkerspot, 102–3, 105–6, 108
bear, 2, 34–35, 59–60
bearded saki monkey, 176
Beluga sturgeon, 95
Benton's cave crayfish, 117
Bigfoot (Sasquatch), xi
big lupine plant, 105
bison, 57–58, 70, 160, 194
black-footed ferret, 58
black rhino, 52, 53, 54–55, 105, 160
black-tailed prairie dog, 59

Bluebuck, 33
Boiga irregularis, 68
Bolyeria multocarinata, 87, 90
bonobo, 42–45, 47, 65
Boophis williamsi, 85
Bornean orangutan, 41–42
Bowerbird, ix, x
Bramble Cay rat, 17, 33, 234, 236
Brassica oleracea, 135
Brosimum alicastrum, 134
brown bear, 12, 35
brown capuchin, 176
brown tree snake, 68
Buddleia, 110
Burmese python, 36
butterfly, 11, 99, 101–103, 104, 105–6,
   107, 108, 110–11, 114–15, 117, 146, 167,
   193, 211–12, 214–16, 225, 226, 237
butterfly bush, 110

C. (Cacatua) sulphurea, 76
C. (Corepgonus) reighardi, 91
C. (Crocodylus) intermedius, 88
C. (Crocodylus) siamensis, 88
Cacatua haematuropygia, 76
California condor, 79, 210, 232–33, 234,
   239
California plantain, 104
California rose, 9–10
Cambarus aculabrum, 117
Campephilus imperialis, 68
Camptorhynchus labradorius, 69
Canary Islands' oystercatcher, 68
Canis familiaris, 9
Canis latrans, 9
Canis lupus, 9
Capsicum annum, 135
caracara, 68
Carex mitchelliana, 107

Caribbean monk seal, 33
Carolina parakeet, 69, 237
Castor canadensis, 64
cat, 33, 36, 37, 73, 76, 178–79, 186, 224,
   228–29, 233, 237
Catarina pupfish, 92
Cebus apella, 176
Celestus occiduus, 87
Cepaea, 122
Cercopithecus kandti, 40
checkerspot butterfly, 102–3, 104,
   105–6, 108, 111, 214, 237
chili, 135
Chinese giant salamander, 85
Chinese sturgeon, 81, 92, 95
Chinook salmon, 93–94, 95
Chiropotes utahicki, 176
chytridiomycosis, 81, 82, 84, 85, 86, 179
Cissa thalassina, 79
Clostridium difficile, 154, 196
Coccidioides posadasii, 150
Coccidoides immitis, 150
cockatoo grass, 78
Coleoptera, 115
common alpine, 106
Compsilura concinnata, 109
Conuropsis carolinensis, 69
Conus, 122, 123
coral, xvi, 6, 17, 74, 120, 122, 127–29,
   196
Coregonus johannae, 91
corn, 99, 112, 167, 209
coyote, 9, 35, 36, 37
Craugastor omoaensis, 82
Crocodylus midorensis, 88
Crotalus stejnegeri, 88
crow, 68
Culex quinquefasciatus, 179
Cylindraspis indica, 87

Mission blue butterfly, 225
Mitchell's sedge, 107
Moho braccatus, 68
mold, 125, 201, 204
monarch butterfly, 106, 108, 109, 110
mountain gorilla, 39
mountain mist frog, 82
Mustela nigripes, 58

Neomonachus tropicalis, 33
Nesiota elliptica, 142
Nile perch, 181
North American beaver, 64
North American bison, 57–58, 70, 160, 194
North Atlantic right whale, 59, 243
Northern gastric-brooding frog, 82
Norway rat, 168

Oarisma poweshiek, 112
Oncorhynchus tshawytscha, 93
Ophrys, 112
orange-bellied parrot, 76, 77
orangutan, 41–42, 48, 160; of Sumatra, 42, 160; of Tapanuli, 41
Orinoco crocodile 89
Oryza sativa 200
Ostrea angasi 120

Pacific pocket mouse, 211
Pacific rat, 14
Palm cockatoo, 76
Pan (chimpanzee), 43
panda, 17, 37–38, 143, 145
Passenger pigeon, 17, 18, 69–71, 99, 114, 121, 143, 162, 190, 204, 237, 239, 241
Passer montanus, 200
Penicilium notatum, 204
Perameles notina, 33

Pharomachrus mocinno, 192
Philippine cockatoo, 76
Philippine crocodile, 88
photosynthesizing algae, 127
Phytophthora infestans, 150
pigeon, xiv, 14–15, *180, 181*, 241;
    Passenger, 17, 18, 69–71, 99, 114, 121,
    143, 162, 190, 204, 237, 239, 241
Pinguinus impennis, 68
pipevine swallowtail, 103
Plantago erecta, 104
Plebejus icarioides, 106
Plebejus icarioides missionenesis, 225
Podilymbus gigas, 69
Podilymbus podiceps, 69
polar bear, 34–35, 59–60, 61
Polyborus luctuosus, 68
Pongo abelli, 41
Pongo pymaeus, 41
Pongo tapanuliensis, 41
Potosi pupfish, 92
Poweshiek skipperling, 112
prairie dog, 57–58, 160, *193, 194*
Probosciger aterrimus, 76
Procambarus fallax, 118
Prolagus sardus, 33
Pseudophilautus adsperus, 82
Pteropus tokudae, 33
pterosaur, 20
Ptilinopus insularis, 74

queen conch, 123
Quercus robur, 200
quetzal, 192

R. *(Rheobatrachus) silus*, 82, *83*
Rafetus swinhoei, 88, 89
Rainbow burrowing frog, 85

Velociraptor, 20, 237
Vibrio cholerae bacteria, 146
Vini stepheni, 75

warbler, xiii, xiv, 71, 73, 192
Wels catfish, 179, *180*, 181
West African black rhino, 52
Western black rhino, 54
Western gorilla, 41
whale, 36, 59–61, 145, 180, 243
wheat, 98, 136, 167, 200
white abalone, 126
white rhino, 51, 52, 53, 54
whitebark pine, 195
wild dog, 202
wild domestic dog, 32

Williams' Boophis frog, 85
winter-run Chinook, 93–94
wolf, 9, 32, 35, 36, 37, 47, 160, 227, 234, 243
wooly rhinoceros, 53, 56
woolly monkey, 175, 176
Worthen's sparrow, 193

Yangtze softshell river turtle, 88, 89, 234
yellow-crested cockatoo, 76
yellow-headed parrot, 76

Zapornia atra, 75
Zea mays, 200
zebra swallowtail, 110
zooxanthellae, 120, 127, 128
Zostera marina, 125

Subject Index

archaeans, 13, 148, 153
Arctic National Wildlife Refuge, 220-21, 223
arthropods, 12-13, 161
Asia, ix, xv, 1, 22, 32, 56, 78, 79, 81, 82, 118, 160, 176, 203; species and population declines/extinctions, 32, 52, 55-56, 79, 104, 160, 186; wildlife markets, 78, 151
asteroid and meteorite collisions, xvi, xxi, 21-23, 24, 29-30, 175, 240-41
atmosphere: formation and composition, 6-7, 132; geoengineering of, xvii
atolls, 16-17, 75, 89-90
Australia, 6, 86-87, 115, 179, 197, 218, 225-26, 231-32; species and population declines/extinctions, ix-x, 16-18, 31-33, 76-78, 77, 82, 173, 197, 228, 229-30, 231; wildfires, 173-74, 229
Australopithecus, 87
avian malaria, 179

bacteria, 6, 11, 13, 147-48, 155-56; of human gut microbiome, 146, 147, 152-53, 155-56, 196; nitrogen-fixing, 146
Balkan, Joel, 187
Barnosky, Anthony, 27-28
bats, 33, 63-64, 151-52, 175, 199
bears, 37; brown, 11-12, 34-35; giant pandas, 17, 37-38; grizzly, 11-12, 34-35; polar, 34-35, 60, 61
Beattie, Andrew, 146-47
beavers, 35-36, 64-65, 83, 127
Beebe, William, 29
bees. *See* honey bees

beetles, 12-13, 199; dung beetle, 115
behavior, 42, 43-44, 45, 46-47, 102, 103, 170, 243
Beyond Words (Safina), 50-51
Biden, Joseph, 223
Big Myth, The (Oreskes and Conway), 188
biodiversity, xvii, xxii, 28-30, 65, 99, 165, 166, 169, 182-83, 184; Cambrian explosion, 7-8, 132; classification and measurement, 9-13; "conservation," 146; dinosaur-dominated, 21; evolution, 5-8; mammals, 45, 52; marine, 127; microbes, 145-56; plants, 131-44; production, 147, 149, 152, 155-56; reptiles, 86; of the soil, 125, 149
biodiversity loss, xvii, xxii, 30, 146, 160-61, 165, 168, 169, 173, 185; as ecosystem services loss, 190-204; as existential crisis, 29, 144, 166; rate, 127, 191. *See also* defaunation; drivers of extinction; mass extinctions; population extinction crisis
biodiversity protection, 217-41; individual actions for, 243; policy and education approaches, 206-13; in reserves/protected areas, 217-25; science and education approaches, 213-17
biogeography, 74, 110, 125, 147-48, 223
biological annihilation, xxi, 32, 160-61, 204
biological control, 63, 108, 121, 147, 199-200
biological corridors, 164, 226-27
biomass, 116, 133, 134, 158, 174-75, 185, 191
bioprospecting and biopiracy, 123, 147

biosphere, 4, 144, 158, 204, 217–18, 220–21, 244

bioturbation, 126

birds, 22, 67–79, 122, 168, 174, 192; ancestors/evolution, 19, 26, 73; breeding habitats, xiii, 36, 70, 73, 74, 76, 77; building and wind turbine collisions, 73, 224; diseases, 68, 79, 179; ecosystem services, 190, 191, 192, 194–95, 199–200; extinction drivers, 174, 183; predator control programs, 227–32, 234; predators, 14, 36, 68, 73, 74–75, 76, 78, 179–81; seabirds, shorebirds, and water-birds, xii, 74, 178, 180–81, 184–85; songbirds, xiii–xiv, 73–75, 76, 78–79, 146, 224; species and population declines/extinctions, 5, 14–17, 18, 28, 67–68, 69–71, 73–75, 161–62, 210. *See also individual bird species*

bison, North American, 57–58, 70, 160, 194

bivalves, 118–20

Blight (Monosson), 150–51

bluebuck, 33

Bolsonaro, Jair, 242

bonobos, 42–45, 46, 65

Botanic Gardens Conservation Initiative, Threat Search, 138, 139

breeding grounds, xii–xiv, 60–61, 73, 74, 76–77, 88, 99, 107

Breedlove, Dennis, 106

Brewis, Harriet, 173

bubonic plague, 163, 165

Burkle, Laura, 214

bushmeat, 40–41, 44, 47, 48, 70–71, 126, 175–76, 193, 200, 203, 217

butterflies, 11, 83, 99–112, 167; caterpil-lars, 105–6, 107, 108, 111, 167, 199, 216;

checkerspots, 72, 102–6, 110, 214, 237; conservation programs, 211–12, 225, 237; monarch, 106–9; Paul Ehrlich's research with, 101–7, 110, 214–15; species and population declines/extinctions, 99, 101, 102–6, 103, 214; square-bashing census, 111, 214–15, 216

caecilians, 81

Cambrian explosion, 7–8, 132

camera traps, 158–59, 186

Cannery Row (Steinbeck), 174

captive breeding programs, x–xi, 38, 56, 62, 76, 86, 88, 89, 94, 117, 120, 124, 126, 177, 232, 233, 234

carbon dioxide, 6–7, 26, 59, 123, 167, 221–22

Carboniferous period, 25–26

carbon sequestration, 38, 59, 197–98

catfish, Wels, 179–81

cats: Arabian sand cat, 186; domestic/house, 33, 36, 37, 73, 76, 228–29, 230; feral/wild, 36, 73, 178–79, 186, 223–24, 228, 230

Ceballos, Gerardo, xx, xxvi, 28, 29, 38–40, 45–46, 54–55, 192, 193, 204, 209, 245

cetaceans, 59–62

chemical pollution, 101, 117–18, 172

chimpanzees, 42–44, 47

China, 53, 88–89, 141, 176–77, 199–200, 218–19; Belt and Road Initiative, 170–71, 242; as COVID-19 origin site, 151; giant pandas, 37–38; ivory trade, 49, 51, 235; sturgeons, 81, 92–93

chytridiomycosis fungal disease, 81–84, 85, 150, 179

Cichlidae, 94

dams: beaver-constructed, 35–36, 65; human-constructed, 92–93, 119, 171, 227–28

Darwin, Charles, 21, 33, 125, 132–33, 204, 215; *On the Origin of Species*, 10

Dasgupta, Partha, 166

decomposition, 115, 116, 117, 122, 148, 198, 232

deer, 108, 109, 152, 163, 175

de-extinction, 237–41; *Jurassic Park* (film) about, 19, 237–38, 240

defaunation, xxi, xxii, xxiv, 21–23, 157–65, 168, 178, 181, 184–85, 188–89, 217; aftermath, 23–27; differential, 23–25, 151, 160–65, 188–89; toxicant-driven, 184–85

deforestation, xvii, 15, 26, 71, 74, 109, 137–38, 172, 178, 197, 229; rates, 137–38, 157, 161; tropical forests, 84–85, 195, 242

desert species, 80, 90, 92, 186, 214, 237

Devonian period, 5, 24–26

Digirinana, Francois, 40

dingoes, x, 32, 228–29

dinosaurs, xi–xii, xvi, xxi, 5, 8, 19–21, 20–24, 26–27, 30, 86, 175

diprotodonts, x

Dirzo, Rodolfo, xx, xxvi, 29–30, 75, 133, 134–35, 142, 157, 192

diseases, 35, 80, 172; of bats, 63; of birds, 68, 77, 79; control, 191; of domestic animals, 46, 146; of fishes, 92, 93; human, 182; human, animal models, 82, 84; human, transmission to animals, 43; introduced, 33, 77, 81; relation of extinctions to, 70–71, 77; vectors, xvi, 103, 152, 163, 165, 179, 240; zoonoses, 151–52, 165

dodo, xiv, 14–18, 68, 69

dogs, 32, 37, 38, 76, 202, 230, 233

dolphins, 62

domesticated animals, 51, 113, 188, 191, 200, 259

drivers of extinction, 133–34, 166–89; interactions, 188–89; proximate, 171–87, 205-6; socio-political, 187–88; ultimate, 166–71, 187–89. *See also* climate change impacts; habitat loss and degradation; human population growth / overpopulation

droughts, 84, 93–94, 98, 164, 191, 200

ducks, 69

Earth, age, 18

earthworms, 124–25, 147

echidnas, 175

ecosystem engineers, 115, 124–25, 194. *See also* beavers; corals and coral reefs; oysters; prairie dogs

ecosystems, xvii, 3, 34, 72, 115, 147, 163, 213, 215, 230; Leuser, 42, 53; plant endemism and, 219; protection, 217–25; redundancy, 129–30; savannas, 37, 48, 54, 77, 160, 195; wolf-elk, 35–36. *See also* microbiomes; temperate regions/ecosystems; tropical species/ecosystems

ecosystem services, xx–xxi, 44, 47–48, 70–71, 84, 165; cultural, 109, 190–91, 192–93, 202–3; invertebrates, 115, 118, 122, 126–27; mammals, 35–36, 38, 47, 58–59, 63–65, 84, 128; medicinal, 123; population extinctions-related loss, 191–204; provisioning, 190–91, 196, 200–202; regulating, 190–91, 196–99, 202; soil organisms, 124–25; supporting, 124–25, 190–91, 193–96, 202. *See also* pollinators

Ecuador, Yasuni Biosphere Reserve, 220–23, 241
Edwards, William Henry, 109–10
eelgrass, 125–26
eggs: bird, xii–xiii, 14–15, 82, 85, 122, 183, 232, 233; fish, xiv, 94; frog, 83, 85; insect, 99, 104, 105–6, 107, 167; turtle, 88
Ehrlich, Anne, xix–xx, 43
Ehrlich, Lisa, 43
Ehrlich, Paul, 28, 43, 46, 60, 77, 79, 101–7, 101–11, 110, 120, 133, 169, 192, 193, 204, 214–15, 216
elephants, 31, 40, 48–51, 48–52, 133, 160, 171, 175, 176, 192, 202, 208, 235–36; ivory trade, 48–51, 168, 235, 241
empty forest syndrome, 157
endangered species, 33–34; critically endangered, 34, 41, 53, 88, 96, 117–18, 131; giant pandas, 37–38; international protections, 207–9; legal national protection, 209–13; population extinction rates, 5. *See also* threatened species
Endangered Species Act, 115, 209–13; distinct population segments (DPSs), 210–12
endemic species: animals, 40, 46, 74–75, 90, 91, 93–94, 229; plants, 140, 142–43, 218–19
End of the Game, The (Beard), 31–32
energy production and consumption, xxi, 59–61, 128, 142, 143, 166, 168, 190, 220–23, 241
Erwin, Terry, 12–13
ESA. *See* Endangered Species Act
Eskimo curlews, 67–68, 74
eukaryotes, 13
Europe, 11–12, 32, 56, 79, 165, 183, 215

European exploration and colonization, xii, xiv–xv, 1–2, 57, 64, 65, 70, 139, 140, 142–43, 178, 228
European Union, 208, 225
evolution, 6–8, 7, 18, 19, 21; adaptive radiation, 7–8, 94; after mass extinctions, 22–25, 27, 240–41; of cognitive skills, 44; Darwinian theory, 10; of humans (*Homo sapiens*), xv, 5, 14, 25, 27, 42–44, 47, 87; of plants, 131–33, 219; population extinctions and, xiv, xxi, 204, 210, 228
evolutionary radiation, 24
extinction, xvii, 29; mass extinction threat, 5–6; risk, 32; of top predators, 34–36. *See also* defaunation; deforestation; drivers of extinction
extinction rates, 3, 4, 18, 27–28, 31–32, 215, 240; background, 4, 18–19, 27–28, 159, 240. *See also* mass extinctions
extinct species: de-extinction, 237–41; number of, xxi, 32. *See also* species and population declines/extinctions *under individual species*

family size limits, 169, 243
ferns, 131, 132
ferrets, black-footed, 58
fertilizers, 149, 164, 183
fires/wildfires, x, 26, 71–73, 77–78, 82, 107, 141, 172, 173–74, 185, 191, 197
fishes, 8, 184; cichlid, 94, 181; extinction drivers, 92–93, 119, 174, 176, 186–87; freshwater, 80, 91–94, 168, 179–81; "living fossil," xi, 81; marine, 94, 96–97, 126, 174, 196; as pets, 76; species and population declines/extinctions, xxi, 5, 28, 80, 91–94

fishing/fisheries, 126, 127, 145, 147, 172, 174, 181, 195–96; commercial, 60–62, 93, 96, 112, 124, 217

flies, 81, 106, 109, 114–15; fruit *(Drosophila)*, 114, 155

flooding, 65, 93, 191, 196–97; coastal, 172, 196, 203, 221–22, 240

food chains, 74, 97–98, 133, 182, 191, 228

food webs, 2–3, 21, 75, 91, 133

Foraminifera, 22, 24–25

forestry / forest management, 193–94, 198, 220–21; logging, 38, 41, 44, 47, 48, 68–69, 82, 138, 168, 194, 212, 217, 221, 242

Fossey, Dian, 39

fossil fuels, xvi, xxi, 206, 223, 243

fossil record, x–xi, 6, 8, 14, 18, 19, 20–21, 23–26, 28, 74, 89–90, 131

foxes, 33, 36, 178, 228–29, 230, 231–32

Frankenstein (Shelley), 239

Frisch, Karl von, 202

frogs, 64, 82–86, 183; fungal disease, 81–84; gastric-brooding, xiv, 82–84, 204

fungal diseases, 63, 81–84, 85, 136–37, 138, 146, 150–51, 179

fungi, 5, 11, 53, 141, 148, 153, 201, 204; Operational Taxonomic Units, 148; root-associated (mycorrhizal), 148, 149

Galápagos Islands, giant tortoises, 89–90

geese, 125–26

genetic diversity, 39, 135–37, 147, 161, 196, 204

genome, human-chimpanzee comparison, 42–43

genomics, 149, 238

geographic range, 41, 57, 61, 77, 112, 117, 118, 159–60, 186, 194; contraction, 58, 107, 127, 134–35, 139, 143, 159, 160–62, 185

gharials, 88

giraffes, 48, 208

Gleick, Peter, 166

Global Diversity Hotspots, 140, 218–19

Global Invasive Species Database, 179

goats, 142–43

Goodall, Jane, 43

gorillas: Eastern, 41; mountain, 38–41; Western, 41

grasshoppers, 98–99, 100, 199

Great Auk, xii–xiii, 68

greenhouse gases, 26, 59, 128, 169, 221–22, 224, 239

Gruber, T., 42–43

guano, 75

Gymnosperms, 132, 139

Habitat Conservation Plans (HCPs), 212, 225–26

habitat fragmentation, 33, 38, 46, 51, 68, 76, 83, 115, 143, 163, 171, 178, 205–6, 226

habitat loss and degradation, 15, 32–33, 163, 169, 172–74, 205–6; amphibians, 80, 84–86; birds, 15, 68–69, 71, 73–74, 80; under ESA, 212; fishes, 80, 92; implication for de-extinction initiatives, 239; invertebrates, 101, 105, 107–8, 109, 116–17, 121, 124, 126; mammals, 51, 56, 58, 60, 64, 163, 164; primates, 41–44, 46; reptiles, 80, 87, 90

habitat recovery initiatives, 38, 57–58, 71, 73, 77–78, 86, 90–91

habitat requirements, 166–67

Haddad, Nick, 112
Haldane, J. B. S., 115
Hammer, Bill, 102
Hanski, Ilkka, 103, 215
Harvesting the Biosphere (Smil), 158
Hawaii, 68, 121, 179
hawks, 69–70
Hayes, Tyrone, 183
Henderson fruit-dove, 74–75
herbivores, x, 34–35, 37–38, 52, 142–43, 149, 161, 163, 199
hippopotamuses, 160
home ranges, 23, 50
honey bees, 102, 112–14, 146, 168, 199, 202–3, 214; colony collapse disorder, 113
hormones, 182, 183, 184
human population growth / overpopulation, xxi, 5–6, 34, 38, 77, 128, 150–51, 158, 166–70, 188, 205
humans *(Homo sapiens)*: diet and nutrition, xvii, 113, 119, 124, 126, 154–55, 182, 196, 199, 203; evolution and ancestors, 5, 14, 25, 27, 42–44, 87; extinctions, xiv–xv, 4; family size limits, 168–69, 243; gut microbiome, 146, 147, 152–56, 196; as invasive species, 25; as sixth extinction cause, 4, 5–6, 19, 27–30, 157, 168–69
hunting (by humans), as extinction driver, 2–3, 15, 32, 34, 68–69, 158, 172, 174–75; birds, 73; crocodilians, 88; mammals, 162, 175; market hunting, 70; primates, 41–42, 44, 47, 175–76; tortoises and turtles, 87–90; trophy hunting, 39, 231; ungulates, 175; whales, 59–61; wolves, 35. *See also* poaching

Hurricane Mitch, 197
hybridization, 35, 55, 69

ice ages, xii, 25, 56, 158
inbreeding, 32, 39, 163, 234
Indigenous people, xii, 123, 176; agricultural practices, 135–36, 137; Australian aboriginal, xiv–xv, 32, 76–77; exterminations, xiv–xv; Inuit, 60, 61; Native Americans, xv, 57, 70
inequity, social and economic, xvii, 170, 171, 178, 187, 197, 203, 205, 241, 244, 245, 252
infrastructure development, 44, 87, 133–34, 138, 170–71, 220, 225
In Search of the Golden Toad (Crump), 84
insects, 20, 83, 94, 97–116, 99, 176; ecosystem services, 97–98, 199; plant-eating, 105–6; species and population declines/extinctions, 5, 8, 114, 115, 244. *See also names of individual insect species*
intelligence, 39, 43, 44, 124, 236
Intergovernmental Platform on Biodiversity and Ecosystem Services, 173, 191
International Union for Conservation of Nature (IUCN), 34, 43, 54, 92, 95, 175, 185, 207–8; Red List, 43, 138, 139, 179, 207
invasive and introduced species, 25, 68, 76, 80, 91–92, 94, 101, 118, 138, 142–43, 178–81, 199, 205–6, 227–28, 230; fishes, 69, 85, 179–81, 216; humans as, 14, 25, 123, 228; invertebrates, 109, 113, 121, 123, 138; mammals, 15, 33, 65, 87, 90, 178–79, 228–32, 233, x; pathogens, 82, 179

invertebrates, 8, 97–130, 147; conservation protections, 211; marine, 125, 186–87; species and population declines/extinctions, 5, 99, 186–87, 211. *See also* insects; mollusks

ivory trade, 48–51, 61, 168, 235, 241

jaguars, 34, 36, 160, 163, 164
Janzen, Dan, 143
Javan green magpie, 79
jellyfish, 6, 127
Jurassic Park (film), 19, 238, 240
Jurassic period, 5, 19–21

kangaroos and wallabies, ix–x, 33, 173–174, 231
kelp, 2–3, 8–9
keystone species, 53, 58, 72, 194
Kigame, Paul, 39–40
King, Mary-Claire, 42–43
koalas, 173–74, 229

landscape connectivity, 111, 223
land-use change, 172–74. *See also* agriculture; deforestation; fires/wildfires; habitat fragmentation; urbanization
Last of the Curlews (Bodsworth), 67
Last Rhino, The (Lawrence), 51–52
lemurs, 47
Lepidopterists' Society, 102, 110–11
life: classification, 8–11; origin, 6–8, 19
light pollution, 101
limpets, eelgrass, 125–26
Linnaeus, Carl, 140; *Systema Naturae*, 8–10
lions, 34, 145, 167, 202, 227, 231

livestock and livestock grazing, 34, 35, 38, 58–59, 76, 77, 87, 99, 105, 121, 138, 172, 215, 229
"living dead" species, 143, 163, 173, 215
lizards, 80–81, 87–88, 90
locusts, 49, 67, 98–99, 101, 103
longevity, 23, 141, 143
lupine, 105–6
Lyme disease, 70–71

macaws, 76, 162
Madagascar, 85, 90, 179, 218–19
mammals: body size, ix–x, 20, 24–25, 29, 79, 107, 161, 175, 176, 228–29, 231; endangered, 34; evolution, 19–20, 26, 27, 240–41; extinction drivers, 174, 185–86; extinct species, 32–33; flying, 63–64; identification of new species, 12; post-Chicxulub Impact recovery, 24–26; scaled, 56–57; species and population declines/extinctions, 5, 27–28, 32–33, 229, 230. *See also individual mammal species*
mammoths, 8, 34, 237, 239
marine species: differential defaunation, 24; impact of plastics pollution on, 181–82; invertebrates, 125–26; mammals, 34, 36, 60–62, 181–82, 243; mass extinctions, 26; number of species, 13; predators, 26, 36, 60–61
marmots, 214
marsupials, ix–x, 32–33, 173–74, 175, 231
mass extinctions, 18–23, 161; biodiversity changes after, 24–25, 27, 240–41; fifth (Cretaceous-Tertiary), 5, 21–23, 26–27; first (Ordovician-Silurian), 5, 25; fourth (Triassic-Jurassic), 5, 26;

rates, 18-19; second (Devonian), 5, 24-26; sixth, 4, 5-6, 19, 27-30, 157, 168-69; third (Permian-Triassic), 5, 26

McKenna, Malcolm, 22

Mearns, Edward, 58

medicine: biomimetics, 147; derived from natural products, 123, 147, 204; traditional/folk medicine, 41, 52, 55-56, 62, 87-88, 89, 104, 120, 171-172, 192

megafauna, 29

megapodes, 232

MERS, 152

mesosaurs, 26

Mesozoic era, 19-21, 86

meta-populations, 103-4, 211-12, 215, 223

methane, 224

Mexico, 6, 90, 109, 163, 219; Chicxulub Impact, xxi, 21-24, 29-30, 175, 240-41; conservation initiatives, 62, 209; plant extinctions, 142; species and population declines/extinctions, 92, 142

mice: deer, 70-71; house, 168; Pacific pocket, 211

Michener, Charles "Mich," 102

microbes, 5, 13, 145-56; evolution, 6-7; plastics-borne, 182; soil, 149; species and population declines/extinctions, 5, 13; species richness, 13. *See also* bacteria; microbiomes

microbiomes, 147, 150-56, 196

migration, in response to climate change, 226-27

migratory species: beluga whales, 60; birds, xiii, 64, 67, 74, 76, 77, 101, 224; butterflies, 106-7; effect of light

pollution on, 101; fishes, 92-93, 95; grasshoppers, 98-99, 100

milkweed, 107-8

Millennium Ecosystem Assessment, 190-91

mining, 44, 48, 85, 117, 138, 142, 217, 242

Mojave Desert, CA, 80

molecular biology, 155, 236, 238

molecular genetics, 12-13, 148

mollusks, 26, 46-47, 69, 94, 118-24, 125, 126, 176

mongooses, 87, 90

monkeys, 40, 45-47, 175-76

moral hazard, 240

mosquitoes, 103, 167, 179, 191

moths, 53, 101, 102-3, 109

mussels, freshwater, 118-19

Myers, Norman, 218-19

National Geographic, 75

National Marine Fisheries Service, 210

national parks, 39, 42, 55, 57, 104, 217-18, 221, 241

natural selection, 121-22

nature reserves, 38, 42, 56, 163, 173, 217-18, 220-21, 223, 236, 245

nematodes (round worms), 125, 147

New Guinea, ix-x, xiv, 15, 17, 32, 33

nitrogen, 6, 105, 164, 198

nitrogen-fixing bacteria, 146

noise pollution, 60

Norscia, I., 42-43

North America, xi, 1, 11-12, 37, 109, 111, 195, 221, 226, 237, 239; species and population declines/extinctions, 35, 57, 63, 69-73, 98-100, 108, 119, 121, 125-26, 139-40, 160, 161-62, 229, 232-33

nuclear war, xv, 344
nutrient cycling, 25–26, 58–59, 181, 191,
193–94

oceans: acidification, 26, 94, 96, 123,
127–29, 172, 185; ecosystem services,
195–96; evolution of life in, 6–8;
nutrient cycling, 25, 26, 59; plastic
pollution, 74, 88, 181–82, 184–85;
pollution, 94, 96; temperature, 22,
25, 127–29, 185; toxification, 25, 26,
94, 96
octopi, 46–47, 124
offsets, 187, 225
oil and gas industry, 128, 168, 190,
220–23
olive trees, St. Helena, 142–43
opossums, 36
orangutans, 41–42, 160
Ordovician-Silurian period, 5, 25
oryx, Arabian, 232, 234
overexploitation/overharvesting, 163,
168, 172, 174–78, 205–6, 229; aquatic
species, 2–3, 59, 91, 92, 117–18, 119–20,
123, 128; crayfish, 117–18; mammals,
2–3, 59; marine life, 2–3, 59, 128;
mollusks, 119–20, 123; plants, 138,
168. See also fishing/fisheries;
hunting (by humans)
owls, 69–70
oxygen, 6–7, 26, 126, 132, 167, 191
oysters, 119–20
ozone layer, xvii, 7

Pacific islands: bird populations,
74–75; extinctions, xiv, 14–15, 33, 68,
74, 120–21, 226; human colonization,
xiv, 14, 74, 233; invasive/introduced
species, 14–15, 68, 74, 226, 233

Paleozoic era, 24
pandas, giant, 17, 37–38
pandemics, 19, 151, 152, 163, 165
Pangea, 21
pangolins, 56–57, 89, 171, 175, 176, 178,
192
parasites and parasitic diseases, 109,
113, 120, 125, 146, 167, 219
Parker, Martin, 188
parrots, 76–78, 233–34
parthenogenic species, 118
Patagonia, 67
perfluorinated compounds (PFAS),
183–84
Permian period, 26
Permian-Triassic period, 5, 26
persistent organic pollutants (POPs),
182
pest control, 49, 63, 98, 99, 100, 105–6,
109, 116, 121, 147, 199–200
pesticides, 149, 183, 224–25; DDT,
183, 233; neonicotinoid, 101, 109,
112
pet trade, 42, 46, 76, 81, 82, 87–88, 89,
217
Phanerozoic period, 132
Phillips, John, 20
philopatric species, 89
photosynthesis, 6–7, 127, 133
phytoplankton, 75
pigeons, 14–15; Passenger, 17–18, 69–71,
99, 114, 121, 143, 162, 190, 204, 237,
239, 241
pigs, 78, 90, 124, 230, 236
Pitcairn Islands, 74–75
plantain, California, 104
plants, 131–44; as aquatic deoxygenation cause, 26; biodiversity, 132–35,
138, 143; chemical compounds, 105,

123, 155, 201; climate change threat, 74; colonization of land, 131–32; defense mechanisms, 105, 123; diseases and parasites, 125, 150; ecosystem services, 200–201; evolution, 19, 131–32; flowering (Angiosperms), 19, 132, 134; Global Diversity Hotspots, 140, 218–19; human impact on, 34; insect interactions with, 105–6; introduced species, 75; "living dead," 215; medicinal value, 105, 123, 168, 194, 200–1; overgrazing on, 35–36; pollinator interactions, 112–14, 140, 214–15; scientific and common names, 9–10; species and population extinctions, 3–5, 18, 137–43, 157; vascular, 131–32, 134, 139, 140

plastic pollution, 74, 181–82; micro-plastics, 182

platypuses, 175

Pleistocene era, 29, 232–33

plesiosaurs, xi–xii, xvi, 26

Poached (Nuwer), 178

poaching, 38, 39, 43, 53, 56, 62, 104, 158, 172, 176–78, 216–17, 232, 234. *See also* wildlife trade, illegal

polar regions, 26, 32, 149

political and corporate opposition, to conservation initiatives, 170, 187–88, 190, 220–21, 225, 243

pollinators, 97–98, 109, 183, 191; bees, 102, 112–14, 145–46, 168, 199, 202–3, 214; plant interactions, 112–13, 140, 214–15

pollution, 80, 205–6. *See also* air pollution; plastic pollution; water: pollution

Polynesian people, x, 14, 74

population(s), 3–4; definition, 11; number of, 13

population extinction crisis, 3–6, 27–30. *See also* biodiversity loss; mass extinctions: sixth

population extinction crisis solutions. *See* conservation initiatives

population extinctions, 3–6, 4, 13, 158–59, 191; climate change-related, 186–87; relation to species extinctions, 3–6, 27–28, 28–29, 71, 159–60, 210. *See also* species and population declines/extinctions *under individual species*

porpoises, vaquitas, 61–62, 163

potato blight fungus, 136–37, 146, 150

prairie and grasslands ecosystems, 57–59, 149, 194

prairie dogs, black-tailed, 58–59, 160, 193, 194

Precambrian eon, 6–8

predators: climate change-induced extinction, 34–35; control programs, 227–32, 234; ecosystem balance role, 35–36; mesopredator release, 35; predator-prey relationships, 2–3, 20, 228; prey species declines, 32, 34; smaller, 35, 227–32; top (large), 34–37, 227–28

predator saturation strategy, 69–70, 239

primates (non-human), 38–48; genetic relationship to humans, 42–43, 44, 46; importance to humans, 47

productivity, xvi, 126, 136, 149, 191, 195–96, 199, 204, 223; net primary, 133–34; primary, 133–34, 204

prokaryotes, 13

protected areas, 217–25
protozoa, 108, 153
Pteridophytes, 132
pterodactyls, 8, 26
pterosaurs, 20
pumas/cougars, 36, 163, 226
pupfishes, 92, 237

rabbits, 36
raccoons, 36
rails (flightless birds), x–xi, 68, 74
rainforest, 56, 82, 127, 134–35, 148, 173,
    176, 215, 242. *See also* Amazon
range contraction, 139, 159, 160–62,
    185
rats, 47, 168, 179; Bramble Cay, 15, 17–18,
    33, 234, 236–37; eradication programs,
    74–75; as introduced species, 14, 15,
    33, 74–75, 87, 90, 121, 178, 226, 233;
    Pacific, 14
Raven, Peter, 216
Redford, Kent, 157
remote sensing technology, 137–38,
    140, 157
reproduction: breeding grounds,
    xii–xiv, 60–61, 73–74, 76–77, 88, 99,
    107; rate, 23; systems, 83–84. *See also*
    captive breeding programs
reptiles: conservation initiatives, 90–91;
    endangered species, 87; extinction
    drivers, 87, 174; giant, 19–20; preda-
    tors, 87, 228, 230; species and popula-
    tion declines/extinctions, 5, 28, 80,
    87–89
resource consumption / overcon-
    sumption: human overpopulation
    and, xxi, 166–70; inequities in, xvii,
    170, 178, 187, 197, 203, 241, 244, 245,
    252; per capita, 38, 170

rhinoceroses, x, 31, 48, 51–56, 89, 114
    120, 146, 160, 171, 175, 176, 178, 202;
    illegal horn trade, 52, 54–55, 217
rinderpest, 146, 179
rivet popper hypothesis, 129–30
rodents: as disease vectors, 163, 165.
    See also beavers; mice; rats
Royal Society for the Protection of
    Birds (RSPB), 74–75
Rwanda, Virunga National Park, 38–41

salamanders, 80–82, 85
salmon, 91, 211; Chinook, 93–94, 95
saola, 177
sapsuckers, 71, 72
SARS, 151–52, 163, 165
SARS-2 virus, 152
savanna ecosystem, 37, 48, 54, 77, 160,
    195
scallops, 118–19
Schaller, George, 39
scientific community, response to
    biodiversity loss, 131, 138, 140, 159,
    166–67, 213–15, 218, 230, 233, 241, 244
scientific names, 9–10
sea ice, 35, 59–61
sea level rise, xvi, 17, 26, 236, 240
seals, 33, 35, 158
"sea monsters," 24
sea otters, 2–3
sedge, Mitchell's, 83
seeds, 132; banks, 141, 215; dispersal,
    44, 98, 141, 195
shark-like fish, 24
sharks, 95, 174, 227
shellfish reefs, 119–20
shells, 24, 89, 120–23, 128, 129, 131
shooting galleries, 17, 70
Silva, da Lula, 242

skunks, 36

snails (gastropods), 5, 121–24, 129

snakes, 33, 36, 68, 87–88, 90

snares, 16, 39–40, 177, 217

soil, 141, 172, 191; fertility and nutrient content, 58, 103, 116, 141–42, 155, 164, 167; organisms, xvii, 122, 124–25, 141, 147–49, 155, 191, 193–94

solar power, 80

South Africa, xi, 33, 51–52, 53, 218, 225–26

South America, xiv, 20, 67, 90; Global Diversity Hotspots, 219

sparrows, 193, 199–200, 224

speciation, 12

species: definitions, 10–12, 159–60, 210; new, 12, 176, 178; number of, 11–13

species extinctions, 3–6, 14–18, 159–60, 191, 207; first recorded, 15, 18; global, 159; most recorded, 121; prehistoric, 17–18; relation to population extinctions, 3–6, 28–29, 160, 210. *See also* species and population declines/ extinctions *under individual species*

spiders, 116

Stanford University, 133; Jasper Ridge Biological Preserve, 102–9, 111, 220

Stegosaurus, 20

Steller, Georg W., 2

Steller's sea cow, 2–3, 33, 69

stromatolites, 6

sturgeons, 81, 92–93, 95–96

Sulawesi people, 47

tapirs, 160, 163–64

Tasmania, 32–33, 76–77, 232

Tasmanians, xiv–xv

Tasmanian wolf (thylacine), 32–33

taxonomy, 12

temperate regions/ecosystems, xvi, 105, 161–62, 186, 195, 219, 226–27

Terborgh, John, xiii

termites, 56–57, 78, 149

terrestrial species, 34, 175; evolution, 7; species and population declines/ extinctions, 5, 26, 34

threatened species, xxi, 34, 65–66, 95, 159, 163, 171; amphibians, 81, 179; birds, 73–74, 76–78, 174, 179; disease threats, 77, 81; invertebrates, xxi, 48, 119–24, 119–30; mammals, 34, 38, 44–45, 47–48, 51–56, 59–60, 64, 174, 175, 179, 221, 229; plants, 34, 139; predation threats, 74, 179–81, 227–28, 229, 231; protection, 209–13, 227–28, 231; reptiles, 88–91, 179

"thrush anvils," 121–22

thylacine (Tasmanian wolf), 18, 32–33

tigers, 37, 89, 160, 176

toads, 17, 18, 64, 81–85, 186

topographic heterogeneity, 104

tortoises, 80; giant, 87, 89–90, 121

totoaba fish, 62

tourism, 38, 51, 61, 74, 160, 169, 202; ecotourism/nature, 38, 40, 43, 160, 202, 221

toxification, 80, 87, 94, 96, 182–85, 205–6

traditional/folk medicine, 41, 52, 87–88, 172, 192; Chinese, 55, 56, 62, 89, 104, 120, 171

transpiration, 195, 198

tree of life, 147; mutilation, xxi, 22, 24, 203–4, 205

trees: ecosystem services, xvii, 194–95, 197–99, 200. *See also* deforestation

Triassic-Jurassic period, 5, 26

Triceratops, 20

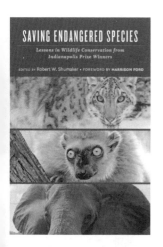
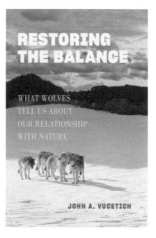
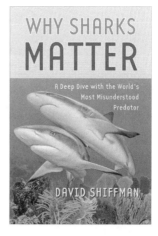
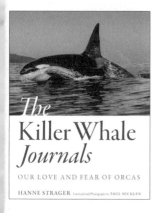
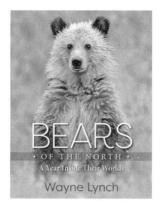
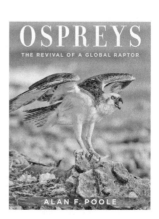